150.00
TSE

*ADVANCES IN*
Chromatography

VOLUME 31

ADVANCES IN Chromatography

VOLUME 31

Edited by

J. Calvin Giddings
Executive Editor
UNIVERSITY OF UTAH
SALT LAKE CITY, UTAH

Eli Grushka
THE HEBREW UNIVERSITY OF JERUSALEM
JERUSALEM, ISRAEL

Phyllis R. Brown
UNIVERSITY OF RHODE ISLAND
KINGSTON, RHODE ISLAND

MARCEL DEKKER, Inc. New York • Basel • Hong Kong

Library of Congress Cataloging in Publication Data
Main entry under title:

Advances in chromatography. v.1-
1965-
New York, M. Dekker
  v. illus. 24 cm.
  Editors: v.1- J.C. Giddings and R.A. Keller.
  1. Chromatographic analysis-Addresses, essays, lectures.
1. Giddings, John Calvin, [date] ed. II. Keller, Roy A., [date] ed.
QD271.A23 544.92 65-27435
ISBN 0-8247-8568-1

Copyright © 1992 by MARCEL DEKKER, INC. All Rights Reserved

Neither this book nor any part may be reproduced or transmitted in any form or by any means, electronic or mechanical, including photocopying, microfilming, and recording, or by any information storage and retrieval system, without permission in writing from the publisher.

MARCEL DEKKER, INC.
270 Madison Avenue, New York, New York 10016

Current printing (last digit):
10 9 8 7 6 5 4 3 2 1

PRINTED IN THE UNITED STATES OF AMERICA

Contributors to Volume 31

S. Shyamali M. DeSilva   Research Associate, Department of Chemistry, The Wichita State University, Wichita, Kansas

Paul L. Dubin   Associate Professor, Department of Chemistry, Indiana-Purdue University, Indianapolis, Indiana

Jiří Gasparič   Professor, Department of Biophysics and Physical Chemistry, Faculty of Pharmacy, Charles University, Hradec Králové, Czechoslovakia

Georges A. Guiochon   Distinguished Scientist, Chemistry Department, The University of Tennessee, Knoxville, Tennessee

Kunihiro Kamata   Pharmaceutical Sciences Department, Tokyo Metropolitan Research Laboratory of Public Health, Tokyo, Japan

Anita M. Katti*   Chemical Engineer, Department of Chemistry, University of Tennessee, Knoxville, Tennessee

Martha Knight   President, Peptide Technologies Corporation, Washington, D.C.

Current affiliation: Senior Development Engineer, Mallinckrodt Specialty Chemicals, St. Louis, Missouri

Roger Meyer   Senior Scientist, Analytical Development, Herbert Laboratories, Irvine, California

Noboru Motohashi   Professor, Department of Medicinal Chemistry, Meiji College of Pharmacy, Tokyo, Japan

Ram P. Singhal   Professor, Department of Chemistry, The Wichita State University, Wichita, Kansas

Contents of Volume 31

Contributors to Volume 31   *iii*
Contents of Other Volumes   *ix*

1. Fundamentals of Nonlinear Chromatography: Prediction of Experimental Profiles and Band Separation   1

   Anita M. Katti and Georges A. Guiochon

       I. Introduction
      II. Basic Equations and Models
     III. Results and Discussion
     IV. Conclusion
          References

2. Problems in Aqueous Size Exclusion Chromatography   119

   Paul L. Dubin

       I. Introduction
      II. Commercial Packings
     III. Electrostatic Interactions
     IV. Hydrophobic Interactions

V. Universal Calibration
VI. SEC of Proteins
VII. SEC of Micelles
References

3. Chromatography on Thin Layers Impregnated with Organic Stationary Phases   153

   Jiří Gasparič

   I. Introduction
   II. The Partition Solvent System
   III. Applications
   IV. Conclusions
   References

4. Countercurrent Chromatography for the Purification of Peptides   253

   Martha Knight

   I. Introduction
   II. Early Instrumentation for the Countercurrent Chromatography of Peptides
   III. Theory of Countercurrent Chromatography
   IV. Methods
   V. Use of the Horizontal Flow-Through Coil Planet Centrifuge
   VI. Multilayer Coil Planet Centrifuge
   VII. Multicoil Countercurrent Chromatograph (MC-CCC)
   VIII. Potential Instruments for CCC of Peptides
   IX. Summary
   References

5. Boronate Affinity Chromatography   293

   Ram P. Singhal and S. Shyamali M. DeSilva

   I. Introduction
   II. Boronate Complex Formation
   III. Conditions to Enhance Complex Formation
   IV. Boronate Ligands Used for Affinity Chromatography
   V. Chemistry of Activation and Coupling of Ligands
   VI. Affinity Matrices Used for Immobilization of Different Ligands
   VII. Methods for Studying Complex Formation
   VIII. Examples of Applications of Boronate Affinity Chromatography

IX. Concluding Remarks
References

6. Chromatographic Methods for Determining Carcinogenic Benz(c)acridine   337

    Noboru Motohashi, Kunihiro Kamata, and Roger Meyer

    I. Introduction
    II. Column Chromatography
    III. Paper Chromatography
    IV. Thin-Layer Chromatography
    V. High-Performance Liquid Chromatography
    VI. Gas-Liquid Chromatography
    References

Index   383

Contents of Other Volumes

Volume 1

Ion-Exchange Chromatography *F. Helfferich*
Chromatography and Electrophoresis on Paper and Thin Layers: A Teacher's Guide *Ivor Smith*
The Stationary Phase in Paper Chromatography *George H. Stewart*
The Techniques of Laminar Chromatography *E. V. Truter*
Qualitative and Quantitative Aspects of the Separation of Steroids *E. C. Horning and W. J. A. VandenHeuvel*
Capillary Columns: Trials, Tribulations, and Triumphs *D. H. Desty*
Gas Chromatographic Characterization of Organic Substances in the Retention Index System *E. sz. Kováts*
Inorganic Gas Chromatography *Richard S. Juvet, Jr., and Franjo Zado*
Lightly Loaded Columns *Barry L. Karger and W. D. Cooke*
Interactions of the Solute with the Liquid Phase *Daniel E. Martire and Luigi Z. Pollara*

Volume 2

Ion-Exchange Chromatography of Amino Acids: Recent Advances in Analytical Determinations *Paul B. Hamilton*
Ion Mobilities in Electrochromatography *John T. Edward*

Partition Paper Chromatography and Chemical Structures *J. Green and D. McHale*

Gradient Techniques in Thin-Layer Chromatography *A. Niederwieser and C. C. Honegger*

Geology—An Inviting Field to Chromatography *Arthur S. Ritchie*

Extracolumn Contributions to Chromatographic Band Broadening *James C. Sternberg*

Gas Chromatography of Carbohydrates *James W. Berry*

Ionization Detectors for Gas Chromatography *Arthur Karmen*

Advances in Programmed Temperature Gas Chromatography *Louis Mikkelsen*

Volume 3

The Occurrence and Significance of Isotope Fractionation during Analytical Separations of Large Molecules *Peter D. Klein*

Adsorption Chromatography *Charles H. Giles and I. A. Easton*

The History of Thin-Layer Chromatography *N. Pelick, H. R. Bolliger, and H. K. Mangold*

Chromatography as a Natural Process in Geology *Arthur S. Ritchie*

The Chromatographic Support *D. M. Ottenstein*

Electrolytic Conductivity Detection in Gas Chromatography *Dale M. Coulson*

Preparative-Scale Gas Chromatography *G. W. A. Rijnders*

Volume 4

R_F Values in Thin-Layer Chromatography on Alumina and Silica *Lloyd R. Snyder*

Steriod Separation and Analysis: The Technique Appropriate to the Goal *R. Neher*

Some Fundamentals of Ion-Exchange-Cellulose Design and Usage in Biochemistry *C. S. Knight*

Adsorbents in Gas Chromatography *A. V. Kiselev*

Packed Capillary Columns in Gas Chromatography *István Halász and Erwin Heine*

Mass-Spectrometric Analysis of Gas-Chromatographic Eluents *William McFadden*

The Polarity of Stationary Liquid Phases in Gas Chromatography *L. Rohrschneider*

Volume 5

Prediction and Control of Zone Migration Rates in Ideal Liquid-Liquid Partition Chromatography *Edward Soczewiński*

Chromatographic Advances in Toxicology *Paul L. Kirk*
Inorganic Chromatography on Natural and Substituted Celluloses
   R. A. A. Muzzarelli
The Quantitative Interpretation of Gas Chromatographic Data
   H. Wilson Johnson, Jr.
Atmospheric Analysis by Gas Chromatography *A. P. Altshuller*
Non-Ionization Detectors and Their Use in Gas Chromatography
   J. D. Winefordner and T. H. Glenn

Volume 6

The Systematic Use of Chromatography in Structure Elucidation of
   Organic Compounds by Chemical Methods *Jiři Gasparič*
Polar Solvents, Supports, and Separations *John A. Thoma*
Liquid Chromatography on Lipophilic Sephadex: Column and Detection
   Techniques *Jan Sjövall, Ernst Nyström, and Hero Haahti*
Statistical Moments Theory of Gas-Solid Chromatography: Diffusion-
   Controlled Kinetics *Otto Grubner*
Identification by Retention and Response Values *Gerhard Schomburg*
   (translated by Roy A. Keller)
The Use of Liquid Crystals in Gas Chromatography *H. Kelker and
   E. von Schivizhoffen*
Support Effects on Retention Volumes in Gas-Liquid Chromatography
   Paul Urone and Jon F. Parcher

Volume 7

Theory and Mechanics of Gel Permeation Chromatography *K. H. Altgelt*
Thin-Layer Chromatography of Nucleic Acid Bases, Nucleosides,
   Nucleotides, and Related Compounds *György Pataki*
Review of Current and Future Trends in Paper Chromatography
   V. C. Weaver
Chromatography of Inorganic Ions *G. Nickless*
Process Control by Gas Chromatography *I. G. McWilliam*
Pyrolysis Gas Chromatography of Involatile Substances *S. G. Perry*
Labeling by Exchange on Chromatographic Columns *Horst Elias*

Volume 8

Principles of Gel Chromatography *Helmut Determann*
Thermodynamics of Liquid-Liquid Partition Chromatography *David
   C. Locke*
Determination of Optimum Solvent Systems for Countercurrent Distri-
   bution and Column Partition Chromatography from Paper Chroma-
   tographic Data *Edward Soczewiński*

Some Procedures for the Chromatography of the Fat-Soluble Chloroplast Pigments *Harold H. Strain and Walter A. Svec*

Comparison of the Performances of the Various Colunm Types Used in Gas Chromatography *Georges Guiochon*

Pressure (Flow) Programming in Gas Chromatography *Leslie S. Ettre, László Mázor, and József Takács*

Gas Chromatographic Analysis of Vehicular Exhaust Emissions *Basil Dimitriades, C. F. Ellis, and D. E. Seizinger*

The Study of Reaction Kinetics by the Distortion of Chromatographic elution Peaks *Maarten van Swaay*

Volume 9

Reversed-Phase Extraction Chromatography in Inorganic Chemistry *E. Cerrai and G. Ghersini*

Determination of the Optimum Conditions to Effect a Separation by Gas Chromatography *R. P. W. Scott*

Advances in the Technology of Lightly Loaded Glass Bead Columns *Charles Hishta, Joseph Bomstein, and W. D. Cooke*

Radiochemical Separations and Analyses by Gas Chromatography *Stuart P. Cram*

Analysis of Volatile Flavor Components of Foods *Phillip Issenberg and Irwin Hornstein*

Volume 10

Porous-Layer Open Tubular Columns—Theory, Practice, and Applications *Leslie S. Ettre and John E. Purcell*

Resolution of Optical Isomers by Gas Chromatography of Diastereomers *Emanuel Gil-Av and David Nurok*

Gas-Liquid Chromatography of Terpenes *E. von Rudloff*

Volume 11

Quantitative Analysis by Gas Chromatography *Josef Novák*

Polyamide Layer Chromatography *Kung-Tsung Wang, Yau-Tang Lin, and Iris S. Y. Wang*

Specifically Adsorbing Silica Gels *H. Bartels and P. Prijs*

Nondestructive Detection Methods in Paper and Thin-Layer Chromatography *G. C. Barrett*

Volume 12

The Use of High-Pressure Liquid Chromatography in Pharmacology and Toxicology *Phyllis R. Brown*

Chromatographic Separation and Molecular-Weight Distributions of
  Cellulose and Its Derivatives *Leon Segal*
Practical Methods of High-Speed Liquid Chromatography *Gary J.
  Fallick*
Measurement of Diffusion Coefficients by Gas-Chromatography Broadening Techniques: A Review *Virgil R. Maynard and Eli Grushka*
Gas-Chromatography Analysis of Polychlorinated Diphenyls and Other
  Nonpesticide Organic Pollutants *Joseph Sherma*
High-Performance Electrometer Systems for Gas Chromatography
  Douglas H. Smith
Steam Carrier Gas-Solid Chromatography *Akira Nonaka*

Volume 13

Practical Aspects in Supercritical Fluid Chromatography *T. H. Gouw
  and Ralph E. Jentoft*
Gel Permeation Chromatography: A Review of Axial Dispersion Phenomena, Their Detection, and Correction *Nils Friis and Archie
  Hamielec*
Chromatography of Heavy Petroleum Fractions *Klaus H. Altegelt and
  T. H. Gouw*
Determination of the Adsorption Energy, Entropy, and Free Energy of
  Vapors on Homogeneous Surfaces by Statistical Thermodynamics
  Claire Vidal-Madjar, Marie-France Gonnord, and Georges Guiochon
Transport and Kinetic Parameters by Gas Chromatographic Techniques
  Motoyuki Suzuki and J. M. Smith
Qualitative Analysis by Gas Chromatography *David A. Leathard*

Volume 14

Nutrition: An Inviting Field to High-Pressure Liquid Chromatography
  Andrew J. Clifford
Polyelectrolyte Effects in Gel Chromatography *Bengt Stenlund*
Chemically Bonded Phases in Chromatography *Imrich Sebestian and
  István Halász*
Physicochemical Measurements Using Chromatography *David C. Locke*
Gas-Liquid Chromatography in Drug Analysis *W. J. A. VandenHeuvel
  and A. G. Zacchei*
The Investigation of Complex Association by Gas Chromatography and
  Related Chromatographic and Electrophoretic Methods *C. L.
  de Ligny*
Gas-Liquid-Solid Chromatography *Antonio De Corcia and Arnaldo
  Liberti*
Retention Indices in Gas Chromatography *J. K. Haken*

xiv / Contents of Other Volumes

Volume 15

Detection of Bacterial Metabolites in Spent Culture Media and Body Fluids by Electron Capture Gas-Liquid Chromatography *John B. Brooks*

Signal and Resolution Enhancement Techniques in Chromatography *Raymond Annino*

The Analysis of Organic Water Pollutants by Gas Chromatography and Gas Chromatography-Mass Spectrometry *Ronald A. Hites*

Hydrodynamic Chromatography and Flow-Induced Separations *Hamish Small*

The Determination of Anticonvulsants in Biological Samples by Use of High-Pressure Liquid Chromatography *Reginald F. Adams*

The Use of Microparticulate Reversed-Phase Packing in High-Pressure Liquid Chromatography of Compounds of Biological Interest *John A. Montgomery, Thomas P. Johnston, H. Jeanette Thomas, James R. Piper, and Carroll Temple Jr.*

Gas-Chromatographic Analysis of the Soil Atmosphere *K. A. Smith*

Kinematics of Gel Permeation Chromatography *A. C. Ouano*

Some Clinical and Pharmacological Applications of High-Speed Liquid Chromatography *J. Arly Nelson*

Volume 16

Analysis of Benzo(a)pyrene Metabolism by High-Pressure Liquid chromatography *James K. Selkirk*

High-Performance Liquid Chromatography of the Steroid Hormones *F. A. Fitzpatrick*

Numerical Taxonomy in Chromatography *Desire L. Massart and Henri L. O. De Clercq*

Chromatography of Oligosaccharides and Related Compounds on Ion-Exchange Resins *Olof Samuelson*

Applications and Theory of Finite Concentrations Frontal Chromatography *Jon F. Parcher*

The Liquid-Chromatography Resolution of Enantiomers *Ira S. Krull*

The Use of High-Pressure Liquid Chromatography in Research on Purine Nucleoside Analog *William Plunkett*

The Determination of Di- and Polyamines by High-Pressure Liquid and Gas-Chromatography *Mahmoud M. Abdel-Monem*

Volume 17

Progress in Photometric Methods of Quantitative Evaluation in TLO *V. Pollak*

Ion-Exchange Packings for HPLC Separations: Care and Use *Fredric M. Rabel*

Micropacked Columns in Gas Chromatography: An Evaluation *C. A. Cramers and J. A. Rijks*
Reversed-Phase Gas Chromatography and Emuslifier Characterization *J. K. Haken*
Template Chromatography *Herbert Schott and Ernst Bayer*
Recent Usage of Liquid Crystal Stationary Phases in Gas Chromatography *George M. Janini*
Current State of the Art in the Analysis of Catecholamines *Anté M. Krstulovic*

Volume 18

The Characterization of Long-Chain Fatty Acids and Their Derivatives by Chromatography *Marcel S. F. Lie Ken Jie*
Ion-Pair Chromatography on Normal- and Reversed-Phase Systems *Milton T. W. Hearn*
Current State of the Art in HPLC Analyses of Free Nucleotides, Nucleosides, and Bases in Biological Fluids *Phyllis R. Brown, Anté M. Krstulovic, and Richard A. Hartwick*
Resolution of Racemates by Ligand-Exchange Chromatography *Vadim A. Danankov*
The Analysis of Marijuana Cannabinoids and Their Metabolites in Biological Media by GC and/or GC-MS Techniques *Benjamin J. Gudzinowicz, Michael J. Gudzinowicz, Joanne Hologgitas, and James L. Driscoll*

Volume 19

Roles of High-Performance Liquid Chromatography in Nuclear Medicine *Steven How-Yan Wong*
Calibration of Separation Systems in Gel Permeation Chromatography for Polymer Characterization *Josef Janča*
Isomer-Specific Assay of 2,4-D Herbicide Products by HPLC: Regulaboratory Methodology *Timothy S. Stevens*
Hydrophobic Interaction Chromatography *Stellan Hjertén*
Liquid Chromatography with Programmed Composition of the Mobile Phase *Pavel Jandera and Jaroslav Churáček*
Chromatographic Separation of Aldosterone and Its Metabolites *David J. Morris and Ritsuko Tsai*

Volume 20

High-Performance Liquid Chromatography and Its Application to Protein Chemistry *Milton T. W. Hearn*
Chromatography of Vitamin D_3 and Metabolites *K. Thomas Koshy*

High-Performance Liquid Chromatography: Applications in a Children's Hospital *Steven J. Soldin*
The Silica Gel Surface and Its Interactions with Solvent and Solute in Liquid Chromatography *R. P. W. Scott*
New Developments in Capillary Columns for Gas Chromatography *Walter Jennings*
Analysis of Fundamental Obstacles to the Size Exclusion Chromatography of Polymers of Ultrahigh Molecular Weight *J. Calvin Giddings*

Volume 21

High-Performance Liquid Chromatography/Mass Spectrometry (HPLC/MS) *David E. Grimes*
High-Performance Liquid Affinity Chromatography *Per-Olof Larsson, Magnus Glad, Lennart Hansson, Mats-Olle Månsson, Sten Ohlson, and Klaus Mosbach*
Dynamic Anion-Exchange Chromatography *Roger H. A. Sorel and Abram Hulshoff*
Capillary Columns in Liquid Chromatography *Daido Ishii and Toyohide Takeuchi*
Droplet Counter-Current Chromatography *Kurt Hostettmann*
Chromatographic Determination of Copolymer Composition *Sadao Mori*
High-Performance Liquid Chromatography of K Vitamins and Their Antagonists *Martin J. Shearer*
Problems of Quantitation in Trace Analysis by Gas Chromatography *Josef Novák*

Volume 22

High-Performance Liquid Chromatography and Mass Spectrometry of Neuropeptides in Biologic Tissue *Dominic M. Desiderio*
High-Performance Liquid Chromatography of Amino Acids: Ion-Exchange and Reversed-Phase Strategies *Robert F. Pfeifer and Dennis W. Hill*
Resolution of Racemates by High-Performance Liquid Chromatography *Vadium A. Davankov, Alexander A. Kurganov, and Alexander S. Bochkov*
High-Performance Liquid Chromatography of Metal Complexes *Hans Veening and Bennett R. Willeford*
Chromatography of Carotenoids and Retinoids *Richard F. Taylor*
High Performance Liquid Chromatography of Porphyrins *Zbyslaw J. Petryka*
Small-Bore Columns in Liquid Chromatography *Raymond P. W. Scott*

Volume 23

Laser Spectroscopic Methods for Detection in Liquid Chromatography *Edward S. Yeung*
Low-Temperature High-Performance Liquid Chromatography for Separation of Thermally Labile Species *David E. Henderson and Daniel J. O'Connor*
Kinetic Analysis of Enzymatic Reactions Using High-Performance Liquid Chromatography *Donald L. Sloan*
Heparin-Sepharose Affinity Chromatography *Akhlaq A. Farooqui and Lloyd A. Horrocks*
Chromatopyrography *John Chih-An Hu*
Inverse Gas Chromatography *Seymour G. Gilbert*

Volume 24

Some Basic Statistical Methods for Chromatographic Data *Karen Kafadar and Keith R. Eberhardt*
Multifactor Optimization of HPLC Conditions *Stanley N. Deming, Julie G. Bower, and Keith D. Bower*
Statistical and Graphical Methods of Isocratic Solvent Selection for Optimal Separation in Liquid Chromatography *Haleem J. Issaq*
Electrochemical Detectors for Liquid Chromatography *Ante M. Krstulović, Henri Colin, and Georges A. Guiochon*
Reversed-Flow Gas Chromatography Applied to Physicochemical Measurements *Nicholas A. Katsanos and George Karaiskakis*
Development of High-Speed Countercurrent Chromatography *Yoichiro Ito*
Determination of the Solubility of Gases in Liquids by Gas-Liquid Chromatography *Jon F. Parcher, Monica L. Bell, and Ping J. Lin*
Multiple Detection in Gas Chromatography *Ira S. Krull, Michael E. Swartz, and John N. Driscoll*

Volume 25

Estimation of Physicochemical Properties of Organic Solutes Using HPLC Retention Parameters *Theo L. Hafkenscheid and Eric Tomlinson*
Mobile Phase Optimization in RPLC by an Iterative Regression Design *Leo de Galan and Hugo A. H. Billiet*
Solvent Elimination Techniques for HPLC/FT-IR *Peter R. Griffiths and Christine M. Conroy*
Investigations of Selectivity in RPLC of Polycyclic Aromatic Hydrocarbons *Lane C. Sander and Stephen A. Wise*

Liquid Chromatographic Analysis of the Oxo Acids of Phosphorus
   Roswitha S. Ramsey
HPLC Analysis of Oxypurines and Related Compounds   *Katsuyuki Nakano*
HPLC of Glycosphingolipids and Phospholipids   *Robert H. McCluer, M. David Ullman, and Firoze B. Jungalwala*

Volume 26

RPLC Retention of Sulfur and Compounds Containing Divalent Sulfur
   Hermann J. Möckel
The Application of Fleuric Devices to Gas Chromatographic Instrumentation   *Raymond Annino*
High Performance Hydrophobic Interaction Chromatography
   Yoshio Kato
HPLC for Therapeutic Drug Monitoring and Determination of Toxicity
   Ian D. Watson
Element Selective Plasma Emission Detectors for Gas Chromatography
   A. H. Mohamad and J. A. Caruso
The Use of Retention Data from Capillary GC for Qualitative Analysis: Current Aspects   *Lars G. Blomberg*
Retention Indices in Reversed-Phase HPLC   *Roger M. Smith*
HPLC of Neurotransmitters and Their Metabolites   *Emilio Gelpi*

Volume 27

Physicochemical and Analytical Aspects of the Adsorption Phenomena Involved in GLC   *Victor G. Berezkin*
HPLC in Endocrinology   *Richard L. Patience and Elizabeth S. Penny*
Chiral Stationary Phases for the Direct LC Separation of Enantiomers
   William H. Pirkle and Thomas C. Pochapsky
The Use of Modified Silica Gels in TLC and HPTLC   *Willi Jost and Heinz E. Hauck*
Micellar Liquid Chromatography   *John G. Dorsey*
Derivatization in Liquid Chromatography   *Kazuhiro Imai*
Analytical High-Performance Affinity Chromatography   *Georgio Fassina and Irwin M. Chaiken*
Characterization of Unsaturated Aliphatic Compounds by GC/Mass Spectrometry   *Lawrence R. Hogge and Jocelyn G. Millar*

Volume 28

Theoretical Aspects of Quantitative Affinity Chromatography: An Overview   *Alain Jaulmes and Claire Vidal-Madjar*

Column Switching in Gas Chromatography   *Donald E. Willis*
The Use and Properties of Mixed Stationary Phases in Gas Chromatography   *Gareth J. Price*
On-Line Small-Bore Chromatography for Neurochemical Analysis in the Brain   *William H. Church and Joseph B. Justice, Jr.*
The Use of Dynamically Modified Silica in HPLC as an Alternative to Chemically Bonded Materials   *Per Helboe, Steen Honoré Hansen, and Mogens Thomsen*
Gas Chromatographic Analysis of Plasma Lipids   *Arnis Kuksis and John J. Myher*
HPLC of Penicillin Antibiotics   *Michel Margosis*

Volume 29

Capillary Electrophoresis   *Ross A. Wallingford and Andrew G. Ewing*
Multidimensional Chromatography in Biotechnology   *Daniel F. Samain*
High-Performance Immunoaffinity Chromatography   *Terry M. Phillips*
Protein Purification by Multidimensional Chromatography   *Stephen A. Berkowitz*
Fluorescence Derivitization in High-Performance Liquid Chromatography   *Yosuke Ohkura and Hitoshi Nohta*

Volume 30

Mobile and Stationary Phases for Supercritical Fluid Chromatography   *Peter J. Schoenmakers and Louis G. M. Uunk*
Polymer-Based Packing Materials for Reversed-Phase Liquid Chromatography   *Nobuo Tanaka and Mikio Araki*
Retention Behavior of Large Polycyclic Aromatic Hydrocarbons in Reversed-Phase Liquid Chromatography   *Kiyokatsu Jinno*
Miniaturization in High-Performance Liquid Chromatography   *Masashi Goto, Toyohide Takeuchi, and Daido Ishii*
Sources of Errors in the Densitometric Evaluation of Thin-Layer Separations with Special Regard to Nonlinear Problems   *Viktor A. Pollak*
Electronic Scanning for the Densitometric Analysis of Flat-Bed Separations   *Viktor A. Pollak*

1
Fundamentals of Nonlinear Chromatography: Prediction of Experimental Profiles and Band Separation

Anita M. Katti* and Georges A. Guiochon *The University of Tennessee, Knoxville, Tennessee*

I.	INTRODUCTION	2
II.	BASIC EQUATIONS AND MODELS	7
	A. The Mass Balance Equation	7
	B. Relationship Between Concentrations in the Two Phases of the Chromatographic Column	8
	C. Initial and Boundary Conditions	9
	D. The Multicomponent Case and the Competitive Equilibrium Isotherms	10
	E. The Ideal Model	13
	F. The Semi-ideal Model	13
	G. Kinetic Models	14
III.	RESULTS AND DISCUSSION	14
	A. Single-Component Problems	14
	B. Two-Component Problems	43
	C. Three-Component and Multicomponent Problems	89
IV.	CONCLUSION	109
	REFERENCES	110

---

*Current affiliation: Mallinckrodt Specialty Chemicals, St. Louis, Missouri

I. INTRODUCTION

The theory of chromatography should answer the following question: What is the response of the chromatographic column to the various possible perturbations caused at its inlet, as a function of the thermodynamic and kinetic parameters of the system and the experimental conditions? This is a basic problem whose complexity results from the involvement of several major areas of physical chemistry and chemical engineering.

This problem was stated first in the 1940s by Wilson [1] and studied by DeVault [2], Glueckauf [3-5], Goldstein [6], and Thomas [7]. Various analytical solutions were proposed but a general discussion was impossible because the theory of partial differential equations was insufficiently developed and computers were not available to give numerical solutions of the general problem. Seminal contributions were made in the 1950s and 1960s by Giddings [8], Houghton [9], Haarhoff and Van der Linde [10], Lapidus and Amundson [11], and Van Deemter et al. [12]. Considerable progress was made around 1970, when preparative gas chromatography raised high hopes as a new separation method, while the theory of characteristics became available to the chemical engineer through the work of Aris and Amundson [13]. The work performed by Jacob, Valentin, and Guiochon [14-17] and that by Conder and Purnell [18] permitted a much deeper understanding of the chromatographic process with the advance of the theory of partial differential equations. Analytical solutions were derived by Rhee et al. [19,20] and Helfferich and Klein [21]. Finally, in the late 1980s most of the remaining problems were solved through a collective effort making use of the advances made in the theory of nonlinear partial differential equations as well as of the huge computer power now available. Recently, the most significant work is by the groups of Byers [22], Carta [23], Guiochon [24], Horvath [25-27], Knox [28], and Wang [29,30].

Nonlinear chromatography can be categorized after the nature of the perturbation at the column inlet. The most common perturbations are the introduction of a concentration step and the injection of a narrow rectangular pulse. The first perturbation is known in mathematics as the Riemann problem. To chromatographers, this is frontal analysis. The second perturbation is the Dirac problem, also called elution chromatography. Both can be performed with a pure component or a multicomponent sample. An additive can be added to the mobile phase, either at constant concentration (system peaks) or as a concentration ramp (gradient elution). When the additive competes with the sample components for the stationary phase, a perturbation in the equilibrium concentration of the additive occurs. This is the origin of system peaks. In gradient elution, the retention of the sample components decreases progressively, either because their solubility in the mobile phase increases or because the strength of their interaction with the stationary phase decreases. The result, compared to isocratic analysis, is more rapid elution. Finally, the injection of a sample pulse may be followed by a concentration step of a compound more strongly retained than any sample component. This is displacement chromatography. Other combinations are possible (e.g., the successive injection of pulses of two compounds) but they have not yet found practical interest.

For the sake of convenience and clarity, this presentation of the theory of nonlinear chromatography has been divided into sections according to the total number of components in the system studied. The mobile phase is not counted as a separate component unless it contains additives which can compete with the sample components for interaction with the stationary phase. The one-component problems encompass input perturbations made of either a step or a narrow impulse injection function. This corresponds to frontal analysis and elution, respectively. Two-component problems include frontal analysis and elution of a binary mixture. They also include displacement chromatography of a single-component sample, gradient elution with a single-component sample, and the simplest system peak problem, occurring when a pure component is injected in a mobile phase with one strongly adsorbing additive. Three-component problems primarily involve the separation of a binary mixture in the displacement mode, in gradient elution, or by isocratic elution under the conditions where system peaks form. Most multicomponent problems involving a larger number of compounds can be considered as the superimposition of several simpler problems. For example, the profiles in overloaded elution with three components can be understood qualitatively as two successive two-component problems.

Several assumptions have to be made regarding the thermodynamics of the chromatographic system considered (are the equilibrium isotherms linear or nonlinear?) and the mass transfer kinetics in this system (are they rapid or not?). Solutions for calculating the band profiles in linear chromatography have been derived. Closed form solutions under linear conditions are available even when the mass transfer kinetics are slow. In the case where the kinetics are fast, we can consider that the column is always near equilibrium and we can use ideal or semi-ideal models to determine the individual band profiles. Assuming equilibrium, we need to know only the competitive equilibrium isotherms of the components involved in the chromatographic system and the column efficiency at infinite dilution. If the kinetics are slow, we need to use a nonequilibrium or kinetic model. The sources of diffusion/dispersion and the rate of transport of molecules from the mobile to the stationary phase must be assumed or determined in addition to the competitive equilibrium isotherms.

Finally, there are several implementations of chromatography, depending on the retention mechanism. The most important ones are liquid-solid chromatography (LSC), which includes normal phase chromatography (NPC), reversed phase chromatography (RPC), and hydrophobic interaction chromatography (HIC); ion exchange chromatography (IXC); liquid-liquid chromatography (LLC); size exclusion chromatography (SEC); and biospecific affinity chromatography (AFC). The present work deals primarily with those modes where consideration of the mass balance of each component and of a simple isotherm equation is sufficient to model the chromatographic process. This includes obviously all LSC modes, LLC, and AFC (a case where a kinetic model is almost always needed). The extension to SEC is ambiguous. In the case of IXC, a charge balance may be needed, in which case a more complex model is required. In many cases, however, an excess of a convenient, unretained counterion is used to maintain electroneutrality everywhere along the column and the theory presented here applies straightforwardly. Finally, gas and dense gas chromatography raises the complex issue of the progressive decrease of the mobile phase density along the column, making necessary the use of a mass balance equation for the mobile phase [14-18]. The sorption effect is a result of the difference in partial molar volume of the analytes in the two phases of the chromatographic system. These phenomena are not discussed here.

This chapter is divided into four major sections. The first section derives and discusses the basic equations of nonlinear chromatography and explains the origin, advantages, and drawbacks of the different models used to account for the chromatographic process. The other three sections review the literature concerning the one-, two-, three-, and multicomponent problems, respectively. In each of these four sections, the solutions pertaining to elution, frontal analysis, displacement, system peaks, and gradient elution are presented successively. In each case where they are available, we give the results of the ideal, semi-ideal, and kinetic models. As often as possible, illustrations have been selected among experimental results rather than among theoretical schematics or calculated profiles. Prominent contributions to the field are described with emphasis on nonlinear chromatography. For the sake of easy reference, Tables 1 and 2 summarize the main equations which are referenced in the discussion. The various model simplifications presented in these tables will be discussed in detail in Section II. Although symbols are defined when they are used for the first time, a glossary of terms is provided.

Table 1 Summary of the Mass Balance Equations

I. General Form of the Mass Balance Equation

$$\frac{\partial C}{\partial t} + F \frac{\partial q}{\partial t} + u_0 \frac{\partial C}{\partial z} = D_m \frac{\partial^2 C}{\partial z^2} \qquad (1)$$

where:
  C and q are the concentrations of the component in the mobile and stationary phases, respectively,
  z and t are the column length and time, respectively,
  F is the phase ratio $(1 - \varepsilon)/\varepsilon$, with ε porosity of the column packing,
  u_0 is the mobile phase velocity, and
  D_m is the axial diffusion coefficient of the component in the column.

II. Relationship Between C and q

In the ideal and semi-ideal models:

$$q = Q = f(C) \qquad (2)$$

with $Q = f(C)$ equilibrium isotherm of the component in the chromatographic system.

In kinetic models:

$$\frac{\partial q}{\partial t} = g(C, Q, q) \qquad (3)$$

with $g(C,Q,q)$ kinetic equation.

III. Mass Balance Equation for the Ideal Model

$$\frac{\partial C}{\partial t} + F \frac{\partial q}{\partial t} + u_0 \frac{\partial C}{\partial z} = 0 \qquad (4)$$

with

$$q = Q = f(C)$$

IV. Mass Balance Equation for the Semi-ideal Model

$$\frac{\partial C}{\partial t} + F \frac{\partial q}{\partial t} + u_0 \frac{\partial C}{\partial z} = D_{ap} \frac{\partial^2 C}{\partial z^2} \qquad (5)$$

with

$$q = Q = f(C)$$

and

$$D_{ap} = \frac{HL}{2t_0} \qquad (6)$$

Table 2 Summary of Equilibrium and Nonequilibrium Equations

I. Single-Component Isotherms

Langmuir equation:

$$Q = \frac{aC}{1 + bC} \tag{1}$$

where Q and C are the concentrations in the stationary and mobile phase at equilibrium, respectively, and a and b are numerical coefficients. When C becomes very large, Q tends toward a limit, $Q_s = a/b$.

BiLangmuir equation:

$$Q = \frac{a_1 C}{1 + b_1 C} + \frac{a_2 C}{1 + b_2 C} \tag{2}$$

where a_1, a_2, b_1, and b_2 are numerical coefficients.

Freundlich equation:

$$Q = aC^n \tag{3}$$

Fowler equation:

$$C = \frac{1}{b} \frac{\theta}{1 - \theta} e^{-\chi\theta} \tag{4}$$

where θ is the surface coverage ratio or ratio q/Q_s of the concentration in the stationary phase to the saturation concentration in this phase, and b and χ are numerical coefficients.

II. Single-Component Rate Equations

Linear driving force rate equation:

$$\frac{\partial q}{\partial t} = K(Q - q) \tag{5}$$

where Q is the equilibrium concentration in the stationary phase and K a rate constant. q is the local concentration in the stationary phase.

Linear kinetics equation:

$$\frac{\partial q}{\partial t} = K_1 C - K_2 q \tag{6}$$

Langmuir rate equation:

$$\frac{\partial q}{\partial t} = K_a C(Q_s - q) - K_d q \tag{7}$$

where K_a and K_d are the adsorption and desorption rate constants, respectively, and Q_s is the saturation concentration in the stationary phase.

III. Binary Competitive Isotherms

Langmuir equation:

$$Q_i = \frac{a_i C_i}{1 + b_1 C_1 + b_2 C_2} \qquad (8)$$

with i equal to 1 or 2.

BiLangmuir equation:

$$Q_i = \frac{a_{1,i} C_i}{1 + b_{1,1} C_1 + b_{1,2} C_2} + \frac{a_{2,i} C_i}{1 + b_{2,1} C_1 + b_{2,2} C_2} \qquad (9)$$

where $a_{i,j}$ and $b_{i,j}$ are numerical coefficients.

Fowler equation:

$$C_i = \frac{1}{b} \frac{\theta_i}{1 - (\theta_1 + \theta_2)} e^{-(\chi_1 \theta_1 + \chi_2 \theta_2)} \qquad (10)$$

II. BASIC EQUATIONS AND MODELS

The most fundamental approach to the general chromatography problem consists of applying the basic principles of physical chemistry. Since there is no reaction in the system, the differential mass balance of each compound is written. This mass balance must be completed by an assumption regarding the mass transfer kinetics and by a set of initial and boundary conditions. After discussing these equations, we present the different models used.

A. The Mass Balance Equation

We make the following assumptions [1,13]:

The mobile phase density is constant. We neglect the compressibility of the mobile phase. It has been shown that the effect is very small in liquid chromatography.

The partial molar volumes of the sample components are the same in both phases. This neglects the sorption effect, which is legitimate in liquid chromatography where the differences between these volumes do not exceed a few percentage points, as compared to gas chromatography, where the partial molar volume in the mobile phase is two orders of magnitude larger than in the stationary phase.

The mass balance for a compound is then written:

$$\frac{\partial C}{\partial t} + F \frac{\partial C_s}{\partial t} + u_0 \frac{\partial C}{\partial z} = D \frac{\partial^2 C}{\partial z^2} \qquad (1)$$

where:

C and C_s are the concentration of the compound in the mobile and stationary phase, respectively,

z and t are the column distance and time, respectively,

F is the phase ratio, equal to $(1 - \varepsilon)/\varepsilon$, where ε is the internal porosity of the column packing,

u_0 is the mobile phase velocity, and

D is the axial dispersion coefficient of the compound in the mobile phase. This includes the contributions to the axial dispersion of the band due to molecular diffusion, to the tortuosity of the packing, and to the nonhomogeneity of this packing, i.e., the terms B/u and $Au^{1/3}$ of the Knox equation, or the terms A and B/u of the Van Deemter equation.

Equation 1 is a partial differential equation of the second order. It contains two functions, C and C_s, and a relationship between them is necessary.

B. Relationship Between Concentrations in the Two Phases of the Chromatographic Column

If the mass transfer kinetics between and across the mobile and stationary phases in the column are very fast, these phases are very close to equilibrium. As a first approximation, we can assume that they are in equilibrium and write:

$$C_s = q = f(C) \qquad (2)$$

where C_s is the instantaneous concentration in the stationary phase and q is the concentration of the component in the stationary phase in equilibrium with the concentration C in the mobile phase. We see later, in Section II.F on the semi-ideal model, how small deviations from equilibrium can be handled while retaining the simplicity of Eq. 2.

When the mass transfer kinetics are slow (e.g., some applications of LSC, IXC, or AFC to the separation of large peptides, proteins, or polynucleotides), we must take their finite rate into account. Three different models have been used:

The Langmuir kinetic model:

$$\frac{\partial C_s}{\partial t} = K_a(q_s - C_s)C - K_d C_s \qquad (3)$$

where:

q_s is the column saturation capacity (in the same units as C_s),

K_a and K_d are the rate constants of adsorption and desorption. This model was used by Thomas [7], Goldstein [6], and Wade et al. [31].

The linear kinetic model:

$$\frac{\partial C_s}{\partial t} = K_1 C - K_2 C_s \tag{4}$$

where K_1 and K_2 are rate constants. This model was used by Lapidus and Amundson [11].

The linear driving force model:

$$\frac{\partial C_s}{\partial t} = K_f(q - C) \tag{5}$$

where K_f is a rate constant. This model was used by Glueckauf and Coates [32], Hiester and Vermeulen [33], Lin et al. [34], Golshan-Shirazi et al. [35], and Yu et al. [29,30].

The properties, advantages, and drawbacks of these models are discussed in Sections II.E-G.

C. Initial and Boundary Conditions

These conditions define mathematically the way the experiment is performed. The initial conditions describe the state of the columns when the experiment begins, i.e., at $t = 0$. For example, in conventional elution (no mobile phase additive), gradient elution, frontal analysis, or displacement, the column is filled with a weak mobile phase which does not participate in the equations (i.e., the reference state of adsorption assumes that the solvent is not adsorbed). Hence, in general the initial condition is:

$$C_i(z,0) = 0 \tag{6a}$$

where C_i is the concentration of the ith component of the feed. In elution when the mobile phase contains a competitive additive, the column contains a constant concentration of this component. The initial condition is:

$$C_a = C_a^0 \tag{6b}$$

The boundary condition describes in mathematical terms what enters the column as specified by the mode of operation of the chromatographic column. In the elution mode, the sample is injected as a pulse. In analytical applications, only a small amount can be introduced as Dirac δ pulses. In preparative applications, the injection lasts a finite amount of time and the boundary condition for elution is:

$$C_i(0,t) = \phi(t) \tag{7}$$

where $\phi(t)$ is the injection profile. If the injection is performed in a mobile phase containing an additive, the boundary condition for this additive is usually:

$$C_a(0,t) = C_a^0 \tag{8}$$

Note that the combination of Eqs. 6b and 8 does not mean that $C_a(z,t)$ is constant over the entire length of the column.

In gradient elution or displacement, the boundary condition given by Eq. 7 applies to the sample component. However, there is an additive, the strong solvent (gradient elution) or the displacer (displacement). In gradient elution the boundary condition for the additive is:

$$C_a(0,t) = C_a^0(z,0) + \phi_a(t) \tag{9}$$

where:
 C_a^0 is the concentration of the additive at the injection time, usually constant over the entire column, and
 $\phi_a(t)$ is the gradient profile, linear, exponential, or other.

In displacement, the boundary condition for the feed components is given by Eq. 7 while the boundary condition for the displacer is:

$$\begin{aligned} C_d(z,t) &= 0 & t &\leq \Delta t_i \\ C_d(0,t) &= C_d^0 & t &> \Delta t_i \end{aligned} \tag{10}$$

where Δt_i is the duration of the sample injection, the elapsed time before the displacer can be pumped into the column.

In frontal analysis, the boundary condition for the sample is:

$$\begin{aligned} C_i(0,t) &= 0 & t &\leq 0 \\ C_i(0,t) &= C_i^0 & t &> 0 \end{aligned} \tag{11}$$

For other experiments, more complicated boundary conditions may be necessary. There usually are no difficulties in translating the experimental procedure into a boundary condition.

D. The Multicomponent Case and the Competitive Equilibrium Isotherms

In the case of a multicomponent sample, we must write the mass balance, the competitive equilibrium isotherm, as necessary, the kinetic equation, and the initial and boundary conditions for each component. Usually, there are only two initial and boundary conditions, one for all the sample components and one for the additives. Besides the mathematical complexity which increases rapidly with the number of partial differential equations, the difficulty of the problem comes from the coupling of these equations. The origin of the coupling is in the competition between the different components (the sample components and the additives) for interaction with the stationary phase. It must be noted, however, that some additives, e.g., acetonitrile used in gradient elution, do not act by competition for retention but by increasing the solubility of the components in the mobile phase. The retention factor (k_0' under linear conditions) of acetonitrile on octadecyl silica in pure water is of the order of 1. Thus, acetonitrile cannot displace in practice any organic compound from a C18 phase in RPLC, although it can elute many components on such a column or act as a regenerant because it is an excellent solvent.

The equilibrium isotherm of most pure compounds in most chromatographic systems is well accounted for by the Langmuir isotherm [36]:

$$q = \frac{aC}{1 + bC} \quad (12)$$

where a and b are numerical coefficients. Depending on the units chosen, the limit retention factor at infinite dilution is either $k_0' = a$ or $k_0' = Fa$. This isotherm indicates that there is a saturation limit, the column saturation capacity q_s, which is proportional to a/b, the coefficient of proportionality being the amount of packing material used to pack the column. Since the Langmuir isotherm model is widely applicable, it is normal to take the column saturation capacity as a unit to measure the degree of column overload achieved, by reporting the actual amount injected as a fraction of the column saturation capacity. The loading factor is defined for each component as the ratio of its amount in the sample to the column saturation capacity for that component:

$$L_{f,i} = \frac{n_i}{q_s} = \frac{n_i b_i}{SLk_{0,i}'} \quad (13)$$

where:

n_i is the amount of the ith component injected,

S and L are the column geometrical cross-section area and length, respectively, and

$k_{0,i}'$ is the retention factor for the component i, at infinite dilution (linear conditions).

Equation 12 has been established on firm theoretical ground for gas-solid adsorption, a case where there is no competition between the adsorbate and the mobile gas phase. On the contrary, in liquid-solid adsorption, there is strong competition between the molecules of any component and those of the solvent for adsorption. Although we can choose a convention canceling the apparent effect of this competition, the conditions of validity of Eq. 12 are not met anymore. These conditions are that (a) the solution be ideal; (b) the solute give monolayer coverage; (c) the adsorption layer be ideal; (d) there be no solute-solute interactions in the monolayer; and (e) there be no solvent-solute interactions. These conditions cannot be valid in liquid-solid adsorption, especially at high concentrations. Nevertheless, Eq. 12 is an excellent approximation for single-compound adsorption equilibrium in LSC. In the case of proteins, however, desorption hysteresis has been reported, which results in a murky situation, not easy to handle theoretically or empirically.

For multicomponent systems, the Langmuir theory of equilibrium isotherm has been extended [37]. For the ith component of the system, the competitive isotherm is written:

$$q_i = \frac{a_i C_i}{1 + \sum_{j=1}^{n} b_j C_j} \quad (14)$$

where n is the number of components in the system. The coefficients a_i and b_i are the coefficients of the single-component Langmuir equilibrium isotherm for the component i (see Eq. 12).

Unfortunately, experimental results available suggest that the column saturation capacity is not the same for the components of a binary mixture, so Eq. 14 does not account accurately for the competitive adsorption behavior of these components. The deviation is significant and more sophisticated models seem necessary. Using the ideal adsorbed solution model [38], LeVan and Vermeulen have derived a competitive binary isotherm equation which accounts for differences in the column saturation capacities for the two components [39]. Statistical thermodynamics suggests that a more accurate prediction could be obtained using the ratio of two second-degree polynomials instead of the ratio of two first-degree polynomials as in Eq. 14. In the case of a binary mixture, the equation obtained is [40]:

$$q_i = \frac{q_s(b_i C_i + b_3 C_1 C_2 + 2\beta_i C_i^2)}{1 + b_1 C_1 + b_2 C_2 + b_3 C_1 C_2 + \beta_1 C_1^2 + \beta_2 C_2^2} \quad (15)$$

This equation contains 6 parameters instead of four. Attempts at reducing the number of independent parameters as well as at

E. The Ideal Model

The simplest chromatography model was formulated by Wilson [1]. It assumes that the column efficiency is infinite. There is no axial dispersion and the two phases are constantly at equilibrium. Then, the mass balance equation (Eq. 1) simplifies to:

$$\frac{\partial C}{\partial t} + F \frac{\partial C_s}{\partial t} + u_0 \frac{\partial C}{\partial z} = 0 \tag{16}$$

This equation neglects completely the influence of the mass transfer kinetics and of the axial dispersion and focuses the attention on the influence of the nonlinear thermodynamics of phase equilibria on the individual band profiles. Equation 16 for a single component can be solved in closed form for many equilibrium isotherms [4,41,42] and for one kinetic expression [31]. For a binary mixture, Eq. 16 is written for each component and can be solved in closed form almost completely only in the case of a Langmuir isotherm [3,43]. In all other cases, numerical solutions have to be calculated, which is difficult for an ideal model but becomes easy for real columns because all numerical methods introduce some degree of numerical diffusion.

The band profiles determined by the ideal model are in excellent agreement with experimental chromatograms at high loading factors. On the contrary, the agreement with experimental results is poor in the case of low efficiency columns *and* when the loading factor is low or moderate [44].

F. The Semi-ideal Model

It has been shown by Giddings [8] and by Haarhoff and Van der Linde [10] that when the mass transfer kinetics is fast but not infinitely fast, the system of the mass balance equation (Eq. 1) and the kinetic equations (Eqs. 2-5) can be replaced by the following equation:

$$\frac{\partial C}{\partial t} + F \frac{\partial C_s}{\partial t} + u_0 \frac{\partial C}{\partial z} = D_{ap} \frac{\partial C}{\partial z} \tag{17}$$

where the apparent diffusion coefficient D_{ap} is given by:

$$D_{ap} = \frac{HL}{2t_0} = \frac{Hu_0}{2} \tag{18}$$

where:

H is the column height equivalent to a theoretical plate for the component considered, and

t_0 is the hold-up time of the column ($t_0 = L/u_0$) or retention time of a mobile phase tracer.

Equation 18 stems from the relationship giving the variance of a Gaussian peak obtained in chromatography, under linear conditions, for a Dirac pulse injection:

$$\sigma^2 = HL = 2D_{ap}t_0 \qquad (19)$$

The semi-ideal model assumes that the diffusion coefficients of the solutes in the mobile phase remain constant, independent of the concentration of the sample components. As concentrations in nonlinear chromatography remain low, rarely exceeding a few percentage points, this is a reasonable assumption. However, for high polymers and especially for proteins, the viscosity of the mobile phase containing the protein bands can be much higher than the viscosity of the pure mobile phase. Fingering may take place, leading to a dramatic decrease in the column efficiency or even to the formation of satellite peaks. Under such circumstances, the semi-ideal model does not remain valid. There is no known solution to the semi-ideal model in closed form. Numerical solutions are easily obtained using the computation methods such as finite differences or finite elements [44-46]. In the former case, extensive results are available on the band profile and how it changes with various experimental conditions [24]. Many comparisons have also been made with experimental data [47-50].

G. Kinetic Models

These models combine the mass balance equation (Eq. 1) with one of the Eqs. 3-5, relating the rate of variation of the concentration of each component in the stationary phase to its concentrations in both phases and the equilibrium concentration in the stationary phase [6-8,11-13,31,33-35].

While in principle these models are more exact, the difference between the individual band profiles calculated using the semi-ideal model or the linear driving force model is negligible when the rate constants are large (i.e., correspond to a column efficiency exceeding a few hundred theoretical plates).

III. RESULTS AND DISCUSSION

A. Single-Component Problems

The solution of the single-component problem does not contribute directly to the understanding of the separation mechanism in chromatography. However, the theoretical analysis of the problem provides some of the fundamental concepts in nonlinear chromatography, such as the notions of velocity associated to a concentration, concentration shocks, shock layers, and diffuse boundaries [13]. It also provides an understanding of the relationship between the shape of the isotherm (i.e., concave up, linear, concave down, or S-shape) and the band profile.

For the single-component problem, the possible boundary conditions which permit solutions are the elution of a pulse by a pure or a mixed solvent (when the strong solvent is not much retained, e.g., acetonitrile in reversed phase HPLC) and frontal analysis. All other chromatographic procedures require two or more solutes. In the case of the solution to the elution problem, the ideal model, the semi-ideal model, and various kinetic models will be discussed.

Elution Mode and the Ideal Model

The mathematical description of the single-component problem was first written by Wilson in 1940 in terms of a differential mass balance [1]. In this work, the ideal model of chromatography was used; thus the column was assumed to have an infinite efficiency (i.e., no axial dispersion, $H = 0$) and the two phases were assumed to be always in equilibrium (Eq. 16 and Table 2, Eq. 7). The Langmuir equilibrium isotherm (Eq. 12 and Table 1, Eq. 4) was used.

In 1943, DeVault made a more rigorous study on the properties of the partial differential equation of chromatography derived by Wilson and demonstrated the formation of concentration shocks (discontinuity) in the cases of a pulse and a step injection [2]. Weiss described in detail the progressive change of the band profile during its migration for a Langmuir (Eq. 12 and Table 2, Eq. 1) and for a Freundlich isotherm (Table 2, Eq. 3) [51].

The theory of chromatography expanded rapidly in the late 1940s. The plate theory [52], the stage model [53], and the solutions of the partial differential equations were developed in parallel [3-5]. Progress in the latter area were hindered, however, by the slow evolution of the theory of hyperbolic systems of partial differential equations. The proper mathematical tools appeared only in the 1950s and were applied to the study of the chromatographic problem much later [13].

Glueckauf [4,5] was one of the early investigators who made major experimental and theoretical contributions in all modes of chromatography. For the single-component problem, he measured the single-component isotherm and showed experimentally and theoretically the development of the band profile in a pure solvent, when the isotherm is concave up, concave down, and sigmoidal. For a convex upward isotherm, the amount at equilibrium in the stationary phase increases less rapidly than the concentration in the mobile phase. The high concentrations are less retained than the low ones and they tend to move faster. Thus, the profiles have a sharp front and a diffuse rear boundary. The converse is true for a convex downward isotherm. The elution profile of the bands have a diffuse front and a sharp rear boundary. For a sigmoidal isotherm, the profile depends on the final concentration with respect to the inflection point. Thus, the single-component problem was solved within the framework of the ideal model as early as 1949. Unfortunately, the method used by Glueckauf and his choice of symbols make his derivation and results extremely difficult to understand and use. For this reason and for the lack of interest in the preparative applications of chromatography in the 40 years which followed, there were few developments in the field.

A solution of the ideal model in the case of a Langmuir isotherm is contained in the book by Aris and Amundson [13]. It was recently reformulated and used for the determination of adsorption isotherms [41]. A simple, general solution of the ideal model, valid for any isotherm was published in 1990 by Golshan-Shirazi and Guiochon [42]. It gives an algebraic equation which can be solved in closed term for the Langmuir isotherm and for a few other isotherms, including the bi-Langmuir isotherm. The diffuse profile of the band is given by the following equation derived by Wilson [1]:

$$t = t_p + \frac{L}{u_0}\left(1 + F\frac{dq}{dC}\right) \qquad (20)$$

where t_p is the width of the injection profile, assumed to be a rectangular pulse. The maximum concentration of the profile C_M is obtained by writing that the peak area is constant and equal to the area of the injected rectangular pulse [42]. It is given by the equation:

$$\left|q(C_M) - C_M\frac{dq}{dC}\right| = \frac{n}{F_v t_0 F} \qquad (21)$$

where F_v is the volume flow rate of the mobile phase. A close form solution of Eq. 21 requires knowing the equilibrium isotherm equation.

Figure 1 describes the profiles for increasing sample sizes calculated as solutions to the ideal model in the case of a Langmuir isotherm [41]. A system of reduced coordinates is used for the presentation of the profiles. It has been shown that in this dimensionless plot the band profiles corresponding to any Langmuir isotherm all have the same rear profile. Only the position of the front shock depends on the value of the loading factor corresponding to the sample size used.

Especially important are the solutions of the ideal model for the Langmuir isotherm, which most pure components seem to follow, and for the bi-Langmuir isotherm. This later model seems justified by

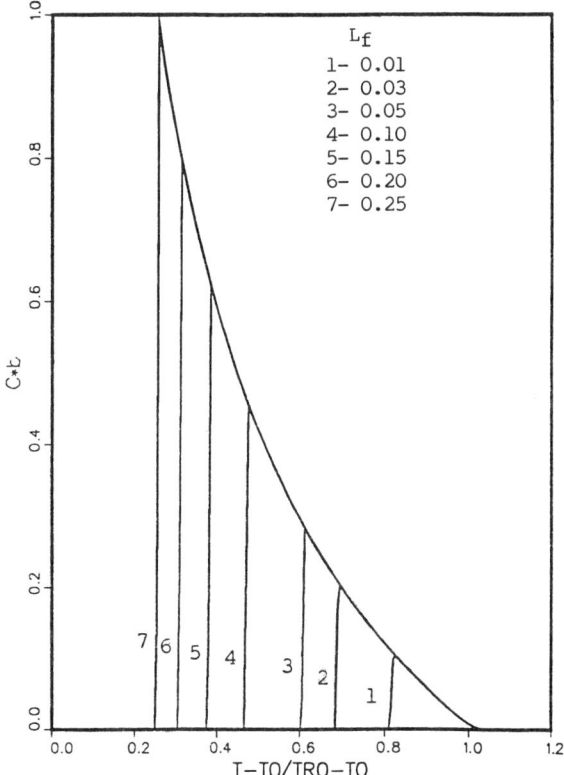

Fig. 1 Dimensionless chromatograms at different loading factors for a Langmuir isotherm. Plots of bC vs. $k'/k_0' = t - t_0/t_{R,0} - t_0$. (Reproduced from Ref. 41 with permission of the American Chemical Society.)

the existence of a bimodal energy distribution on the surface [54,55]. It accounts well for adsorption data determined in reversed phase liquid chromatography, for example. The profiles obtained with this isotherm exhibit a long tail at the end of the large concentration band, a profile which is familiar to many chemists.

Figure 2 shows the comparison made by Wilson between the band profile calculated using the ideal model and an experimental profile recorded for lauric acid eluted on charcoal [1]. This first experimental verification of chromatographic modeling shows a good agreement. Figures 3 and 4 show a comparison between experimental and theoretical band profiles obtained at increasing sample sizes, using high-performance material, in normal and reversed phase chromatography, respectively [49]. The primary difference between

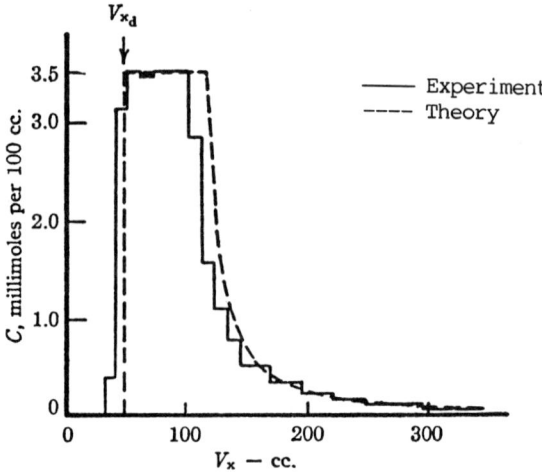

Fig. 2 Comparison of theory and experiment for lauric acid on charcoal. Mobile phase: petroleum ether. Solute concentration, C_0: 0.35 mM. Column: 1 g of carbon. (Reproduced from Ref. 1 with permission of the American Chemical Society.)

experimental and calculated profiles is due to band broadening not accounted for by the ideal model. The solution of the ideal model gives excellent agreement with experimental results when the loading factor exceeds ~3% and the column efficiency is more than about 1000 theoretical plates [44].

Linear Chromatography

Martin was the first to develop the plate theory, which assumes a linear isotherm, divides the column into a series of plates, and writes a mass balance for each plate [52]. In the mid-1940s, Craig developed an apparatus and a method to describe the distribution of solutes in linear chromatography (LLC) based on a multiple-step extraction process where each stage is considered as an ideal mixer [53]. Mobile phase is transferred from one vessel to the next as new mobile phase is added to the first vessel. The Craig machine gives the binomial distribution as the response to a pulse injection in linear chromatography, the same result obtained by Martin. It should be noted that the flow is intermittent in a Craig separation machine, since the content of each vessel must be equilibrated between successive transfers of fluid (mobile phase) from one mixing vessel to the next. The continuous flow model, however, leads to a Poisson-type distribution [56]. For a large number of stages, both the Poisson and the binomial distributions may be approximated

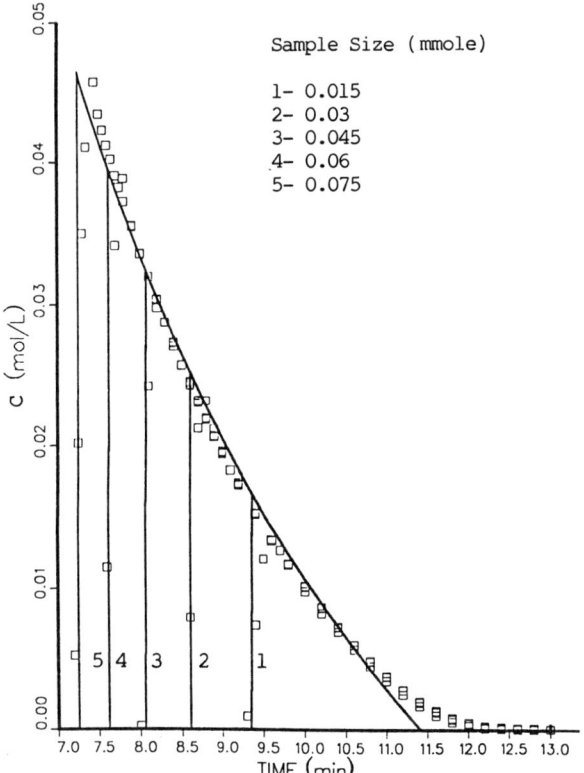

Fig. 3 Comparison of experimental data in reversed phase chromatography and theoretical profiles calculated with the ideal model. Solute: phenol; mobile phase: 20:80 MeOH/water; flow rate: 1 ml/min. Column 25 × 0.46 cm, packed with Vydac ODS silica, 20 μm. (Reproduced from Ref. 49 with permission of the American Chemical Society.)

by a Gaussian distribution function. The difference between these models has been discussed by Klinkenberg and Sjenitzer [57] and by Keulemans [56]. Villermaux also showed the equivalence between the plate and the continuous model [58]. Although the Craig distribution model was used primarily to study linear chromatography, it can easily be extended to nonlinear systems by writing the appropriate isotherm equation [59,60].

The Craig model lumps all contributions to band broadening in a stage. Many other developments were made which describe in more detail the contributions to band broadening [61] and more

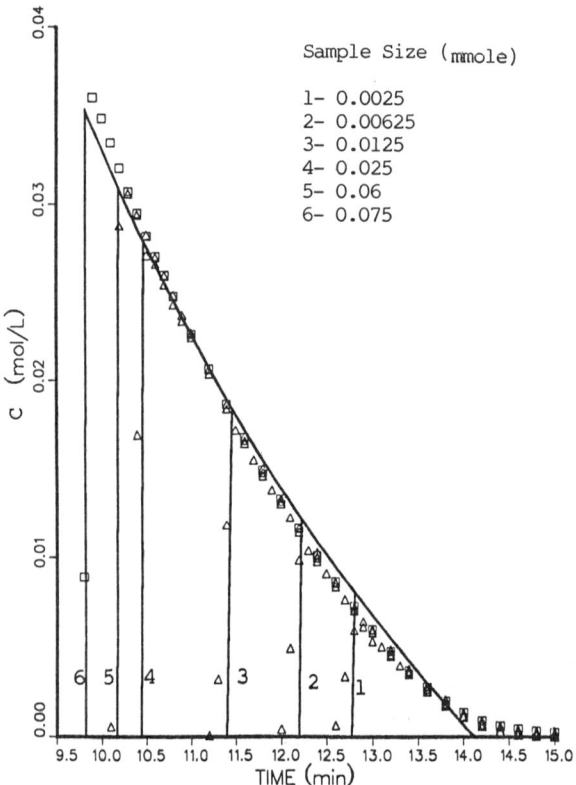

Fig. 4 Comparison of experimental data in normal phase chromatography and theoretical profiles calculated with the ideal model. Solute: benzyl alcohol; mobile phase: 15:85 THF/n-heptane; flow rate: 1 ml/min. (Reproduced from Ref. 49 with permission of the American Chemical Society.)

recently by Horvath and Lin [62] and Huber [63]. The detailed analysis by Giddings [8] cannot be summarized in the limited scope of this work.

In linear chromatography, the variances of the different contributions to band broadening are additive because they are the result of a shift invariant convolution of the thermodynamic (i.e., ideal) band profile [8,64]. In nonlinear chromatography, the convolution becomes shift variant and the variances are no longer additive [64].

Elution Mode and the Semi-ideal Model

Models which assume equilibrium but account for band broadening in a lumped diffusion coefficient are called semi-ideal models [44]. For example, the Craig model is a semi-ideal model. This section presents a review of several semi-ideal models in the literature, and of the solutions proposed by Houghton [9], Haarhoff and Van der Linde [10], Craig [53], and Guiochon et al. [24,41-46]. In these models, we assume that the diffusion coefficients are constant, i.e., independent of the concentration and that the contribution to band broadening can be lumped. Equation 17 and equation 5, Table 2 show the mass balance for the semi-ideal models. A detailed analysis of the mathematical problem and of its different approximate solutions in closed form and numerical solutions was recently published [65].

Houghton simplified the mass balance equation by assuming that the loading factor is small and the equilibrium isotherm can be replaced by its two-term expansion [9]. Using a parabolic isotherm, the mass balance equation becomes:

$$\frac{\partial C}{\partial t} + \frac{\Lambda u C}{(1 + k_0')(1 + \Lambda C)} \frac{\partial C}{\partial \xi} = \frac{D}{(1 + k_0')(1 + \Lambda C)} \frac{\partial^2 C}{\partial \xi^2} \tag{22}$$

with:

$$\Lambda = -2b \frac{k_0'}{1 + k_0'} \tag{23a}$$

$$\xi = L \frac{t_{R,0} - t}{t_{R,0}} \tag{23b}$$

Equation 22 cannot be solved in closed form without a further simplification. Houghton neglected ΛC in the denominator of the last two terms of Eq. 22. This is incorrect to the first order and the modified equation obtained no longer conserves mass [44]. Hence, the area of the Houghton profile is neither proportional to the sample amount injected nor independent of the curvature of the isotherm (coefficient b). Haarhoff and Van der Linde [10] showed the equivalence between the coefficient D in Eq. 22 and the apparent diffusion coefficient (Eq. 18). These authors suggested that the effect of dispersion on the band profile can be calculated at the limit retention time $t_{R,0}$. This corrects the Houghton equation for its lack of mass conservation. The solution proposed by Haarhoff and Van der Linde, also based on a parabolic isotherm, gives a profile whose equation can be written in dimensionless form as:

$$X = \left| \frac{e^{-\tau^2/2}}{\sqrt{2\pi}(\coth m + \mathrm{erf}(\tau/\sqrt{2}))} \right| \tag{24}$$

where X is the dimensionless concentration, τ is the dimensionless time, and m is the dimensionless sample size, defined as follows:

$$X = bC \frac{k_0'}{1 + k_0'} \sqrt{N} \qquad (25a)$$

$$\tau = \sqrt{N} \frac{k_0'}{1 + k_0'} \frac{t - t_{R,0}}{t_{R,0} - t_0} \qquad (25b)$$

$$m = N \left[\frac{k_0'}{1 + k_0'} \right]^2 L_f \qquad (25c)$$

where N is the column efficiency (i.e., number of theoretical plates under linear conditions, determined at low sample size) and L_f is the loading factor defined for the Langmuir isotherm (see Eq. 13) which osculates the parabolic isotherm (i.e., has the same tangent and curvature at the origin). It is worth noting at this stage that

$$X_M = \frac{\tau_M}{2} \qquad (26a)$$

$$k_M' = k_0'(1 + 2bC_M) \qquad (26b)$$

where X_M, τ_M, k_M', and C_M are the reduced concentration, reduced time, retention factor, and concentration of the band maximum, respectively.

Golshan-Shirazi and Guiochon [44] showed that the Haarhoff and Van der Linde profile equation can be derived also by replacing the term ΛC in Eq. 22 by the value derived from the analytical solution of the ideal model (Eq. 20):

$$C = \frac{1}{2b} \frac{t_{R,0} + t_p - t}{t_{R,0} - t_0} \equiv \frac{1}{2b} \frac{t_{R,0} - t}{t_{R,0} - t_0} = \frac{k_0 - k'}{2bk_0'} \qquad (27)$$

Hence:

$$1 + \Lambda C = \frac{1 + k'}{1 + k_0'} = \frac{t}{t_{R,0}} \qquad (28)$$

This simplification is correct to the first order and conserves mass. This amounts to a form of convolution of the dispersion due to the finite column efficiency and of the band profile due to the nonlinear behavior of the isotherm.

Because the parabolic isotherm is acceptable only at low sample sizes, the analytical solution (Eqs. 24 and 25) is valid only for loading factors below ~0.2%. Golshan-Shirazi and Guiochon compared the profiles obtained with the ideal model, the Houghton [9] and Haarhoff-Van der Linde [10] solutions, and the numerical solutions

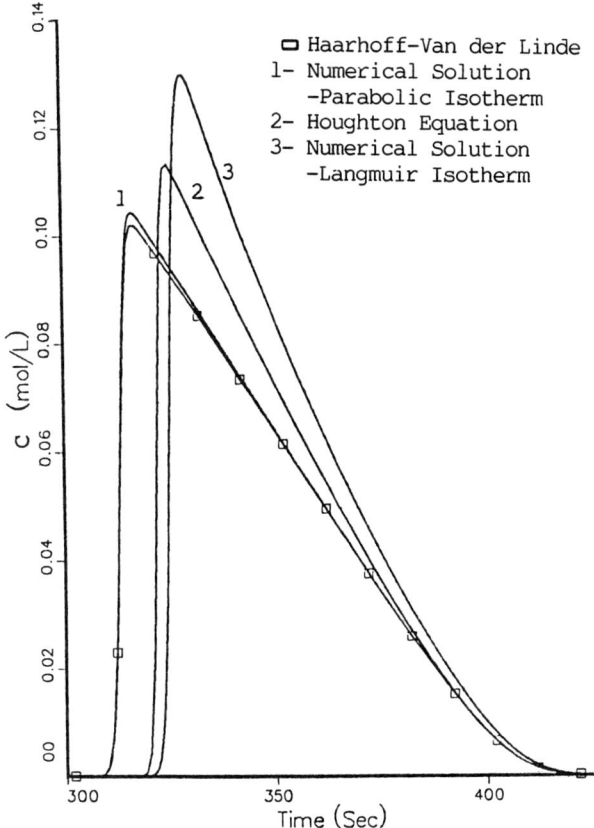

Fig. 5 Comparison of the band profiles predicted by the Haarhoff-Van der Linde and the Houghton equations and by solutions of the Rouchon semi-ideal model of chromatography. The parabolic isotherm is the first two-term expansion of the Langmuir isotherm with parameters $a = 20$, $b = 5$. Column length: 25 cm; phase ratio: 0.25. $t_0 = 200$ sec, $N = 2500$, $L_f = 5\%$. (Reproduced from Ref. 44 with permission of Elsevier.)

of the semi-ideal model under conditions of increasing mass loaded (Fig. 5) (loading factor 5%) [44]. The semi-ideal model for a parabolic isotherm (curve 1) and the Haarhoff and Van der Linde profiles (squares) are nearly coincidental. The difference between the semi-ideal profile for a Langmuir isotherm (curve 3) and curve 1 is due to the fact that the parabolic isotherm approximates the Langmuir isotherm only in a limited concentration range. This explains

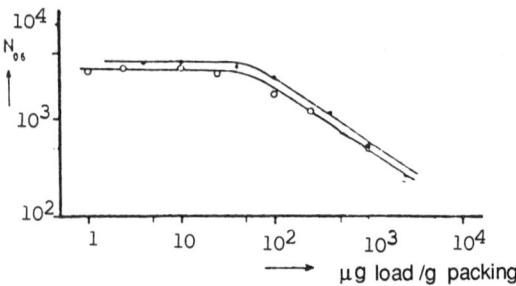

Fig. 6 Plot of the column plate number, $N_{0.6}$, vs. the specific loading of 2,4-dimethylphenol. $N_{0.6}$ is calculated from the band with at 60% of its height. Capacity factor: 0.6; solvent: dichloromethane; columns: o, 100 cm (50 cm × 10.8 mm + 50 cm × 10 mm), packed with 39.5 g of 20-25 μm Si 60. •, 10 cm × 10.8 mm, 4.9 g of 5-8 μm Si 60. (Reproduced from Ref. 66 with permission of Elsevier.)

the limited range of validity of the Haarhoff and Van der Linde profile, as most solutes have a Langmuirian behavior. The Houghton profile (curve 2), because of the simplifications made in the calculations, is not an exact solution for either the parabolic or the Langmuir isotherm. It turns out to be intermediate between the two profiles. It has been shown that the Haarhoff-Van der Linde solution gives a band profile which is reasonably close to the correct one for a Langmuir isotherm for loading factors up to around 1% for very efficient columns and for $bC_M < 0.1$ for all real columns [44]. The Houghton profile is less satisfactory because it does not conserve mass and gives only an approximate band shape.

The contributions to the column HETP have been studied in the nonlinear case [28,41,49,60,66] by lumping all the sources of band broadening, kinetic and thermodynamic, into the column efficiency. These authors have presented correlations between the apparent column efficiency and the sample size. A typical example is given in Fig. 6 [66]. This assumes that kinetic and thermodynamic factors contribute to the total HETP and that the bandwidth can be written [28]:

$$W_{tot}^2 = W_{th}^2 + W_{kin}^2 \tag{29}$$

where:

W_{tot} is the total bandwidth at a certain fractional height, usually the half height,

W_{th} is the bandwidth contribution due to thermodynamics, i.e., to a nonlinear isotherm, and

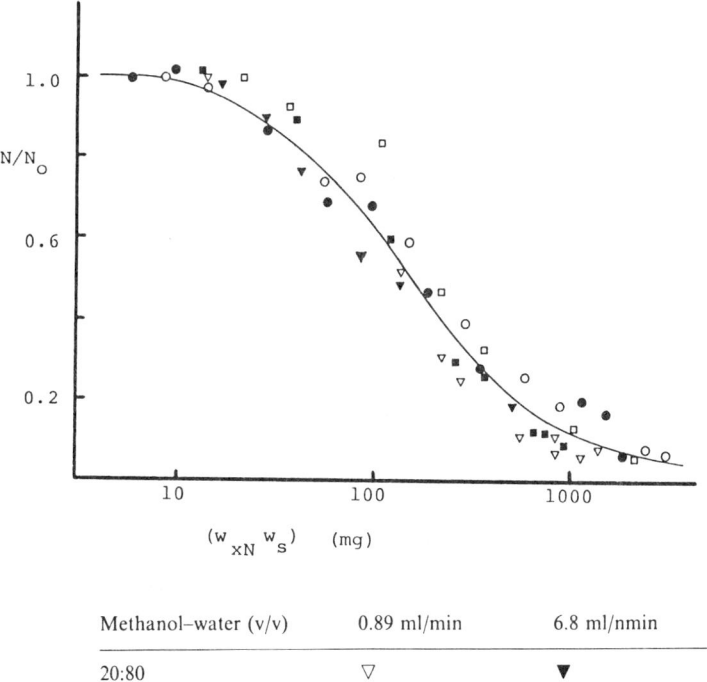

Fig. 7 Dependence of N/N_0 on $w_{xN}w_s$. $w_{xN} = N_0[k_0/(1 + k_0)]^2 (w_x/w_s)$, w_x is the sample size; w_s is the column saturation capacity; $w_x/w_s = L_f$. Correlation curve with w_s = 30 mg. Symbols, experimental data. Solute: benzyl alcohol; column: 5 × 0.46 cm, packed with ODS Zorbax. (Reproduced from Ref. 67 with permission of Elsevier.)

W_{kin} is the bandwidth contribution due to the mass transfer resistances and to axial dispersion.

Equation 29 assumes that the band variance in nonlinear chromatography is the sum of the contributions due to the nonlinear behavior of the isotherm and of the band variance in linear chromatography. This is in fact incorrect as we have shown that the convolution of the nonlinear behavior and the mass transfer resistances is shift variant [64]. As a first approximation, at low values of the loading factor, the approximation remains acceptable.

The difficulty encountered in finding analytical solutions to the semi-ideal model led to the investigation of numerical solutions. Recently, Eble et al. [60] used the Craig model to calculate band profiles in isocratic elution and to develop general correlations between

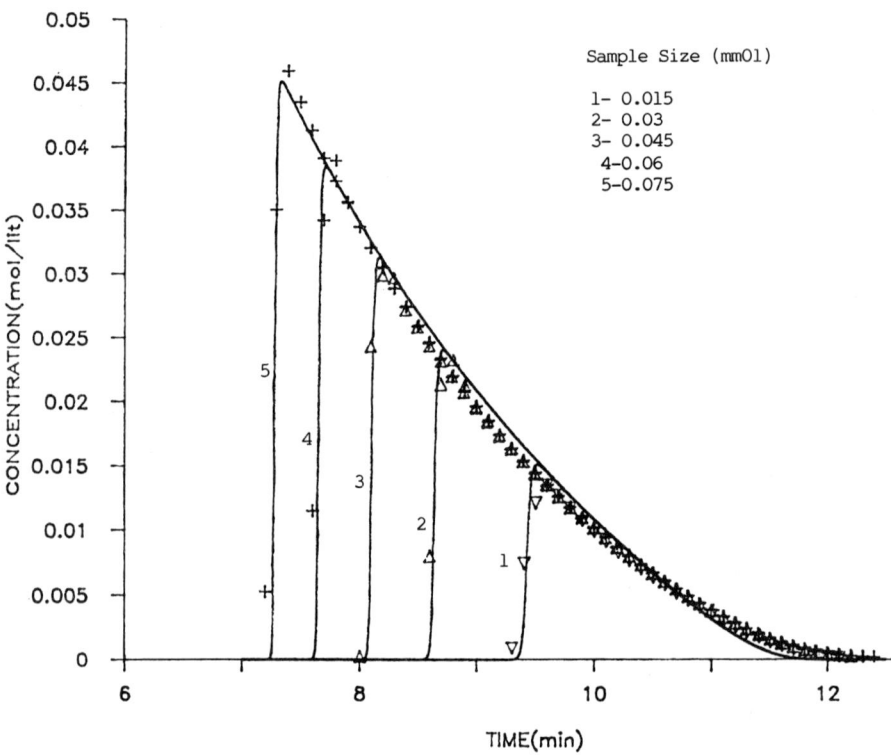

Fig. 8 Comparison of experimental data in reversed phase chromatography and predictions by the semi-ideal model. Solute: phenol; mobile phase: 20:80 MeOH/water; flow rate: 1 ml/min; HETP: 50 μm. Column 25 × 0.46 cm, packed with Vydac ODS silica, 20 μm. (Reproduced from Ref. 48 with permission of the American Chemical Society.)

the capacity factor and the efficiency vs. the sample loading. Experimental data confirm the approximate validity of the relationships obtained [67,68] (Fig. 7). Use of such empirical relationships allowed estimation of the band shape on a personal computer for column efficiencies not exceeding a few hundred theoretical plates.

With the advent of fast computers, more exact numerical solutions have appeared. A numerical solution of the semi-ideal model was developed [45] and used to predict band profiles for different sample size having measured the HETP, flow rate, void volume, extra column volume, column length, and adsorption isotherm. Excellent agreement between the band profiles recorded for large sample amounts and the profiles predicted by the semi-ideal model has

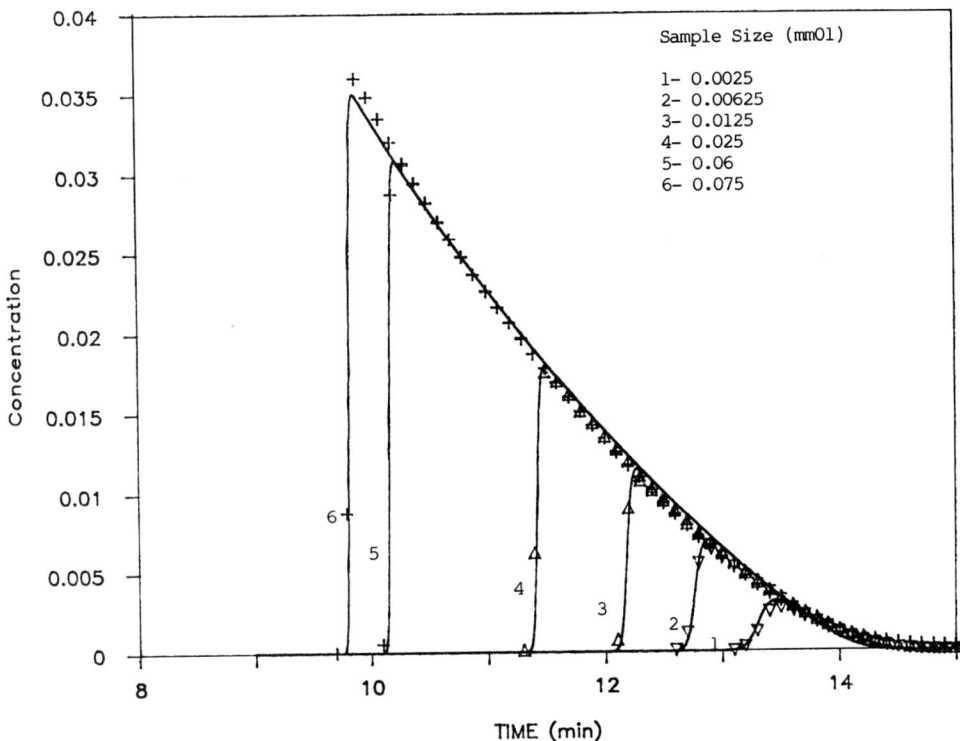

Fig. 9 Comparison of experimental data in normal phase chromatography and predictions by the semi-ideal model. Solute: benzylalcohol; mobile phase: 15:85 THF/n-heptane; flow rate: 2 ml/min; HETP: 50 μm. Column 25 × 0.46 cm, packed with Nucleosil silica, 15-25 μm. (Reproduced from Ref. 48 with permission of the American Chemical Society.)

been reported in many cases, when the equilibrium isotherms are determined with the same column which is used to record the high concentration bands [47-50,60-71]. Three examples are given in Figs. 8-10, in normal and reversed phase chromatography [48,72]. In all these cases, the Langmuir isotherm equation accounts well for the equilibrium isotherm. In contrast with the ideal model, the semi-ideal model accounts virtually exactly for the band broadening of the front and rear boundaries. Figure 11 illustrates the influence of the column efficiency on the elution band profiles by showing a comparison between profiles calculated using the ideal and the semi-ideal model for a 1000-plate column, in the case of a bi-Langmuir isotherm [42].

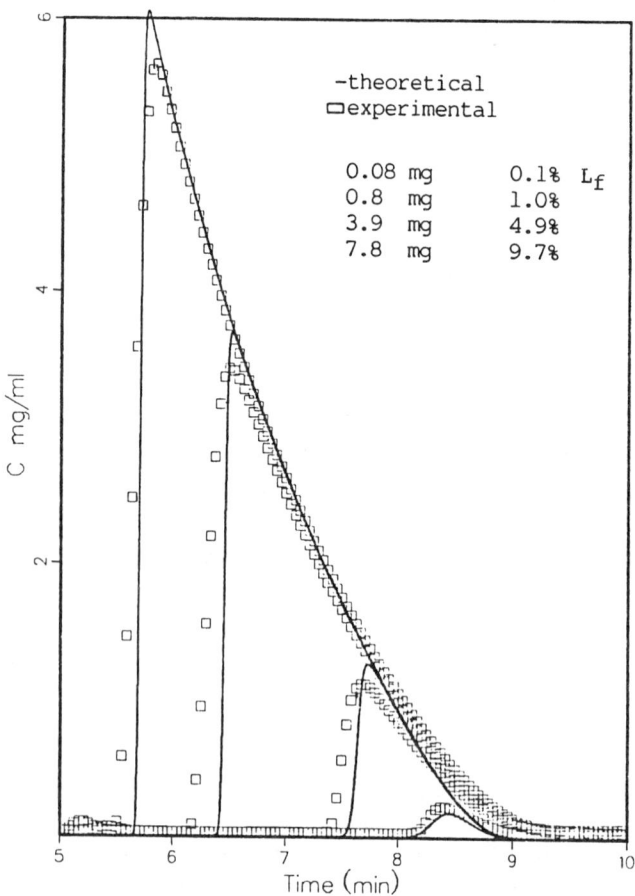

Fig. 10 Comparison of experimental data in reversed phase chromatography and predictions by the semi-ideal model. Solute: 2,6-dimethylphenol; mobile phase: 50:50 MeOH/water; flow rate: 1 ml/min; HETP: 90 µm. Column: 25 × 0.46 cm, packed with Vydac ODS silica 10 µm. (From Ref. 72.)

Less satisfactory results were obtained in the case of chicken albumin and conalbumin eluted on a weak anion exchanger [73]. Deviations were observed between the recorded profiles and those calculated using the semi-ideal model. The influence of the mass transfer kinetics, not properly accounted for by the semi-ideal model, probably explains the differences reported.

The above model was used to describe the profile of a single component in the presence of a binary eluent containing a strongly

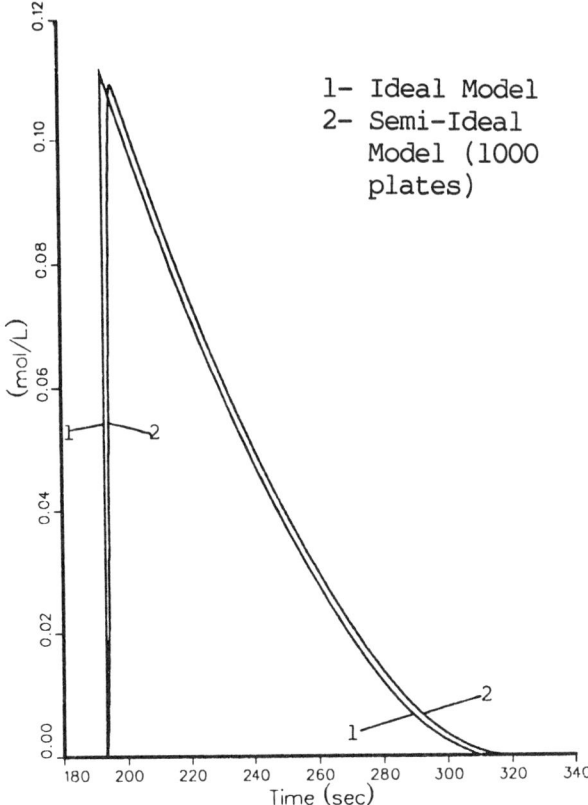

Fig. 11 Comparison of the calculated solutions by the ideal model and the semi-ideal model for a bi-Langmuir isotherm. The calculation with the semi-ideal model assumes 1000 plates. (Reproduced from Ref. 42 with permission of the American Chemical Society.)

adsorbed additive [74]. The appearance of system peaks was observed and the band shapes are in agreement with experimental results [75]. An inversion of the band shape appeared in the presence of a strong solvent. However, since a mass balance must be written for the strong solvent and for the solute, this is considered a two-component problem and is discussed in detail in Section III.C.

The mathematical properties of the solutions of the semi-ideal model are discussed in the next section. Also reported is a discussion of the problems of numerical analysis encountered in the development and use of this algorithm for the computation of the solution.

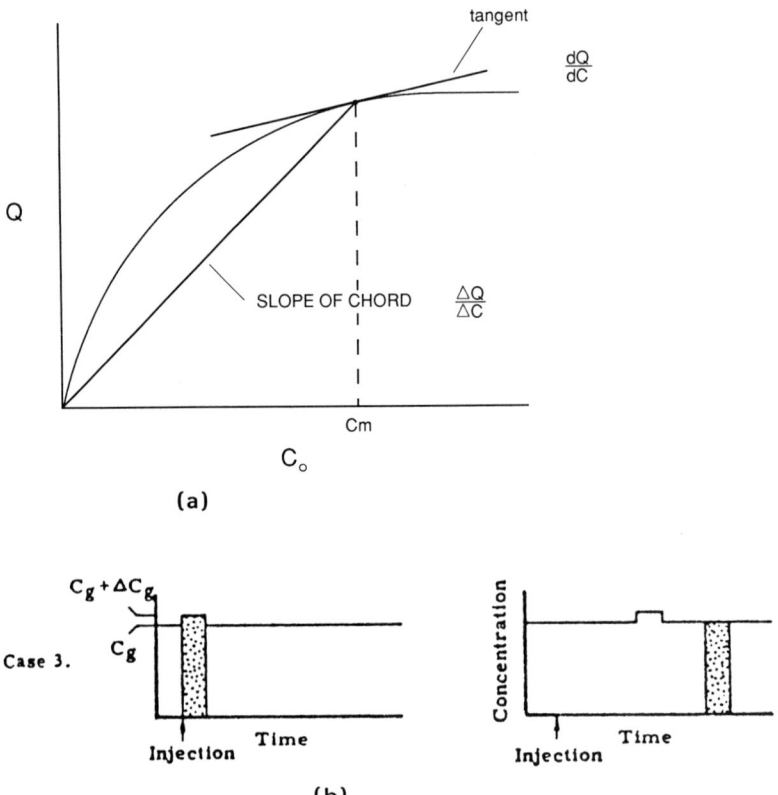

Fig. 12 Schematic describing the velocity of a concentration perturbation and the velocity of molecules. (a) Langmuir adsorption isotherm. The perturbation velocity is related to the slope of the chord (Eq. 32), the molecule velocity to the slope of the tangent (Eq. 31). (b) Elution of tagged molecules (shaded) on a concentration pulse. (Reproduced from Ref. 78 with permission of the American Chemical Society.)

Numerical Analysis of the Ideal and Semi-ideal Model

In practice, elution profiles are better obtained by numerical calculation of solutions of the mass balance equation of chromatography than by using one of the approximate solutions [76]. Writing a computer program which gives both accurate results and a short execution time requires a good understanding of the mathematical properties of the partial differential equation studied.

Equation 16 may be rewritten:

$$\frac{\partial C}{\partial t} + \frac{u_0}{1 + F(dq/dC)} \frac{\partial C}{\partial z} = 0 \quad (30)$$

This equation shows that a velocity u_z is associated with any given value of the concentration of the component C:

$$u_z = \frac{u_0}{1 + F(dq/dC)} \quad (31)$$

Accordingly, the velocity u_z is inversely proportional to the slope of the tangent to the isotherm, illustrated in Fig. 12a. Since the concentration C moves at the velocity u_z, its retention time is equal to L/u_z (Eq. 20). Equation 31 shows that in nonlinear chromatography the velocity associated with a concentration depends on that concentration. In linear chromatography, where $dq/dC = a$ and $u_z = u_0/(1 + k_0')$, the velocity is independent of the concentration. With a Langmuir isotherm, the velocity increases with increasing concentration (see Eq. 31, d^2q/dC^2 is negative), and the front of the band becomes more steep since the high concentrations move faster than the low ones. For the same reason, the rear of the band becomes diffuse. However, the high concentrations cannot pass the low ones. Thus, a concentration discontinuity or shock builds up [13,77]. The velocity of this shock is given by:

$$U_s = \frac{u_0}{1 + F(\Delta q/\Delta C)} \quad (32)$$

where Δq and ΔC are the amplitude of the shock in the stationary and mobile phases, respectively. The velocity of the shock U_s is inversely proportional to the slope of the chord of the isotherm (Fig. 12a). It is faster than the limit velocity of the compound at infinite dilution (i.e., under linear conditions) but slower than the velocity associated to the higher concentration C of the shock on the diffuse rear side of the band profile. Thus, the band broadens during its migration; the shock moves faster than the rear but it also decreases in amplitude. Note that the velocity of a shock, given by Eq. 32, is also the velocity of a molecule.

Figure 12b illustrates how the injection of a plug of labeled molecules on a concentration plateau permits the determination of both velocities in the same experiments [78]. The concentration perturbation moves at the velocity associated with the concentration u_z, while the labeled molecules (shaded area) move at the shock velocity U_s.

In practice, there is no column with an infinite efficiency. The axial dispersion and the mass transfer resistances are finite and thermodynamic equilibrium is never achieved. Since Eq. 17 does not propagate discontinuities, a concentration shock cannot build up, but still the propagation velocity is discontinuous. A steep concentration front is formed. Since it has a finite but narrow thickness, it is called a shock layer [13,77]. The shock layer, however, moves at the velocity of the shock (Eq. 32). Its thickness is a function of the column HETP.

Numerical solutions of Eq. 17 can be calculated using a finite difference algorithm [76]. In this calculation procedure, the continuous z,t space is replaced by a grid ndz,kdt and the concentration at each point of the grid is calculated from the initial and boundary conditions and the concentrations at the previous grid points. The fastest program replaces the partial differential Eq. 16 by the following finite difference [45,76]:

$$C_{z,t} = C_{z-1,t} - \frac{dz}{u_0 dt}[C_{z-1,t} - C_{z-1,t-1}$$
$$+ F(Q_{z-1,t} - Q_{z-1,t-1})] \tag{33}$$

Since the column is sliced into a finite number of segments and the time is divided into a finite number of increments, there is an uncertainty as to where exactly in these elements the actual concentration is located. These errors compound during the calculation which requires a number of loops of the order of the square of the plate number of the column. An analysis of the sources of these errors, replacing the various terms in Eq. 33 by their Taylor expansion, reveals that the error made during the computation is equivalent to a diffusion term, with a numerical diffusion coefficient equal to [80]:

$$D_n = \frac{dzu}{2}(a - 1) \tag{34}$$

where a is the Courant number. The Courant number, which characterizes the stability of the numerical calculation, is equal to $u_z dt/dz$. Its value must be larger than unity. Accordingly, we select the values of the two integration increments, dz and dt, in such a way that:

$$dz = H \tag{35a}$$

$$dt = \frac{2H}{u_z} \tag{35b}$$

Since we define the efficiency at infinite dilution, we choose the value $u_{z,0} = u_0/(1 + k_0')$. Then the Courant number is equal to or larger than 2 and the numerical diffusion coefficient is equal to the apparent diffusion coefficient (Eq. 18). Thus, we obtain solutions

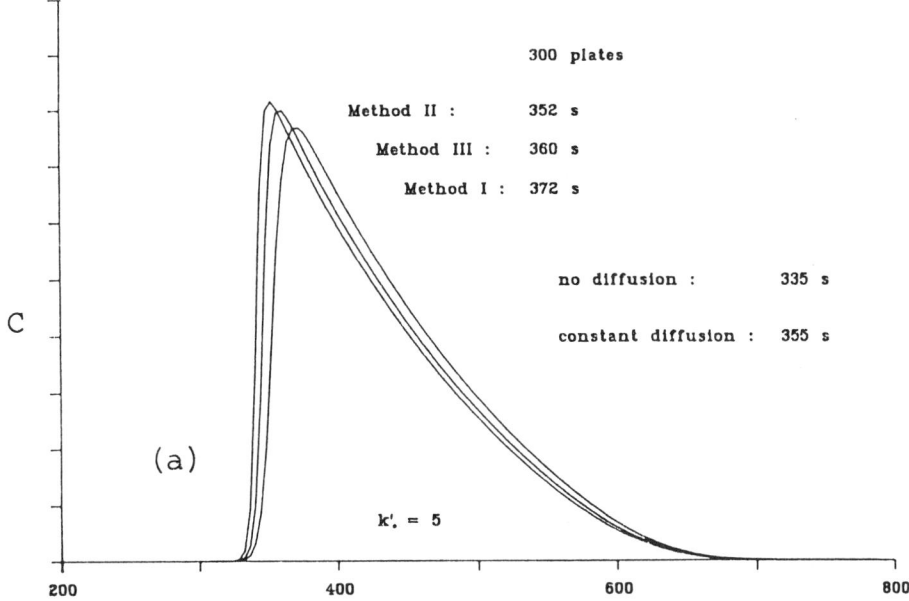

Fig. 13 Comparison of the calculated solution of three different finite difference implementations of the semi-ideal model. $k_0' = 5$, $L_f = 10\%$. (a) $N = 300$ plates. Grid spacing: method I (Rouchon [45,76]) $\Delta z = 500$ μm, $\Delta t = 5$ sec; method II (Craig [80]) $\Delta z = 600$ μm, $\Delta t = 0.4$ sec; method III [80] $\Delta z = 250$ μm, $\Delta t = 1.0$ sec. (b) $N = 5000$ plates. Grid spacing: method I (Rouchon) $\Delta z = 30$ μm, $\Delta t = 0.24$ sec; method II (Craig) $\Delta z = 36$ μm, $\Delta t = 0.024$ sec; method III $\Delta z = 15$ μm, $\Delta t = 0.06$ sec. (Reproduced from Ref. 80 with permission of the American Chemical Society.)

of Eq. 17. This method gives excellent results which are in agreement with experimental band profiles. However, the error analysis shows that there is a third-order error term in the nonlinear case. Nonetheless, this higher order term has a negligible contribution on the single-component band profiles for a column efficiency greater than several hundred theoretical plates [80].

Several different algorithms may be written for the numerical calculation of solutions of Eq. 17 [80]. During the calculation, however, errors propagate differently and the higher order terms are different. Figure 13 illustrates the results obtained with three such algorithms [80]. The profiles obtained are slightly different at low column efficiencies, as shown in Fig. 13a for a 300-plate column, but they are identical at high efficiencies, as shown in Fig. 13b for a 5000-plate column.

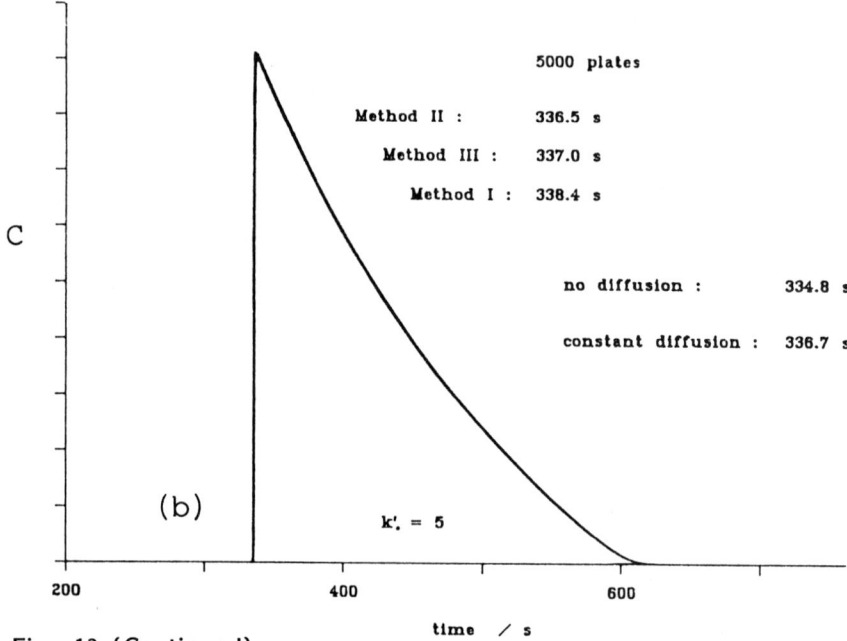

Fig. 13 (Continued)

Elution Mode and the Kinetic Models

Recently, Do and Rice summarized the theoretical forms of many standard models with two parameters [81]. A review of the different kinetic models, comparing their assumptions and the band profiles they predict in different cases, was prepared by Golshan-Shirazi and Guiochon [82].

As mentioned above, attempts were made to account for the contributions of band broadening to the band profiles in nonlinear chromatography. It is more important, however, to derive analytical or even numerical solutions taking into account the different mechanisms of band broadening rather than merely understanding the mechanism of the different band-broadening contributions. The mathematics of linear, nonideal chromatography were discussed by Lapidus and Amundson [11]. Van Deemter et al. showed that when the height of a mixing stage is much smaller than the column length, the elution profile is Gaussian [61]. When the rate constant of the mass transfer kinetics is low, however, the elution profile is not Gaussian, even in linear chromatography. In the case of a linear isotherm, the effect of dispersion (Peclet number) and of the mass transfer coefficient in a linear rate expression on the band profile were investigated by Giddings and Schettler [83,84] and Villermaux

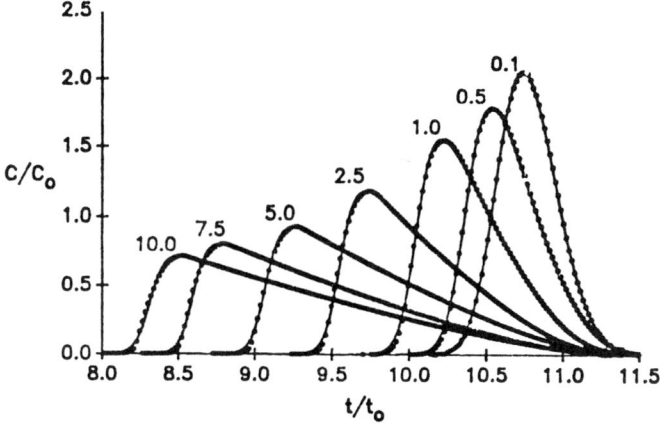

Fig. 14 Comparison of experimental data at various loadings and a kinetic model. Solute: 3-phenyl-1-propanol; mobile phase: 25:75 MeOH/water; temperature: 30.0°C; injection volume: 100 µl; flow rate: 1.0 ml/min. The number above each figure gives the amount of solute in µmoles. (Reproduced from Ref. 89 with permission of Elsevier.)

[85-87]. The appearance of split peaks was shown [85] and methods to extract equilibrium and kinetic parameters from the profiles were proposed [87].

Thomas discussed and solved the Riemann problem in the case of Langmuir kinetics (second-order rate expression, Eq. 3 and Table 2, Eq. 3), with no axial dispersion [7]. Goldstein derived an analytical solution to the mass balance equation for ideal nonlinear chromatography for a pulse injection of any width [6]. Goldstein went further to show that in the limit of fast kinetics the solution obtained is the same as with equilibrium theory [88]. Wade et al. [31] simplified the equations derived by Goldstein for the calculation of single-component overloaded band profiles with a second-order rate expression, assuming a Dirac injection. They also characterized the profiles by calculating the first and second moments and studied their dependence on the amount injected. Kinetic parameters were obtained from experimental band profiles by fitting them to theoretical profiles [31,89]. Figure 14 shows a comparison between experimental profiles and profiles calculated with the kinetic model of these authors [89]. For the system studied (i.e., column efficiency, concentration range), indistinguishable profiles were obtained with the kinetic model, the Houghton and the Haarhoff and Van der Linde equations.

If the rate of mass transfer is slow and equilibrium cannot be assumed, the rate of adsorption and desorption can be modeled with

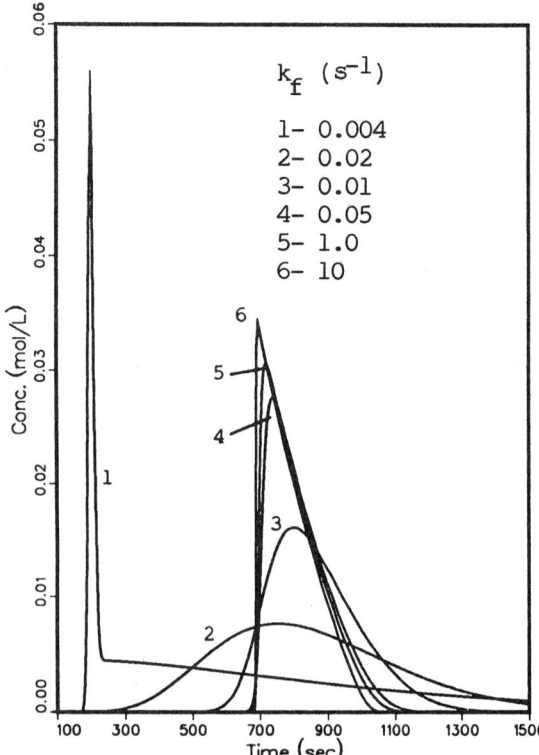

Fig. 15 Effect of the mass transfer coefficient on the overloaded single-component profile. Langmuir isotherm. (Reproduced from Ref. 34 with permission of the American Chemical Society.)

a linear rate expression (Eq. 4 and Table 2, Eq. 6). Carta derived an infinite series solution for a pulse of any width [90]. However, this solution assumes that the concentrations are small enough for linear equilibrium adsorption behavior.

Also using a linear rate expression, Lin et al. studied the effect of the sample size and the kinetic rate constant on the band shape in overloaded elution chromatography [34,35]. Figure 15 illustrates the numerical solution at decreasing values of the rate constant [34]. When the rate constant is very large (curves 5 and 6), the solutions are identical to those provided by the semi-ideal model. When the rate constant decreases (curves 3 and 4), the front of the band becomes less steep and its top becomes rounder. The retention time of the band maximum increases slightly as the local degree of overload (i.e., the deviation of isotherm from linear behavior) decreases with increasing bandwidth. When the rate constant continues to decrease, the smoothing effect on the band profile intensifies (curve 2). One part of the profile (its front with a convex upward isotherm) remains self-sharpening, however, while the other is not and the profile remains non-Gaussian. At low values of the rate constant, the dependence of the retention time on the rate constant reverses, and the retention time decreases with decreasing rate constant. At extremely low values of the rate constant, a bimodal peak appears, with the first mode eluting at the hold-up time (curve 1). The proportion of the component eluted in the first mode increases and tends toward unity as the rate constant tends toward zero.

Frontal Mode

A general analytical solution to the Riemann problem (step injection) for a nonlinear system was first derived by Thomas [7], assuming second-order kinetics and no axial diffusion term. Hiester and Vermeulen derived numerical solutions for Thomas's analytical results [33]. By nondimensionalizing the equations, the entire family of breakthrough curves can be represented with a single curve dependent only on the rate constant. They also showed that the solution calculated using second-order kinetics can be obtained using a linear rate expression. To obtain an exact solution, the rate constant is concentration-dependent but may be approximated by an average constant value.

Similarly, Lapidus and Amundson showed the equivalence between the kinetic model and a semiequilibrium model for fast kinetics in the linear equilibrium case [11]. In the linear case, Fig. 16 illustrates the effect of dispersion on the breakthrough profile. The curves corresponding to different values of the rate constant intersect at the inflection point. Cooney and Lightfoot [91] gave a mathematical proof of the constant pattern behavior for a Langmuir isotherm (i.e., after a certain column length, the breakthrough front does not change shape). In the early 1970s, many analytical and numerical solutions were derived assuming various rate expressions, and either Langmuir, Freundlich, or irreversible isotherms. Ruthven summarized these analytical solutions [92].

Rhee, Aris, and Amundson applied the method of characteristics to the solution of the ideal model and calculated breakthrough profiles [13,19,20,93].

Determination of Isotherm Parameters

The primary use of measurements carried out on single-component elution profiles or breakthrough curves is the determination of single-component adsorption isotherms. Five methods are available for this purpose: frontal analysis (FA) [94,95], frontal analysis by

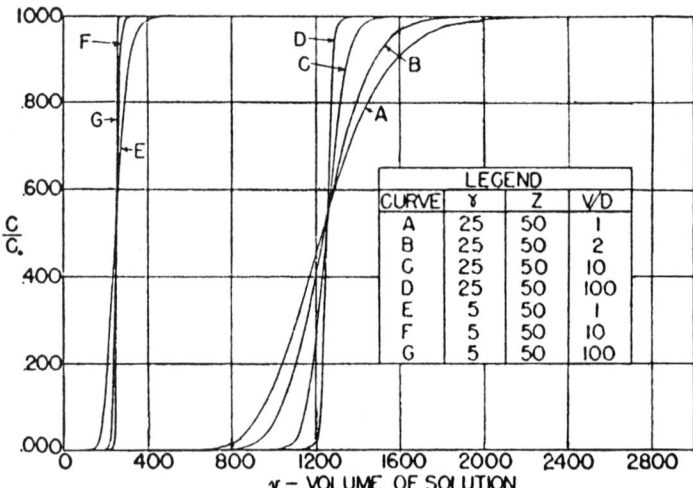

Fig. 16 Effect of the longitudinal diffusion on a column of infinite length in linear chromatography. Local equilibrium is assumed. Initial adsorbate concentration is zero. Inlet concentration is C_0. (Reproduced from Ref. 11 with permission of the American Chemical Society.)

characteristic point (FACP) [96], elution by characteristic point (ECP) [97], elution on a plateau or step-and-pulse method [78], and the retention time method (RTM) [49]. These methods are essentially attempts at solving the general inverse problem of chromatography, given Eqs. 1, 2-6, and the proper set of initial and boundary conditions. The direct and inverse problems can be formulated as follows: The direct problem is to calculate the column response knowing the equilibrium isotherm and the rate constants (or the apparent dispersion coefficient). The column response is either the transient concentration profiles at the column outlet or these concentration profiles in the column at a given time. The inverse problem is knowing solutions of the system of equations (i.e., the band profiles) to determine the isotherm and the rate constants. There is a paucity of mathematical results to guide this last quest.

Frontal analysis was first developed and used independently by James and Phillips [95] and by Schay and Szekely [94] for the determination of adsorption isotherms. Successive abrupt step changes of increasing concentration are performed at the column inlet and the breakthrough curves are determined [98,99]. Figure 17a shows a schematic of a single FA breakthrough curve [100]. Figure 17b shows a typical experimental recording obtained for the determination of the isotherm of phenol between silica and dichloromethane [47]. The area from the dead time to the inflection point of the breakthrough curve gives the amount adsorbed:

$$Q_{i+1} = Q_i + \frac{(C_{i+1} - C_i)(V_{F,i+1} - V_0)}{V_a} \tag{36}$$

where:

Q_i and Q_{i+1} are the amounts of compound adsorbed by the column packing after the ith and the (i + 1)th step, when in equilibrium with the concentrations C_i and C_{i+1}, respectively,

$V_{F,i+1}$ is the retention volume of the inflection point of the (i + 1)th breakthrough curve,

V_0 is the column void volume, and

V_a is the volume of adsorbent in the column.

Frontal analysis has been applied to the determination of isotherms of peptides [100] and proteins [101,102]. In contrast to the other chromatographic methods for the determination of isotherms, detector calibration is unnecessary in FA. However, it can be made concurrently from the plateaus of the step changes in concentration for other uses as needed. Furthermore, the self-sharpening nature of the FA fronts permits accurate determination of the isotherm points even in cases where the mass transfer kinetics is slow. In cases where the fronts are diffuse but the rear part of the profile is self-sharpening (concave upward isotherm), negative concentration steps can be used instead. The main inconvenience of FA is the large amount of material required, although this can be minimized by using microbore technology [101].

Measurement of the single-component isotherm can be performed from the rear boundary recorded in FA or by making a negative step change from a finite concentration to pure mobile phase (e.g., when the column is regenerated after classical FA). This is frontal analysis by characteristic points (FACP) [96]. Figure 17c shows a schematic of FACP [100]. In the last method, elution by characteristic point (ECP) [97], the isotherm can be determined from overloaded elution profiles (Fig. 17d) [100]. In FACP and ECP, the detector response signal must be converted to concentration by calibration of the detector. This can be done by direct injection of the sample into the detector cell at different concentrations. The amount adsorbed is calculated by integration of the area under the peak (using partial sums):

$$Q_C = \frac{1}{V_a} \sum_0^C (V - V_0) dC \tag{37}$$

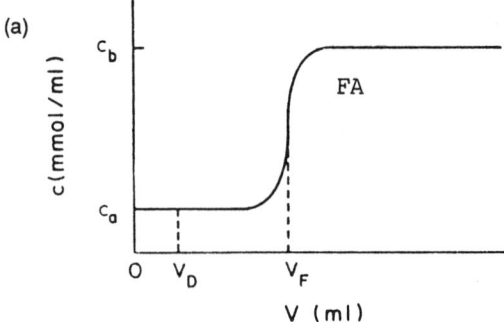

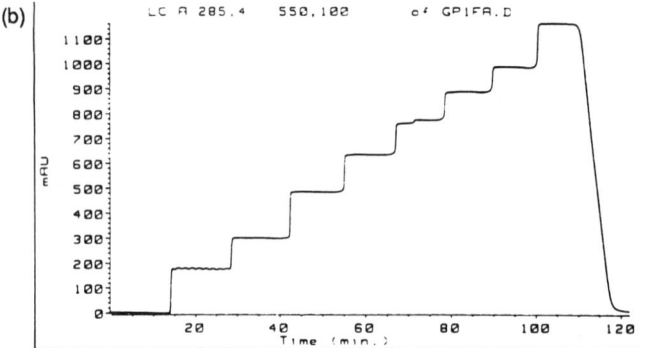

Fig. 17 Schematic of different methods of isotherm determination. (a) Single step in frontal analysis (FA). (b) Typical experimental chromatogram in frontal analysis. (c) Frontal analysis by characteristic point (FACP). (d) Elution by characteristic point (ECP). [Parts (a), (c), (d) are reproduced from Ref. 100 with permission of Elsevier. Part (b) is reproduced from Ref. 47 with permission of the American Chemical Society.]

where:

Q_C is the amount of compound adsorbed by the column packing when in equilibrium with the concentration C,

V_a is the volume of adsorbent in the column, and

V is the retention volume of the characteristic point of the diffuse profile at concentration C.

These latter two methods do not take into account the band broadening due to axial diffusion and mass transfer resistances, which can be significant on the diffuse part of the band profile. This problem is illustrated in Figs. 3 and 4, as ECP is based on

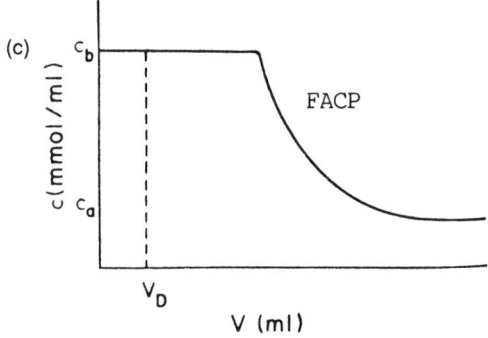

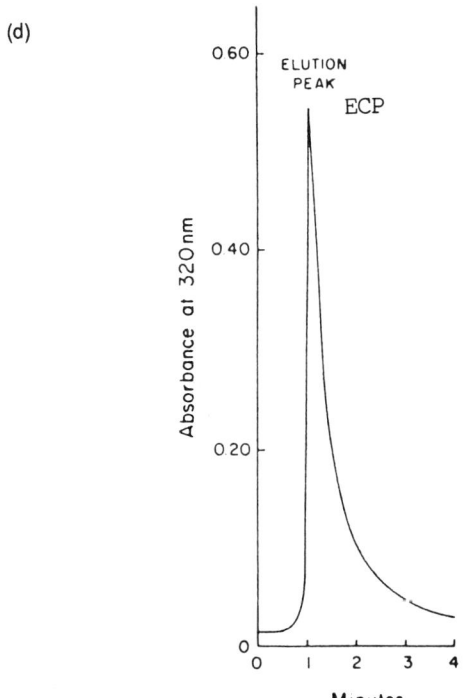

Fig. 17 (Continued)

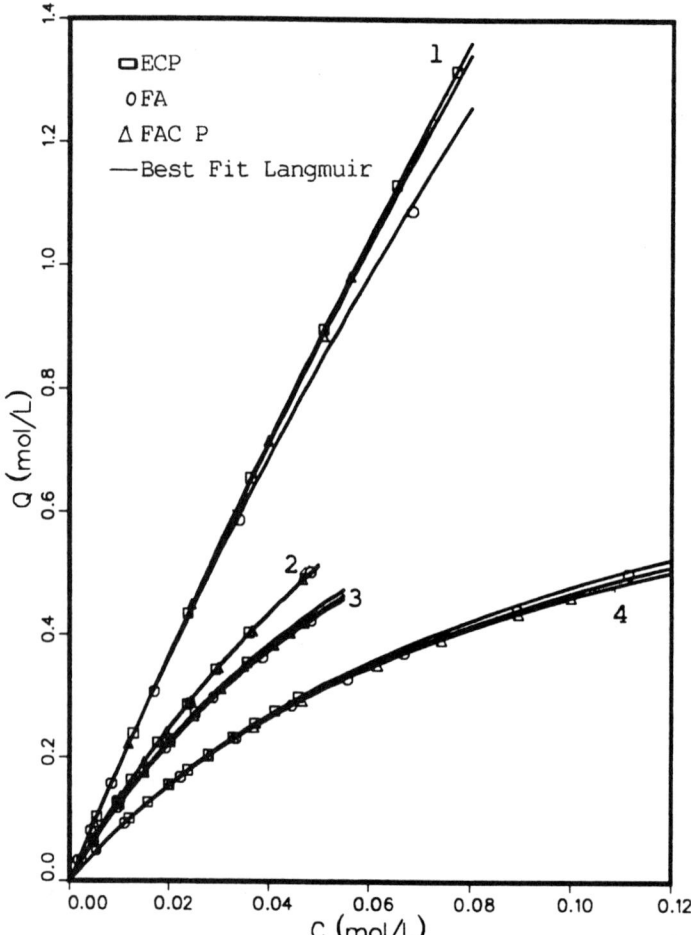

Fig. 18 Comparison of the different methods of isotherm determination in reversed and normal phase chromatography. 1, Acetophenone on silica. Mobile phase: 2.5:97.5 ethyl acetate/n-hexane. 2, Benzyl alcohol on silica. Mobile phase: 15:85 THF/n-hexane. 3, Benzyl alcohol on ODS silica. Mobile phase: 15:85 methanol/water. 4, Phenol on ODS silica. Mobile phase: 20:80 methanol/water. (Reproduced from Ref. 48 with permission of the American Chemical Society.)

the ideal model solution given in solid line. However, experimental data shows band broadening. Thus, FACP and ECP should be used only with high-efficiency columns. The advantage of ECP over FA or FACP is that a much smaller amount of material is needed to determine an isotherm. FACP is the complement to FA and has the major advantage that the comparison of FA and FACP data permit the determination of the possible extent of adsorption hysteresis.

If the isotherm is Langmuirian, the analytical solution of the ideal model can be inverted and a dimensionless plot used to derive the best isotherm parameters from one large and one very small volume injection [49]. Since in most cases the Langmuir model gives an excellent representation of experimental equilibrium data, the method is general. Like ECP, it can be performed with very small amounts of material and is very convenient for most studies involving biochemicals or other costly products.

Other novel approaches for isotherm determination have been developed by systematically guessing values of the parameters of an isotherm model and then minimizing the error between the calculated and the measured peak profiles. The best set of isotherm parameters can be determined rapidly [103-105]. However, with this method an adsorption isotherm model is needed a priori. In the other methods (e.g., FA or ECP) the amount of component adsorbed per unit volume of stationary phase is obtained directly from the experimental data and model fitting is done afterward [104].

It has been shown in many investigations that ECP, FA, and FACP give the same experimental isotherm (Fig. 18) [48]. Among the most recent such studies, we quote Jacobson et al. [100] and Golshan-Shirazi et al. [47-49].

B. Two-Component Problems

With two components, the number of different ways to conduct a chromatographic experiment increases compared to the single component case. Elution and frontal analysis of binary mixtures are the two obvious and most important cases, but we can also have displacement of a single-component band and system peaks (when a single-component pulse is injected in a chromatographic column with a mobile phase containing an additive which is nearly as retained as the compound injected).

Wilson [1] and DeVault [2] both discussed the theoretical statement of the two-component problem and its solution, in the case of the elution of a pulse of a binary mixture. However, it was Glueckauf who first gave a comprehensive analysis of the chromatographic separation process for two solutes following competitive Langmuir adsorption in the elution [106] and the displacement modes [3]. Analytical solutions can be derived in only a few, very specific cases [42,43]. Otherwise, numerical solutions are required [65]. Two basic methods have been used for the numerical integration of partial differential equations: finite differences and collocation on finite elements [65,107].

A review of the theoretical and experimental results obtained when applying the semi-ideal model of chromatography to a variety of two-component problems was presented by Guiochon et al. [24].

Elution Mode and the Ideal Model

In the ideal model, we assume that the column efficiency is infinite; hence, the axial dispersion coefficient is zero. The differential mass balances for the two components are written:

$$\frac{\partial C_1}{\partial t} + F \frac{\partial q_1}{\partial t} + u_0 \frac{\partial C_1}{\partial z} = 0 \tag{38a}$$

$$\frac{\partial C_1}{\partial t} + F \frac{\partial q_2}{\partial t} + u_0 \frac{\partial C_2}{\partial z} = 0 \tag{38b}$$

where C_i and q_i are the concentrations of the component i in the stationary and the mobile phases, respectively (see Eq. 1). The concentrations q_i in the stationary phase are related to the concentrations in the mobile phase by the isotherm equation:

$$q_i = f_i(C_1, C_2) \tag{39}$$

Equation 39 states that the two components compete for interaction with the stationary phase. Since the stationary phase concentration of each component at equilibrium is a function of both concentrations in the mobile phase, the two partial differential equations 38a and 38b are coupled. DeVault showed that with Langmuir competitive isotherms, the two band fronts are self-sharpening, like in the single-component case [2]. He further suggested that the concentration of the first component in the band, where it is pure, is higher than in the original solution, while the concentration of the slower moving solute in the rear band, where it is pure, tends to be smaller than in the mixed band.

Glueckauf was the first to investigate in detail the mathematical properties of the ideal model and give a comprehensive analysis of the chromatographic separation process for two solutes following competitive Langmuir adsorption in the elution mode [106]. Glueckauf [5] calculated the two-component individual band profiles. Coates and Glueckauf determined experimentally the two-component individual band profiles by collecting fractions with an "automatic device" (Fig. 19a) [106]. Except for the severe tailing of the second-component band (which could have been accounted for by a bi-Langmuir

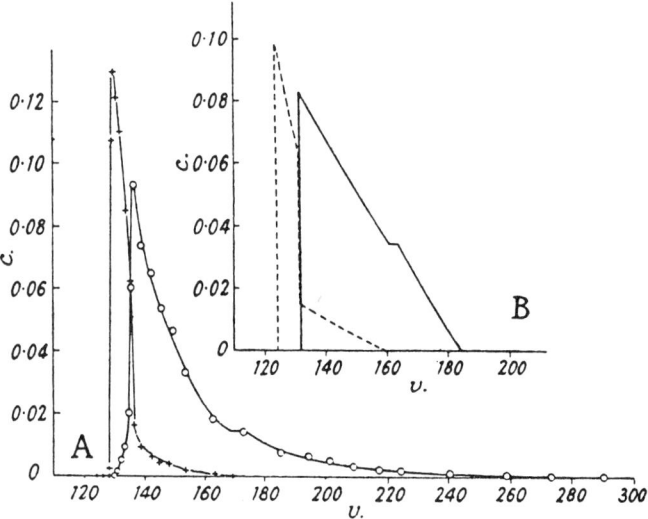

Fig. 19 Comparison of theory and experiment for the separation of a binary mixture in overloaded elution chromatography. (A) Experimental profile. (+) Mn^{2+}, (o) Cu^{2+} on 60 g of 0.01 mm Zeo-karb H. I. Development with 1 N H_2SO_4. Experimental conditions: m_1 = 0.6 meq, m_2 = 2.0 meq, $C_{1,0}$ = 0.1 N, $C_{2,0}$ = 0.4 N. Isotherm constants: a_1 = 2.12, b_1 = 2.3, a_2 = 2.35, b_2 = 2.56. Flow rate (~): 100 ml/24 hr. (B) Theoretical prediction with the ideal model. Dashed line, first component; solid line, second component. (Reproduced from Ref. 106 with permission of the Chemical Society, London, UK.)

isotherm), good agreement was obtained between experimental and theoretical profiles (Fig. 19b). Glueckauf extended these theoretical results to two-component separations with a competitive Freundlich isotherm [4]. Unfortunately, the mathematical derivation is very complex, and the solution is incomplete and obfuscated by a complicated system of symbols. Progress in the theory of system of partial differential equations during the last 40 years has allowed reformulation and clarification of the solution on a more general, more complete, and more rigorous basis.

Helfferich and Klein developed a theory of multicomponent chromatography based on the ideal model of chromatography and using the competitive Langmuir isotherm model [21]. Although primarily applied to the solution of the displacement problem (see Section III.C), this solution can be extended to the elution mode [108,109]. Rhee, Aris, and Amundson discussed in 1970 the chromatographic cycle and the separation of components in the column from a pulse

46 / Katti and Guiochon

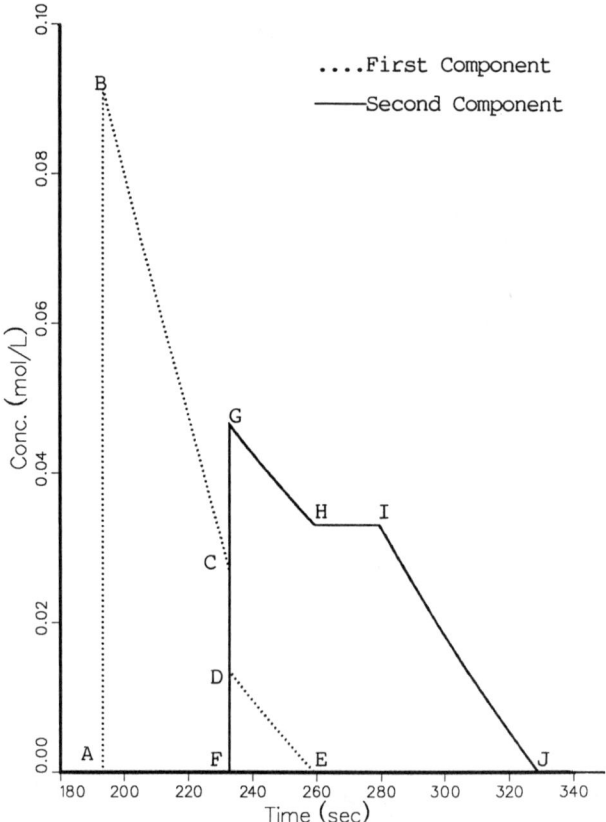

Fig. 20 Schematic of the solution of the ideal model of chromatography. (From Ref. 161.)

injection as well as the separation in time-distance space [20]. The primary purpose of their paper was to explain the mathematics of one-dimensional, isothermal, ideal equilibrium models of chromatography and the theory of simple waves [13]. The theory of simple waves applies to large volume injections where simplifications of the solution occur because there is a constant state (the concentration plateau). This solution is not valid in overloaded elution chromatography when the injection volume is sufficiently small that the plateau has eroded and disappeared by the time the band has eluted from the column. The simple wave method permits the determination of competitive equilibrium isotherms. It has over FA the same advantages that ECP enjoys for the measurement of single-component isotherms, but it does not suffer from the drawback of neglecting the influence of the finite column efficiency (see Section III.B). Kvaalen et al. showed that the system of Eqs. 38a and 38b is strictly hyperbolic [110].

Recently, an analytical solution to the elution problem in the case of a Langmuir competitive equilibrium isotherm was derived by two different methods, both assuming ideal behavior [43,109]. These two solution methods are based on the coherence condition described by Helfferich and Klein [21]. The primary assumption in this derivation is that the concentrations of each of the two components in the mixed zone at a given time and location have the same velocity. Figure 20 illustrates a schematic of the solution to the ideal model. The solution includes three bands, a band of pure first component (ABCF), a mixed band (FGHE), and a band of pure second component (HIJE). Also included are two concentration shocks, one in the front of the first band (AB), the other one between the first and the mixed bands (CD), and two diffuse boundaries for each component, one in the mixed band (DE and GH) and one in the band where each component is pure (BC and IJ) [43]. Lastly, two plateaus are formed. A concentration plateau higher than the concentration of the first component in the feed can appear on the top of the first-component band (displacement effect, not shown in Fig. 20). It is stable as long as the concentration plateau of the injection (in the case of a rectangular pulse injection) has not eroded, but it starts decaying when the injection plateau disappears. Another concentration plateau appears on the rear flank of the second-component band (HI, tag-along effect). The plateau on the rear side of the second-component band (Fig. 20, HI) starts eroding when the separation between the two bands is sufficiently advanced and the maximum concentration of the second component in the mixed band becomes equal to the plateau concentration.

The interaction between the two bands during the separation is due to competition between the component molecules for access to the retention mechanism. For competitive Langmuir-type isotherms, this interaction results in two effects—the displacement effect and the tag-along effect [24,46]. For any component, the stationary phase concentration at equilibrium with a given mobile phase concentration is less in the presence of one or several other components than it would be if it were pure. On the other hand, the velocity associated with a given concentration of, say, the first component (Eq. 31) is higher in the presence of the second component than when the first component is pure. The rapid concentration surge of the second component at the front of its band causes a brutal decrease in the amount of the first component in the stationary phase. The first component is displaced from the stationary phase by the arrival of the second one; this is the origin of the displacement effect. Alternately, we can observe that the concentrations of the first component corresponding to points C and D (Fig. 20) move at the same velocity. This is possible only because in C the first component is pure while in D it is in the presence of the second component, at the concentration G [113].

In chromatography, however, the relative retention of the components of a mixture is often close to 1. This means that for the same mobile phase concentration, the proportion of molecules in the stationary phase at equilibrium is not much larger for the second than for the first component. For $\alpha = 1.2$, for example, there are 20% more molecules of second component than molecules of the first one in equilibrium with a given mobile phase concentration (under linear conditions). When the sample size is large (nonlinear conditions) and the concentration of the second component is much lower than that of the first one, the stationary phase concentration of the second component will be much decreased by the presence of the excess of first component. The front of the second-component band will move faster than the front of a pure second-component band. This is the origin of the tag-along effect. It can be shown [43] that the limit for $C_1 \to 0$ of the velocity associated with a given concentration C_2 of the second component in the presence of the first component is higher than the velocity associated with the same concentration C_2 of the pure second component. This is why the point H moves faster than I (Fig. 20).

The above two paragraphs explain the displacement and the tag-along effects in terms of the velocities associated with concentrations. An alternate, simple physical explanation is presented as follows. The displacement effect is dominant when the amount of second component is greater than the amount of first component. Since the second component is more strongly adsorbed to the stationary phase than the first, it forces the first component to desorb. It is this phenomenon that causes concentration and narrowing of the first-component band. Upon desorption, the first component spends more time in the mobile phase than it would if it were pure. Thus, in nonlinear chromatography the first component elutes faster than in linear chromatography.

The tag-along effect, on the other hand, occurs when the second component is in less amount than the first. Although the second-component band is more strongly adsorbed than the first, the separation factor (i.e., $q_2 C_1 / q_1 C_2$) is not very different from unity (α is of the order of 1.1-1.7 for most practical cases), and the first component is in a larger amount. So the second component cannot readily displace the first one from the stationary phase. Moreover, due to the nature of the isotherm, a smaller fraction of the molecules of second component adsorbs on the stationary phase in the presence of the first component than when the second component is pure. The fraction which remains in the mobile phase is higher and moves faster. Thus, one obtains a dilute, stretched-out band.

The understanding of the displacement and tag-along effects is important for a development of preparative chromatographic applications. The former leads to concentration of the first component. The latter results in the spreading and dilution of the second-component band. The finite column efficiency causes relaxation of the concentration gradients which are created and built up by the nonlinear thermodynamic effects (see next section). Newburger et al. showed how the displacement effect can be used to a profit [111, 112]. Obviously, an excessive tag-along effect must be avoided as much as possible. The intensity of the displacement effect can be measured by the height of the negative concentration shock (line CD, Fig. 20) on the rear of the first-component profile [113]. It increases with increasing ratio of the loading factors, $L_{f,2}/L_{f,1}$, of the two components. The intensity of the tag-along effect can be measured by the length of the plateau (line HI, Fig. 20) on the rear of the second-component profile [113]. It increases with decreasing ratio of the loading factors.

Elution Mode and the Semi-ideal Model

We discussed in the previous section the solution of the ideal model for two components and previously the solution of the semi-ideal model for a single component, in Section II.F. The extension of the solution of the semi-ideal model to a multicomponent problem is straightforward [46]. The system of mass balance equations is written:

$$\frac{\partial C_1}{\partial t} + F \frac{\partial q_1}{\partial t} + u_0 \frac{\partial C_1}{\partial z} = D_{ap,1} \frac{\partial C_1}{\partial z} \tag{40a}$$

$$\frac{\partial C_2}{\partial t} + F \frac{\partial q_2}{\partial t} + u_0 \frac{\partial C_2}{\partial z} = D_{ap,2} \frac{\partial C_2}{\partial z} \tag{40b}$$

Numerical solutions of this system of partial differential equations can be calculated using a finite difference method [46]. The simplest scheme is the same as the one used for calculation of solutions of the semi-ideal model for the elution profile of a single-component band. There is, however, a theoretical difficulty because there can be only one set of integration increments. As is explained in the section on "Gradient Elution" in III.B, the calculation is programmed to give solutions of the ideal model (i.e., with the diffusion coefficient $D_{ap,i}$, equal to 0). However, because of the numerical errors, we obtain profiles which are solutions of Eqs. 40a and b, with $D_{ap,1} = D_{n,1}$ and $D_{ap,2} = D_{n,2}$, where $D_{n,1}$ and $D_{n,2}$ are given by Eq. 34 [79]. If we choose the integration increments to give an HETP equal to H_1 for the first component, then Eqs. 35a and 35b give:

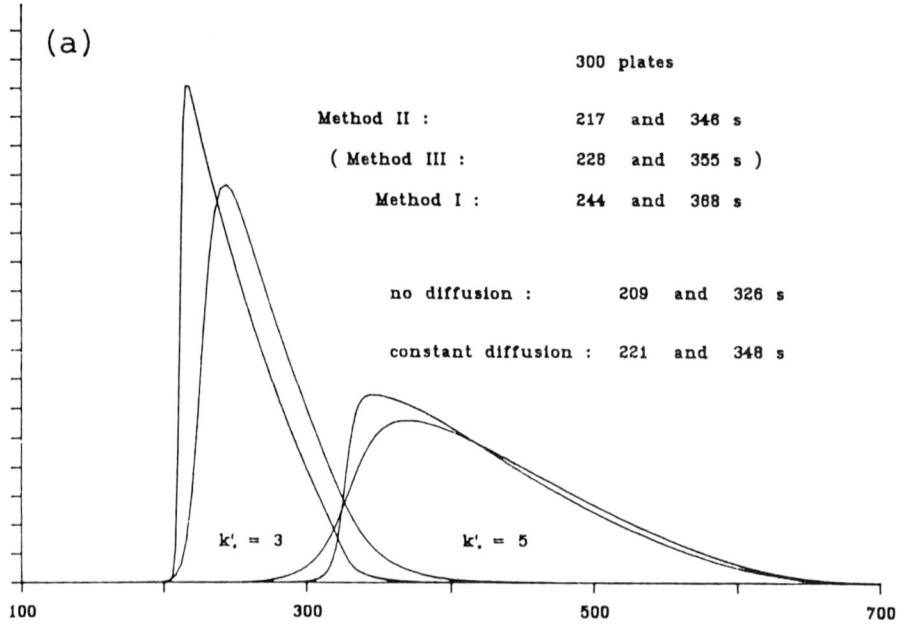
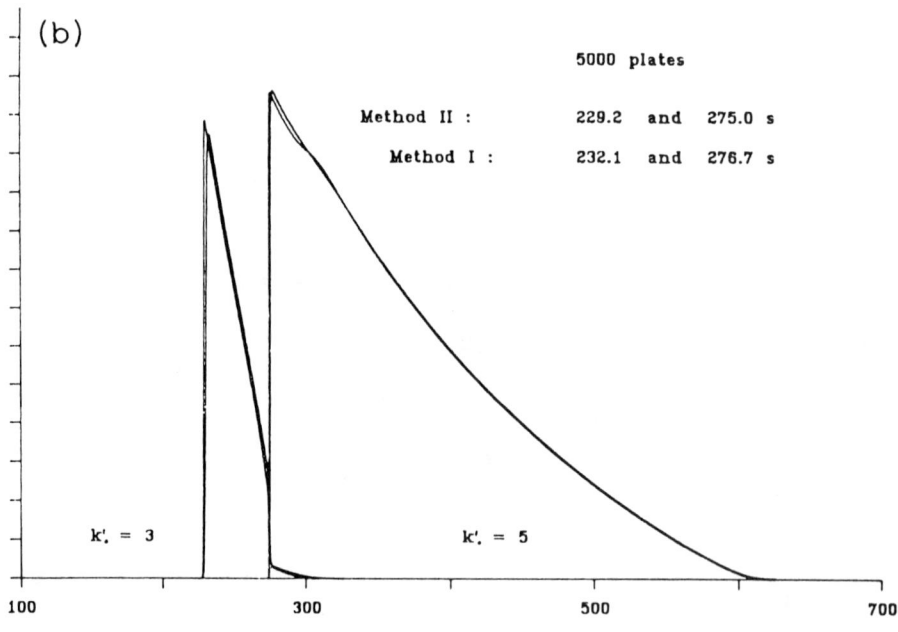

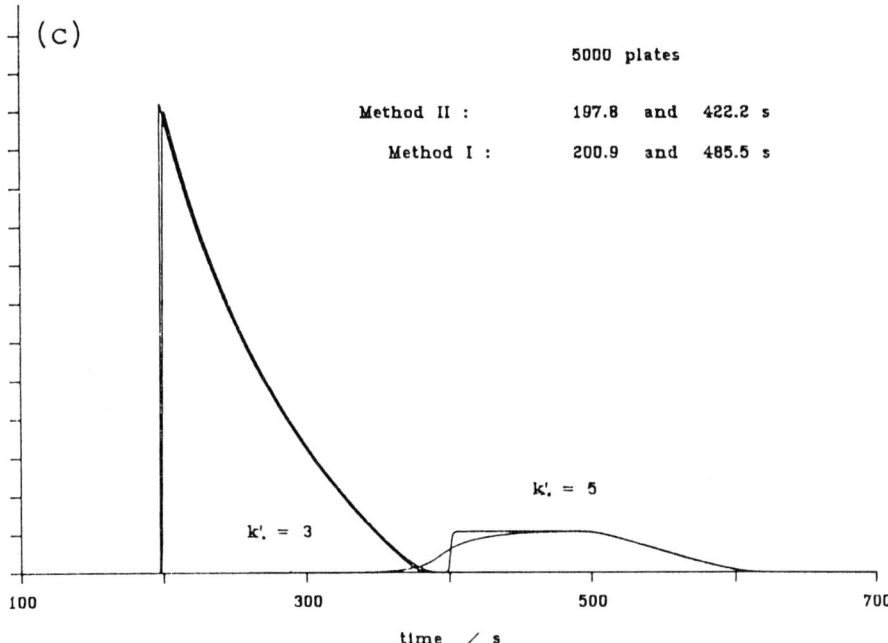

Fig. 21 Comparison of the calculated solutions obtained using two different finite difference implementations of the semi-ideal model. $k_0' = 3$ and 5, $L_f = 10\%$. Grid spacing given in Fig. 13. (a) N = 300 plates, composition ratio: 1:1. (b) N = 5000 plates, composition ratio: 1:5. (c) N = 5000 plates, ratio: 5:1. (Reproduced from Ref. 80 with permission of the American Chemical Society.)

$$dz = H = H_1 \tag{41a}$$

$$dt = \frac{2H_1}{u_0}(1 + k_1') \tag{41b}$$

Accordingly, Eq. 34 gives for the second component:

$$D_{n,2} = \frac{u_0 H_1}{2} \frac{2k_1' - 1}{1 + k_2'} \tag{42}$$

Comparison with Eq. 18 gives:

$$H_2 = H_1 \frac{2k_1' - 1}{k_2' + 1} \tag{43}$$

This relationship is arbitrary and has no physical meaning [80]. Other finite difference algorithms lead to different relations between H_1 and H_2. Comparison of the results obtained with the Craig model and with models based on solution of the semi-ideal model equation (Eq. 1 in Table 2) by using various finite difference techniques showed that the two approaches are equivalent [80,113,115], even though the Craig model is a mechanistic model and finite difference schemes are mathematical approximations. Figure 21 illustrates the band profile obtained using the Rouchon et al. [76] implementation of the semi-ideal model just described (method I) and the Craig model (method II). When the column efficiency is low, differences in the error propagation lead to significant differences in the band profiles (Fig. 21a). At high efficiency, on the other hand, the differences are negligible under conditions of dominant displacement effect because of its self-sharpening nature (Fig. 21b). However, when the tag-along effect is significant, different calculation schemes give significantly different profiles for the front of the second band (Fig. 21c).

The semi-ideal model can be modified so that the amount of dispersion for each component can be controlled independently by adding a dispersion term and using a dispersion coefficient such that the sum of the dispersion term and the numerical errors due to the finite difference scheme gives the total amount of desired band broadening. This leads to more accurate results [80]. The solutions obtained with this method agree very closely with the numerical solutions supplied by a finite element program [107]. Another implementation of the semi-ideal model was made by discretizing the dispersion term directly and using the Lax-Wendroff finite difference scheme to reduce the contribution of numerical errors due to band broadening. With this approach, however, it is much more difficult to obtain a stable solution [79,116] than with the simpler finite difference methods used by Czok and Guiochon [80] and Guiochon et al. [45].

Using the semi-ideal model [46], band profiles have been calculated under numerous sets of experimental conditions in order to investigate the effect of the different parameters on the band interaction (see section on "Optimization of Preparative Chromatography" in III.B). Using the Craig model, Snyder et al. [117] calculated band profiles under heavily overloaded conditions with low-efficiency columns. However, the two-component band profiles calculated are not always in accordance with the theoretical results obtained for a competitive Langmuir isotherm [6,43,46]. Under some sets of conditions, the profiles exhibit multiple maxima. Moreover, the use of an empirical correction factor of 1.8 in order to obtain agreement of the predicted profiles with experimental data was required [117]. These discrepancies are due to the use of an approximate isotherm which is not equivalent to a competitive Langmuir isotherm but presents

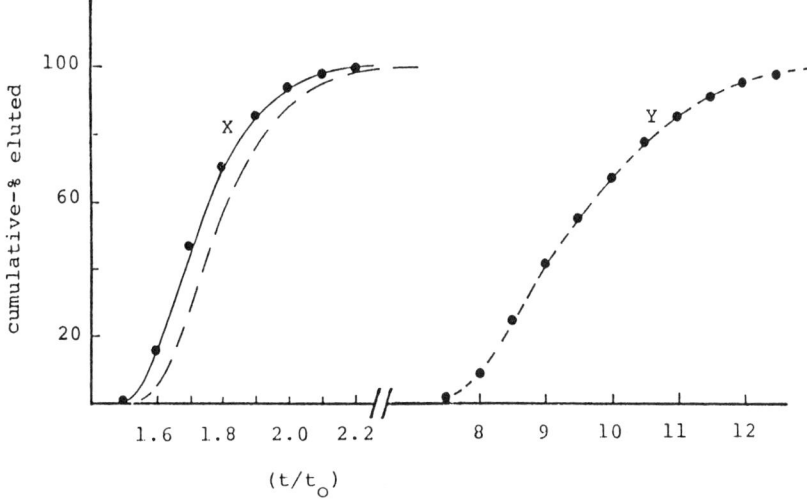

Fig. 22 Cumulative band profiles for a binary mixture. Conditions: number of stage in Craig simulation, n_c = 100, k_0 = 1 and 10 for X and Y, respectively, $w_X = w_Y$ = 6 mg, w_s = 100 mg (L_f = 6%). Symbols, simulation data for Craig two-solute case; dashed lines, simulation data for Craig single-solute case; solid line, simulation data for a shortened column (after blockage, n_c = 94 plus 6 initial stages with no retention). (Reproduced from Ref. 119 with permission of Elsevier.)

unexpected discontinuities. This explains the strange peak shapes obtained [118].

In summary, various procedures permit the numerical integration of the semiequilibrium model of chromatography for multicomponent mixtures. The choice between them is essentially a matter of convenience or rather a compromise between the degree of accuracy required and the amount of CPU time which can be afforded.

The Craig distribution model has been applied to binary mixtures with an isotherm calculated on a scheme based on a nine-term polynomial expansion by Eble et al. [119,120]. They calculated cumulative distribution profiles under slightly overloaded conditions and derived empirical relationships between the capacity factor, the apparent column efficiency, and the loading factor (Fig. 22). These cumulative distribution profiles were also compared with single-solute experimental profiles given by the symbols [120]. Calculations were carried out at very low column efficiencies, which prevented the identification of the displacement and tag-along effects. The displacement effect, which can enhance the separation of two components

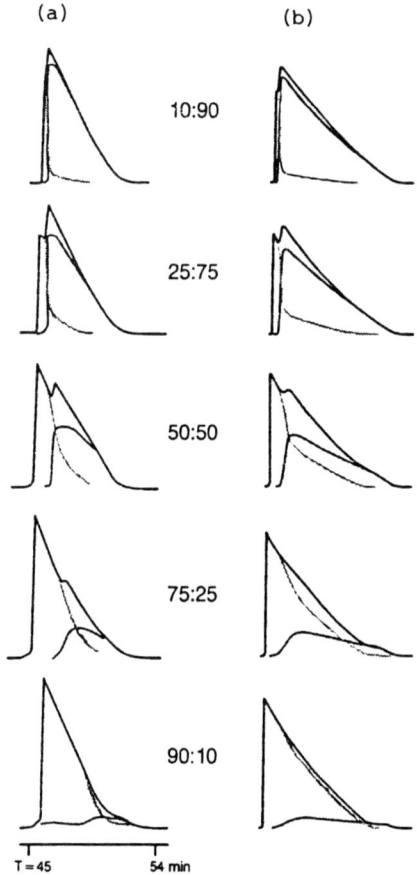

Fig. 23 Qualitative experimental verification of the displacement and tag-along effect at different mixture composition. (a) Experimental results for two epimers methyl cis- and trans-1-methyl-2-oxo-4-(2-propenyl)cyclohexaneacetate ($\alpha = 1.04$), 200-mg injections. (b) Calculation using the semi-ideal model, 300-mg injections. (Reproduced from Ref. 112 with permission of Elsevier.)

compared to the single-solute case, was identified experimentally and reinterpreted as the "blockage effect" [68]. The model is a poor restatement of displacement effects. In view of the long computation times required on personal computers, an attempt was made to combine empirical results from the blockage model with the model of Knox and Pyper [28] to predict the recovery of solutes at increasing sample sizes [121,122]. Because of the combined flaws of the different methods used, the results were not very successful.

Qualitative experimental verification of the displacement and tag-along effects presented in the work of Guiochon and Ghodbane [46] was shown [112] (Fig. 23). A quantitative comparison between theoretical and experimental profiles [50,123] is shown in Fig. 24. Figure 24a presents the predicted band profile as determined using measured single-component isotherm parameters in a competitive Langmuir model and experimental data. Deviations between theoretical and experimental profiles are significant for the second component. Figure 24b shows the predicted profile with the competitive isotherm measured by the method of the hodograph transform (see section on "Determination of Competitive Equilibrium Isotherms" in III.B) [72]. Agreement is now good for the second component but a shift is observed for the first-component profile. Differences are believed to be due to deviations of the isotherms from the competitive Langmuir model. A quantitive comparison of two-component overloaded elution profiles has been made using the N-benzoyl derivatives of D- and L-alanine eluted on immobilized bovine serum albumin [70]. Excellent agreement is observed between the individual profiles determined for 3:1, 1:1, and 1:3 mixtures and those calculated using the equilibrium isotherms measured. The isotherms are accurately fitted by a competitive bi-Langmuir model (Table 2, Eq. 8). This model assumes that the molecular interactions between the two enantiomers and the stationary phase can be classified as chiral-selective and achiral interactions. One term of the bi-Langmuir isotherm corresponds to the nonspecific interactions, the other to those which discriminate between the enantiomers. The column saturation capacities corresponding to each type of site are nearly identical for the respective two components. However, the column saturation capacities for the chiral-selective sites are small and overloading is observed at low concentrations (well below 1 mM). This system represents a most ideal situation and explains in a large part the excellent agreement between experimental results and theoretical predictions.

In an ideal two-component system, the saturation capacity of both components are equal; however, few systems are ideal. The effect of converging and diverging single-component isotherms on the band profile was studied by Cox and Snyder [124] using the Craig model and by El Fallah et al. [125] using the semi-ideal model. This effect on the displacement and tag-along effects has been investigated by Golshan-Shirazi and Guiochon [113] using the ideal model. Figure 25 illustrates the effect of an increase in the ratio of the column saturation capacities, $q_{s,1}/q_{s,2}$, on the band profile. As this ratio increases, the displacement effect decreases and the tag-along

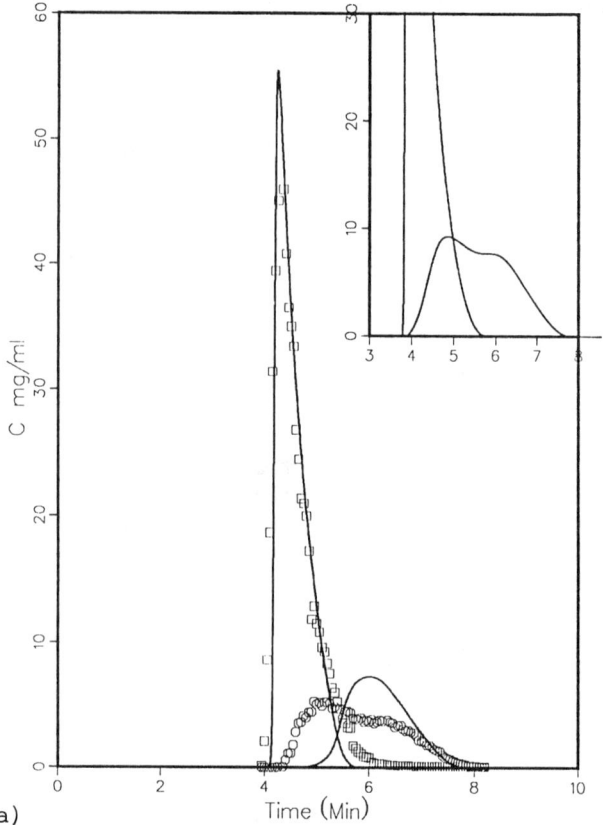

(a)

Fig. 24 Quantitative comparison of experimental and theoretical profiles for a 3:1 binary mixture in overloaded elution chromatography: (1) 2-phenylethanol and (2) 3-phenylpropanol. (a) Predicted profiles are calculated using the semi-ideal model and the competitive Langmuir isotherm model with the parameters obtained by single-component isotherm determination. Insert: chromatogram obtained with a twice larger sample size. (b) Predicted profiles are calculated using the semi-ideal model and parameters for the competitive Langmuir isotherm which are determined by the method of the hodograph transform [72]. (Reproduced from Ref. 50 with permission of Elsevier.)

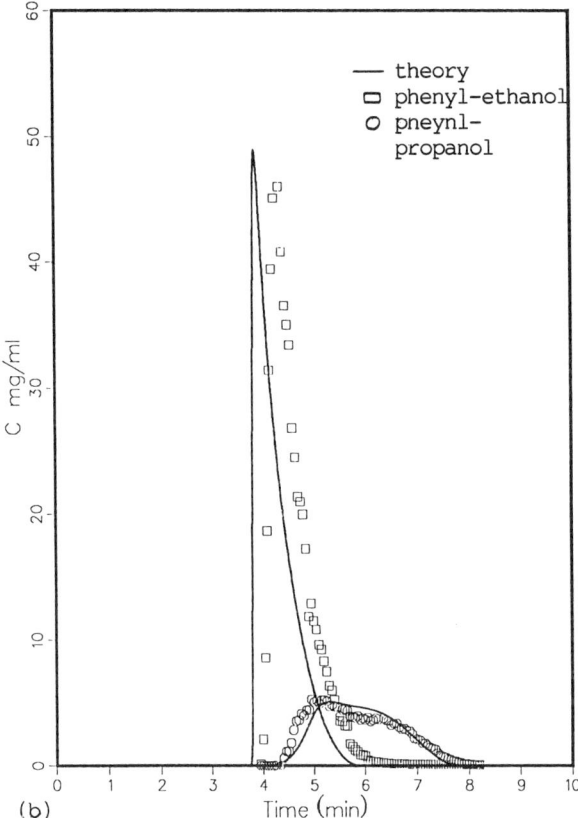

Fig. 24 (Continued)

effect increases. In many cases, however, experimental results using various binary mixtures show an enhancement of the displacement effect with increasing ratio of the column saturation capacities [111,112,124]. This disagreement is explained by the inadequacy of the Langmuir competitive isotherm model.

Elution Mode and the Kinetic Models

Using a forward finite difference method to model Eq. 1 of Table 2 replacing Eq. 2 of Table 2 with a linear kinetic rate expression, and allowing the numerical error to equal the degree of axial dispersion, the effect of the rate constant on two-component overloaded elution separations was studied [35]. The chromatograms calculated showed that a significant decrease in the mass transfer rate constant

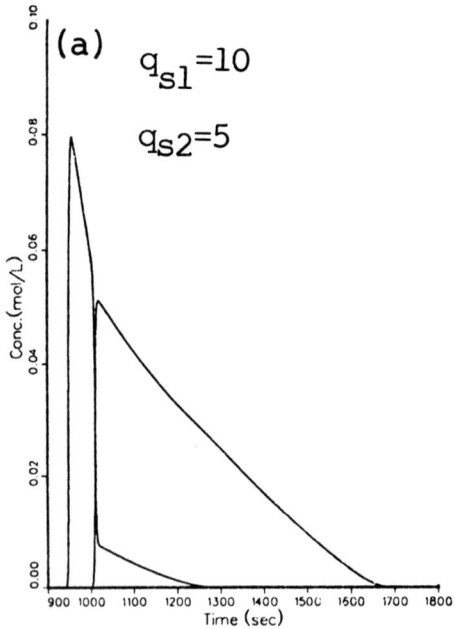

(a) $q_{s1}=10$
$q_{s2}=5$

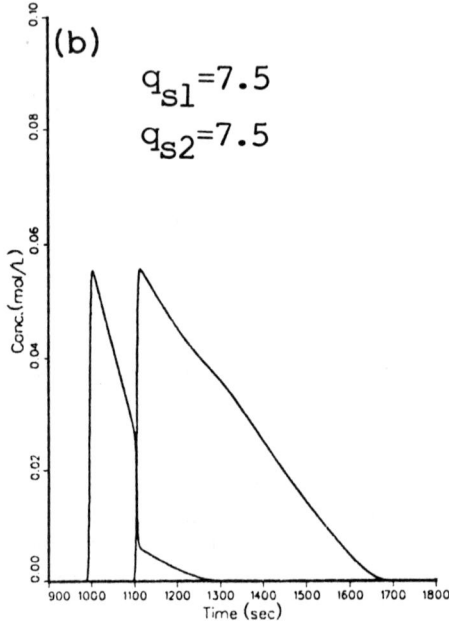

(b) $q_{s1}=7.5$
$q_{s2}=7.5$

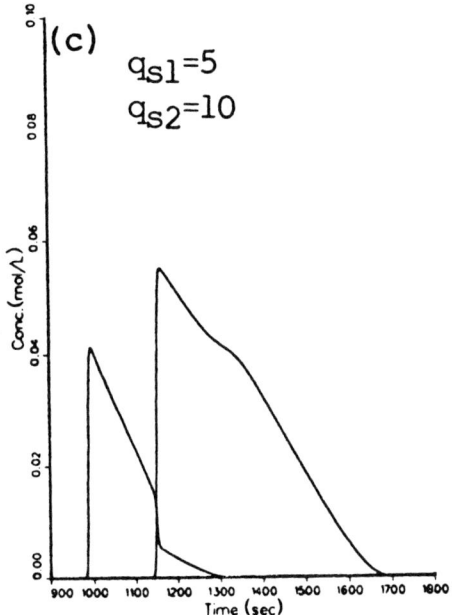

Fig. 25 Effect of the column saturation capacity on the individual profiles of a binary mixture using the semi-ideal model. Column length: 25 cm; efficiency: 5000 plates; phase ratio: 0.25; mobile phase linear velocity: 0.125 cm/sec. Solute relative retention: 1.2, $k'_{1,0} = 6$; feed composition: 1:3; sample size, 0.083 mmol (1) and 0.249 mmol (2). (a) $L_{f,1} = 1\%$, $L_{f,2} = 6\%$, (b) $L_{f,1} = 1.33\%$, $L_{f,2} = 4\%$, (c) $L_{f,1} = 2\%$, $L_{f,2} = 3\%$. (Reproduced from Ref. 113 with permission of the American Chemical Society.)

causes tailing and retention time shifts (Fig. 26). When the mass transfer kinetics is not very slow, the band profiles are controlled by thermodynamic effects and the solutions are identical with those obtained by integrating the equations of the semi-ideal model under the corresponding conditions (Fig. 26a).

When the kinetics becomes slow, the nonlinear thermodynamic effects decrease (Fig. 26b) and eventually vanish (Fig. 26c). The two band profiles become more independent. The retention times of the maximum of the band profiles first increase, then decrease, and eventually bimodal bands appear, with their first maximum eluted with the nonretained compounds (Fig. 26d). When the mass transfer kinetics are very different for the two components, elution profiles are complex. The coupling effect between the two bands disappears rapidly if the component which has the larger concentration

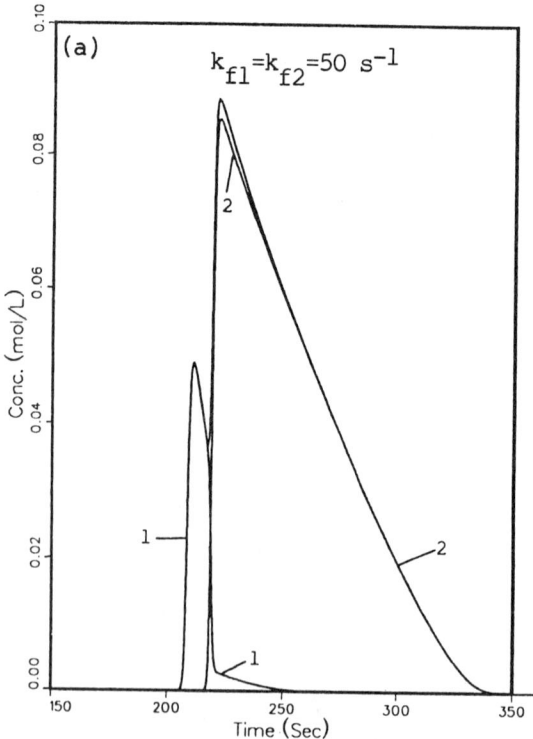

Fig. 26 Effect of the mass transfer coefficient on the profiles of a binary mixture in overloaded elution chromatography. Column length: 25 cm; phase ratio: 0.25; mobile phase linear velocity: 0.625 cm/sec. Components relative retention: 1.2, $k'_{1,0} = 6$; feed composition: 1:9, sample size, 4.15 mmol; $L_{f,1} = 4.5\%$. Competitive Langmuir isotherm parameters: $a_1 = 24$, $b_1 = 2.5$, $a_2 = 29$, $b_2 = 3.0$. 1, Profile of the first component; 2, profile of the second component. (a) Fast kinetics, $k_{f,1} = k_{f,2} = 50$ sec^{-1}. (b) $k_{f,1} = 50$, $k_{f,2} = 1.0$ sec^{-1}. (c) Slow kinetics, $k_{f,1} = 50$, $k_{f,2} = 0.2$ sec^{-1}. (d) Very slow kinetics, $k_{f,1} = 50$, $k_{f,2} = 0.03$ sec^{-1}. (Reproduced from Ref. 35 with permission of the American Chemical Society.)

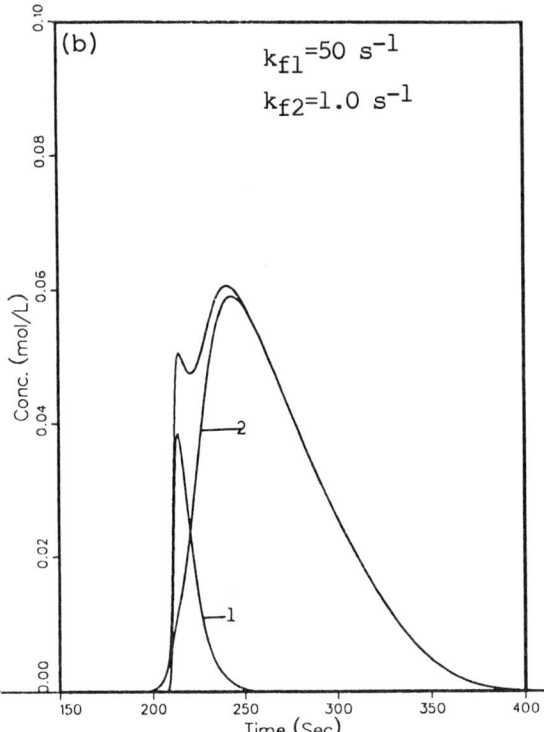

Fig. 26 (Continued)

is the one with the very slow kinetics. The band of a major component with a fast mass transfer kinetics may displace markedly the band of a minor component with a slow mass transfer kinetics. As opposed to the semi-ideal model, the kinetic model permits an accurate prediction of the elution profiles of the components of a mixture when their mass transfer coefficients are different.

Yu et al. [30,126] used orthogonal collocation with gradient-directed moving finite elements in order to solve the mass balance equation including a linear rate expression and inter- and intraparticle diffusion terms. The advantage of this method is that the numerical errors are low compared to finite difference schemes; however, computation times typically require 1-2 hr of CPU to determine the profile on a Gould NPI computer. Computation times for equilibrium models take only a few minutes on a VAX 8600 [114]. The equilibrium and nonequilibrium models have been used to scan the values of the various dimensionless quantities to identify the local

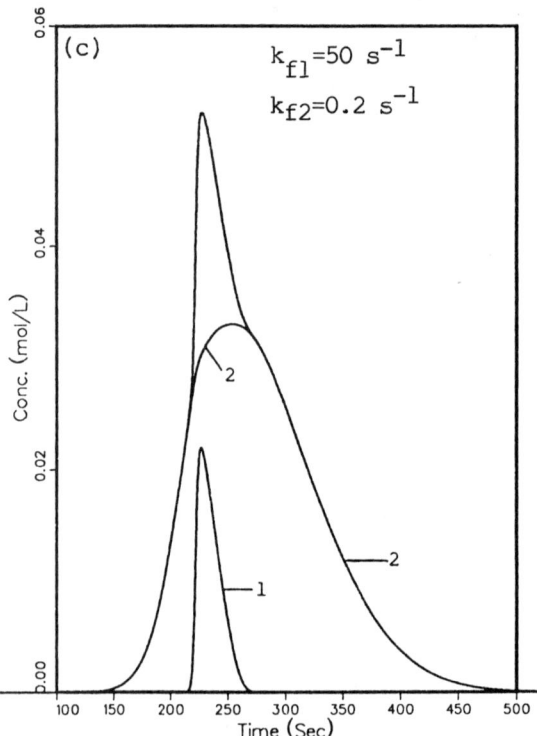

Fig. 26 (Continued)

equilibrium regime and understand the relative importance of film mass transfer, intraparticle diffusion, and axial dispersion for a specific system [29]. Experimental results on proteins such as lysozyme, BSA, and myoglobin show that the equilibrium model fails in predicting accurately the band profiles in ion exchange and a kinetic model must be used. Figure 27 compares experimental band profiles (symbols) of lysine-glutamic acid and lysine-proline with model predictions using the ideal model (dashed line) and a kinetic model (solid line). Kinetic parameters are determined from correlations available in the literature [127].

Gradient Elution

In gradient elution, the composition of the mobile phase is varied continuously during the experiment, in order to increase its eluotropic strength. The concentration of a "strong" solvent is increased. Under such conditions, the simplest gradient elution

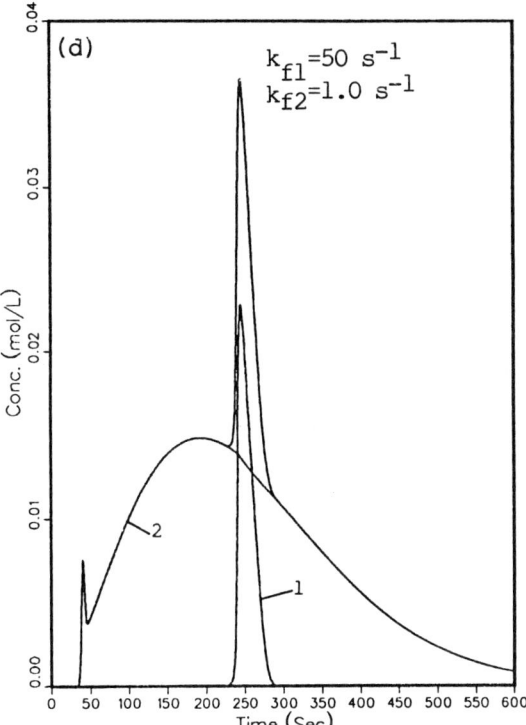

Fig. 26 (Continued)

problem is a two-component problem, with a pure component and the solvent modifier. The effect of the change in the mobile phase composition is due to the combination of two effects. In normal phase chromatography (e.g., on silica), the strong solvent is adsorbed on the stationary phase as strongly as or more strongly than the feed components. The presence of increasing concentration of this additive in the mobile phase accelerates the migration of these components because of the competition for access to adsorption sites. In other cases (e.g., reversed phase chromatography), the strong solvent (e.g., CH_3CN or CH_3OH) is a much better solvent of the feed components than the weak solvent (e.g., water) but it is not very strongly retained by the stationary phase (in pure water, the retension factor k' of acetonitrile on C18 is nearly 1, that of methanol about 11). In this case, gradient elution is not a competitive process and the integration requires merely the use of isotherms whose coefficients are functions of the local mobile phase concentration of this solvent.

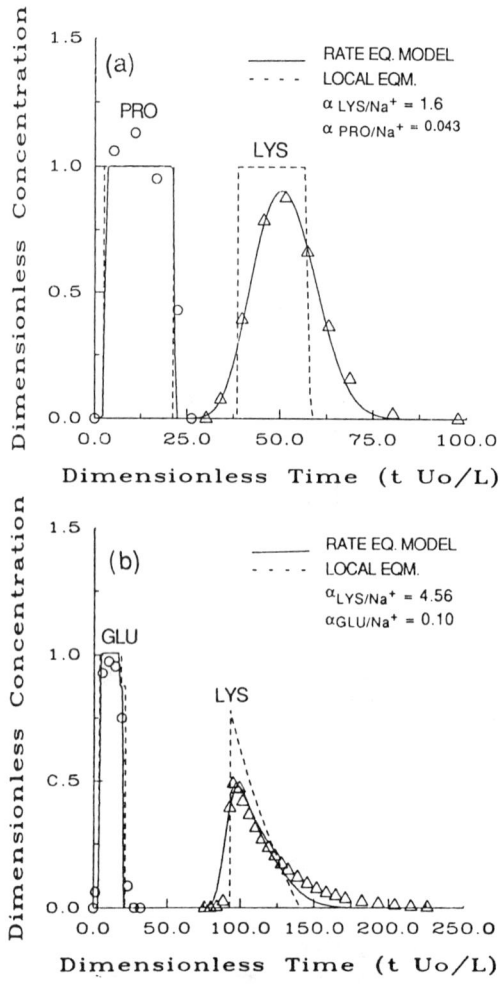

Fig. 27 Comparison of theoretical and experimental band profiles using the kinetic model with a proline/lysine (a) and a glutamic acid system (b). (Reproduced from Ref. 29 with permission of Elsevier.)

Zone migration in gradient elution [128] is more complex especially when the isotherm is nonlinear. Snyder et al. discussed the effect of the sample size on the bandwidth [129] and the band profile [130,131] in gradient elution chromatography. Moreover, they compared overloaded gradient profiles with those obtained in isocratic elution [129]. Using a rigorous model, Antia et al. showed

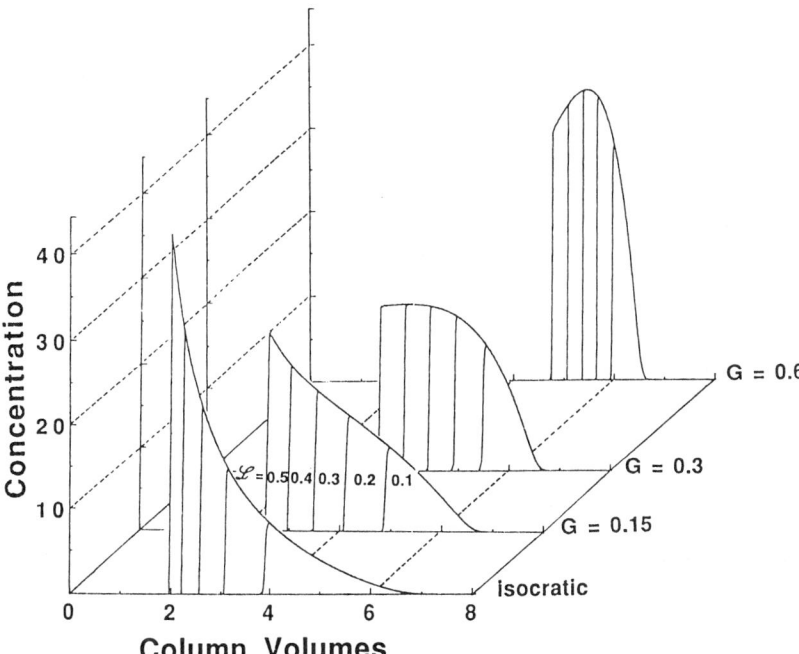

Fig. 28 Calculated single-component profiles in gradient elution as a function of the gradient steepness. (Reproduced from Ref. 132 with permission of Elsevier.)

the effect of the gradient steepness and sample size on the production rate in overloaded gradient chromatography (Fig. 28) [132].

The gradient steepness has a considerable influence on the band profile. This steepness is characterized by the value of $G = S\beta t_0$ (S = slope of the plot of log k_0' vs. the additive concentration, β = slope of the plot of the mobile phase composition vs. time, t_0 = dead time). The velocity of an analytical band (linear conditions) increases exponentially with time. For a concave down isotherm, the elution of the low concentrations of the diffuse rear profile is much more accelerated by the gradient than that of the high concentrations, which move in a less concentrated mobile phase. The concave up rear profile obtained in isocratic elution is altered when the strong solvent reaches the solute and, for shallow gradients ($G = 0.15$), the end of the rear profile appears concave down and forms a lump (Fig. 28). The importance of the lump increases with increasing gradient steepness until the profile assumes the conventional shark fin shape ($G = 0.30$). For a very high steepness ($G = 0.6$), the diffuse part of the profile exhibits a maximum.

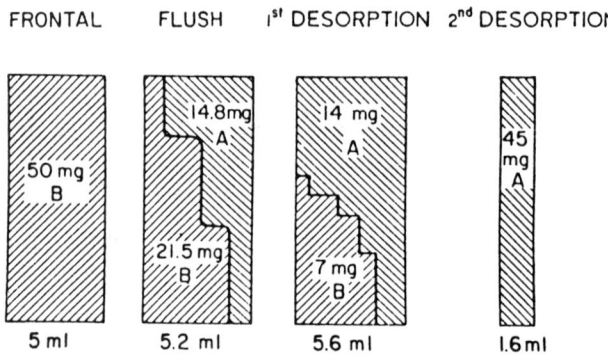

Fig. 29 Experimental results obtained with a step elution technique. (Reproduced from Ref. 133 with permission of Elsevier.)

Gradient operation appears too complex for industrial application because a simpler pumping mechanism is required. The use of step gradient techniques for eluting compounds of widely different capacity factors is a viable alternative. It has been investigated for protein purifications [133-135]. By choosing the appropriate step size and step time, the production rate can be optimized. An example of multistep desorption of β-lactoglobulin A and B is shown in Fig. 29 [133].

Frontal Analysis

The shape of the breakthrough profiles and the effect of axial dispersion on these profile shapes have been studied theoretically [19, 136] and experimentally [23,106,137-140]. The simple wave theory applies to the breakthrough profiles, since eventually the compositions of the feed and of the column effluent become identical. These profiles are characterized by the successive elution of two steep fronts (shock layers) for a binary mixture.

Figure 30a [141] shows calculated curves for two-component frontal analysis with competitive Langmuir isotherms and a column efficiency of 5000 theoretical plates. The dashed and dotted lines are the breakthrough curves of the first and second components, respectively. The solid line is the sum of the two concentrations and is typical of the detector response obtained experimentally [140]. The first step gives a different profile than all the following steps. For this first step, two shock layers signal the successive exit of the lesser and the more retained components. However, after this first step the column is no longer empty and competitive interaction between the two components leads to a different profile for the second-component step. Figure 30b illustrates the typical band shape

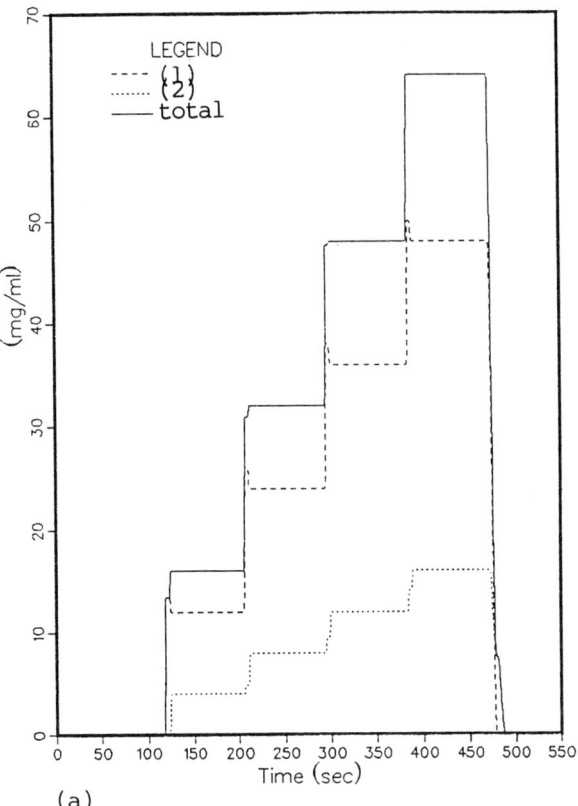

(a)

Fig. 30 Schematic of binary competitive isotherm determination of frontal analysis. Dashed line, concentration profile of the first component; dotted line, concentration profile of the second component; solid line, total concentration profile seen by the detector. (a) Typical experimental chromatogram in two-component frontal analysis. (b) Expansion of one step in (a). (From Ref. 141.)

for the first and second components in each consecutive step. The profile of the first component exhibits an intermediate plateau at a concentration higher than its new feed concentration. This is a consequence of the displacement effect. Then it drops to its injected concentration. On the other hand, the second component proceeds to its plateau concentration via an intermediate plateau which is lower than the injection concentration. Finally, the concentrations of the two components reach simultaneously constant values equal to the feed composition. The first step of a two-component frontal analysis was studied experimentally and theoretically

68 / Katti and Guiochon

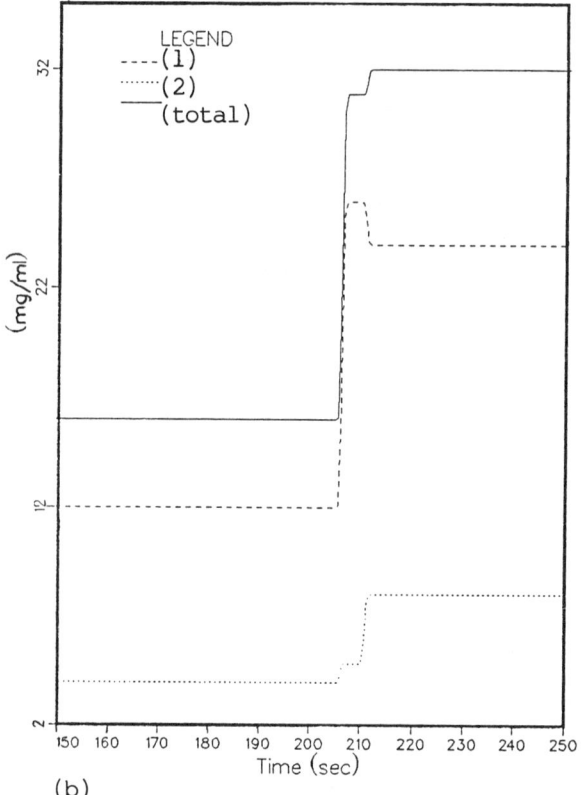

(b)

Fig. 30 (Continued)

by Carta et al. [213] (Fig. 31) for the breakthrough of two amino acids. Note that since the column is initially empty, the second component exhibits only a single front. Excellent agreement is obtained between the experimental and theoretical profiles. This problem is discussed further in the section on "Determination of Competitive Equilibrium Isotherms" in III.C.

The shape of the breakthrough curve has a large impact in affinity chromatography where mass transfer across the column is often slow or where sometimes quasi-irreversible adsorption of some of the feed components of interest may take place. Frontal chromatography has been applied in bioaffinity chromatography [142] and more recently to the preparative purification of a variety of compounds [143]. It has also been used to characterize new adsorbent materials [144].

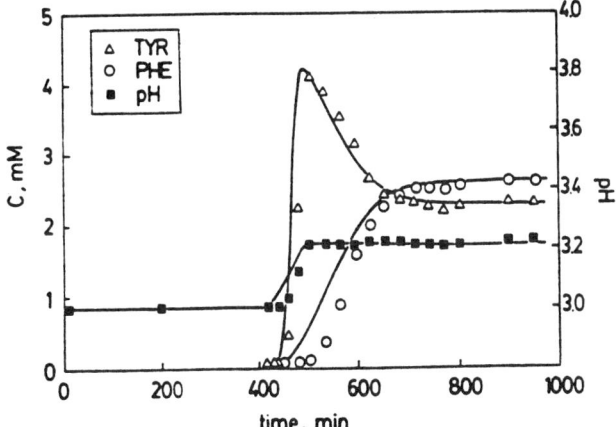

Fig. 31 Breakthrough behavior of phenylalanine and tyrosine for a bed of Amberlite 252 resin. Comparison of model and experimental result. Feed concentrations of phenylalanine: 2.6 mM; of tyrosine: 2.3 mM. Concentration of Cl$^-$, 1 mM. Flow velocity: 0.2 cm/sec. (Reproduced from Ref. 213 with permission of the American Institute of Chemical Engineers.)

Displacement Chromatography

In displacement chromatography, a feed sample is injected in a column saturated by a weak solvent in which the feed components have a high retention. Immediately after the feed injection is completed, a stream of displacer solution is pumped into the column. The displacer is a compound which is more strongly adsorbed by the stationary phase than any sample component. The simplest displacement problem involves one solute and a displacer. Although it is not a realistic separation problem, it can be used to understand the thermodynamic properties of displacement. Basmadjian and Coroyannakis modeled displacement separations neglecting accumulation in the mobile phase and band dispersion [145]. They reexamined some of the pioneering work of Glueckauf [32] by using geometrical concepts to predict adsorption behavior, e.g., by illustrating the significance of the watershed point. They investigated the displacement of a less adsorbed species with a more strongly adsorbed species and vice versa.

Displacement chromatography is discussed in detail in Section III.C as its utility is primarily for the separation of binary and multicomponent mixtures.

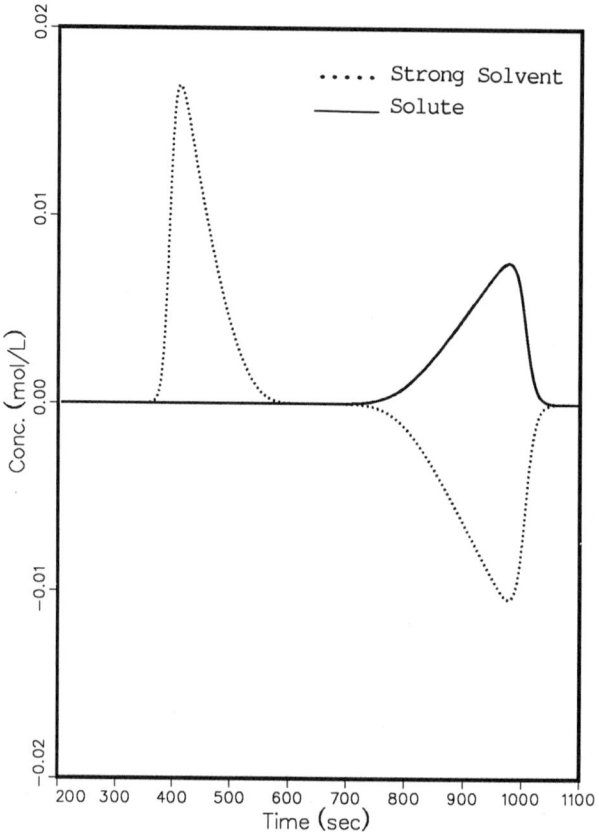

Fig. 32 System peak in overloaded elution chromatography. Band profiles calculated with the semi-ideal model and a Langmuir competitive isotherm for the additive and the sample. (From Ref. 74.)

System Peak with a Pure-Component Sample

The occurrence of system peaks appears as another nonlinear chromatography phenomenon. When a pulse of a pure compound is injected into a mobile phase containing a strongly adsorbing additive, such as an ion-pairing agent or a strong solvent (e.g., an alkanol in normal phase chromatography), system peaks occur [146,147]. The injection of the feed sample causes a perturbation of the additive equilibrium between the two phases in the column. Theory shows that two strong solvent bands appear at the column exit—one positive, the other negative. Figure 32 illustrates the band profile of the strong solvent (dotted line) and the solute (solid line) [74].

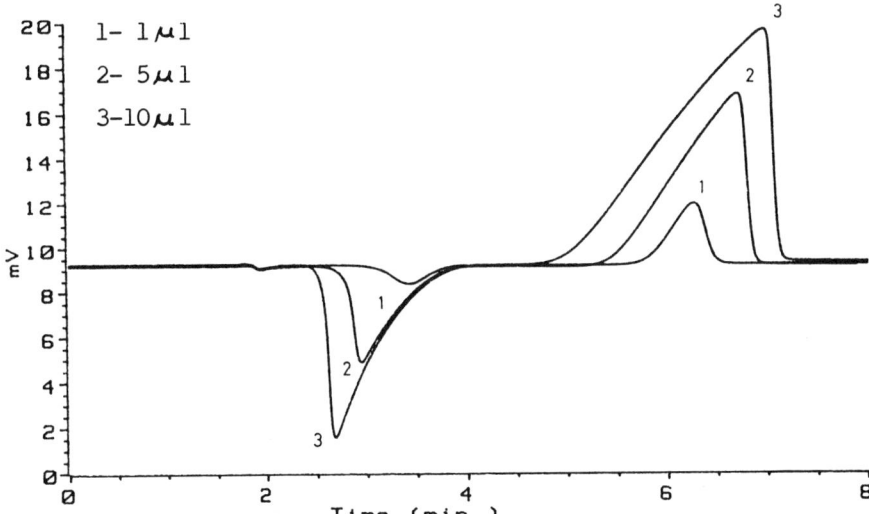

Fig. 33 Influence of the sample size on the band profile of the system peak and of the sample. Experimental profiles obtained with a refractive index detector, a 25-cm-long silica column, dichloromethane with 1% 2-propanol as mobile phase and 3-phenyl-11propanol as sample. (Reproduced from Ref. 75 with permission of Elsevier.)

One of the strong solvent peaks, usually the negative band, coelutes with the compound injected. The profile of the positive strong solvent peak depends on the nature and concentration of the additive. If the sample size is small, its profile is Gaussian. Otherwise, it follows the shape determined by its isotherm.

The effect of the sample size on the shape and size of the perturbation and on the elution profile of the solute is shown experimentally in Fig. 33 [75]. The difference between Figs. 32 and 33 is due to the different signs of the response factors of the strong solvent and the sample with the refractive index detector. Thus, the strong solvent profile in Fig. 32 (dotted line) appears as a negative peak in Fig. 33. Note that under the experimental conditions the shape of the profiles is as if the isotherms were concave up while in fact both single-component isotherms are Langmuir.

The shape of the solvent and solute peaks are strongly influenced by the relative adsorption strength of the strong solvent and the sample [48,74]. This relative strength is measured by the ratio of the retention factors of the compound studied and the additive in the pure weak solvent. When this ratio is smaller than 0.2, the presence of the additive has no effect on the sample band profile. That is, the band has the same shape as would be observed with a

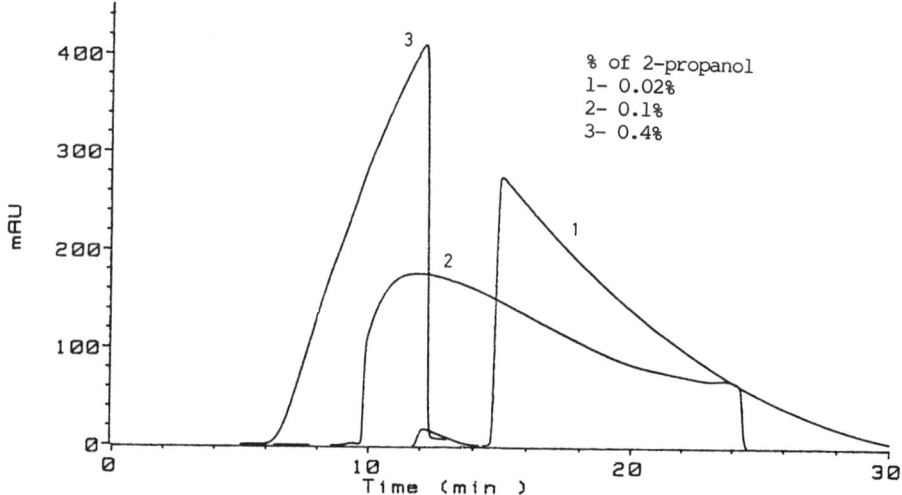

Fig. 34 Change in the shape of the band profile of 3-phenyl-1-propanol eluted on a silica column by dichloromethane with increasing concentrations of 2-propanol. (Reproduced from Ref. 75 with permission of Elsevier.)

(hypothetical) pure solvent S such that the adsorption isotherm of the component between S and the stationary phase would be the same as between the binary mixture and the stationary phase. If the ratio becomes larger, the competition between the sample and the additive becomes more intense.

The nature and concentration of the additive have a profound influence on the elution profile of the sample. Theory and experiment agree in the prediction of this effect [74,75]. Figure 34 shows experimental results which illustrate the influence of the additive concentration [75]. In this example, silica is used as stationary phase, 3-phenyl-1-propanol as the sample, and mixtures of dichloromethane and 2-propanol (0.02-2%) as solvents. The additive and the sample follow the competitive Langmuir isotherm model. At low additive concentration, the profile is Langmuirian (curve 1). At large additive concentrations, an inversion of the band asymmetry is observed and the band profiles appear to correspond to an anti-Langmuir isotherm (Fig. 32, curve 3). In the transition region between the Langmuir and anti-Langmuir behavior, the band profiles are extremely unusual and broad profiles are observed (Fig. 34, curve 2). These profiles have a sharp front, a sharp rear, and sometimes exhibit two maxima. The occurrence of strange band profiles similar to those predicted under the same kind of experimental conditions had already been reported [148,149].

Similarly, increasing the elution strength of the additive, holding all other parameters constant, changes the shape of the band profile from Langmuirian to anti-Langmuirian while its retention time decreases. However, in this case the transition profile is Gaussian [74]. This effect can be understood as the result of a progressive change in the apparent isotherm of the solute in the mobile phase, from concave down to linear to concave up.

Determination of Competitive Equilibrium Isotherms

Several chromatographic methods have been developed for the determination of single-component equilibrium isotherms (Section III.B). Similarly, chromatography can be used for the direct determination of competitive isotherms. Two methods are available: frontal analysis which was originally extended to two-component systems by Jacobson et al. [140] and the method of the hodograph transform developed by Ma et al. [150]. Pulse tracer methods [78] can be used, but severe constraints limit their practical utility.

Frontal analysis is often the method of choice for the determination of single-component isotherms. In two-component FA, a series of binary solutions of constant relative composition of the two components to be studied is prepared. They are pumped through the column in the order of increasing concentration, and the breakthrough curves are recorded (see Fig. 30). Since the breakthrough curves are complex, it is necessary to determine the concentrations at the intermediate plateaus. This is accomplished by sampling the concentrations at specific points, using a second chromatograph. From the breakthrough times and the plateau concentrations, the isotherms of the two components can be determined along a line of constant composition (i.e., the intersection of the isotherm surfaces $q_i(C_1, C_2)$, by the vertical plane C_1/C_2 = const). The experiment is repeated with a series of solutions corresponding to other ratios of the concentrations, in order to determine the entire isotherm surface. This method has been applied to the measurement of the competitive adsorption isotherms where an interchange of the elution order occurs at increasing sample size [151].

Another method was proposed by Ma et al. [150]. It is based on the properties of the simple wave solutions. Large-volume rectangular injections of binary mixtures of various compositions are carried out. The width of these pulses is such that a plateau of significant width is recorded. The individual elution profiles are determined, which requires the collection and analysis of several tens of fractions during the elution of the front and rear of the pulse (Fig. 35a). The hodograph transform (i.e., a plot of the concentration of one component vs. the concentration of the other one) of these two elution profiles contains all the information regarding the competitive equilibrium isotherms. In the case of a

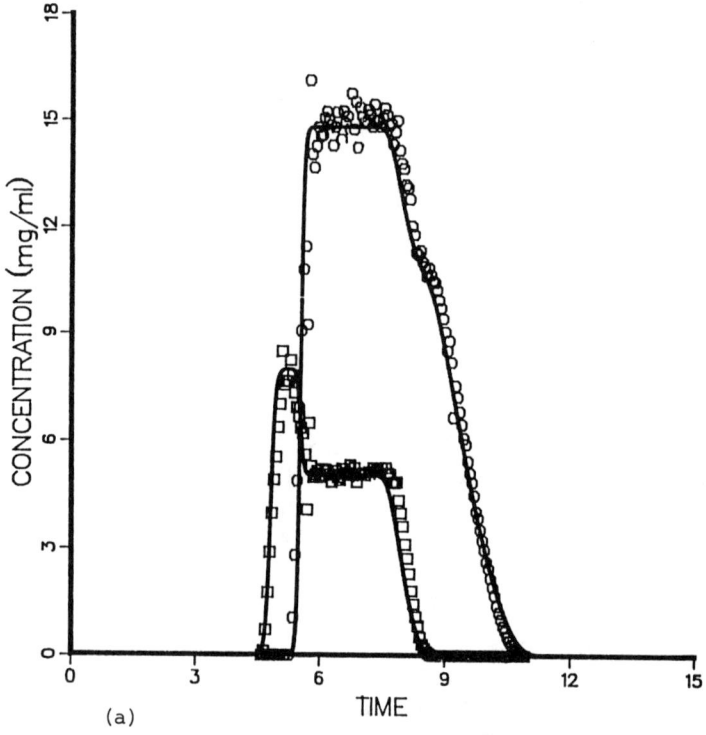

Fig. 35 Comparison between experimental results and model for the simple wave method of competitive isotherm determination. (a) Individual elution profiles of the components of a wide rectangular pulse of a binary mixture. Symbols: experimental results. Solid lines, calculated profiles using a kinetic model and a Langmuir isotherm with the best coefficients derived from the hodograph transform (Fig. 35b). (b) Hodograph transform of the experimental data (symbols) and best linear fit of these data. (Reproduced from Ref. 150 with permission of the American Chemical Society.)

competitive Langmuir isotherm, the transform is made of two straight lines intersection at the coordinates of the plateau (Fig. 35b). The isotherm parameters can be determined by a least-squares fit of the data points to the straight lines.

This second method requires high frequency fraction collection and analysis [50,123]. It is labor-intensive, unless automatic equipment is available. On the other hand, it has the advantage of requiring much smaller amounts of the compounds studied, which permits isotherm determinations with rare biochemicals.

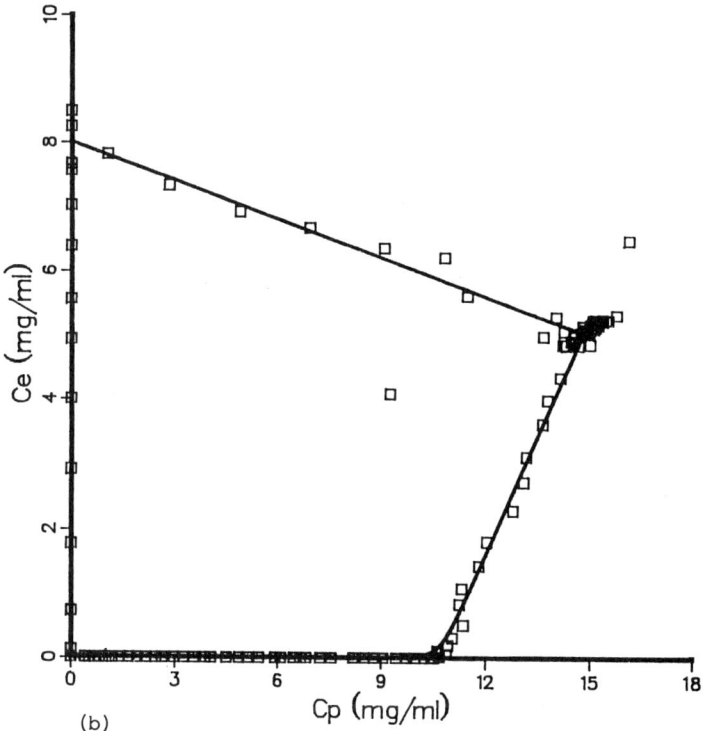

Fig. 35 (Continued)

Optimization of Preparative Chromatography

The term preparative chromatography is primarily associated with the concept of nonlinear chromatography using a finite width pulse injection (i.e., overloaded elution chromatography) [152]. This method is employed primarily by chromatographers and biochemists to produce pure products at the bench scale. However, recently its use has expanded into other fields, essentially in the pharmaceutical industry. Chemical engineers have become increasingly involved in the development of the method. In process scale chromatography, fraction purity, production rates, and recovery yields are more important than the mental comfort or convenience of the chemist. The optimization of experimental conditions for maximum production rate or minimum cost has become a relevant topic of research.

Knox and Pyper [28] developed expressions to calculate the maximum throughput of a chromatographic column at low loading, assuming no competitive interactions between the solute and the stationary

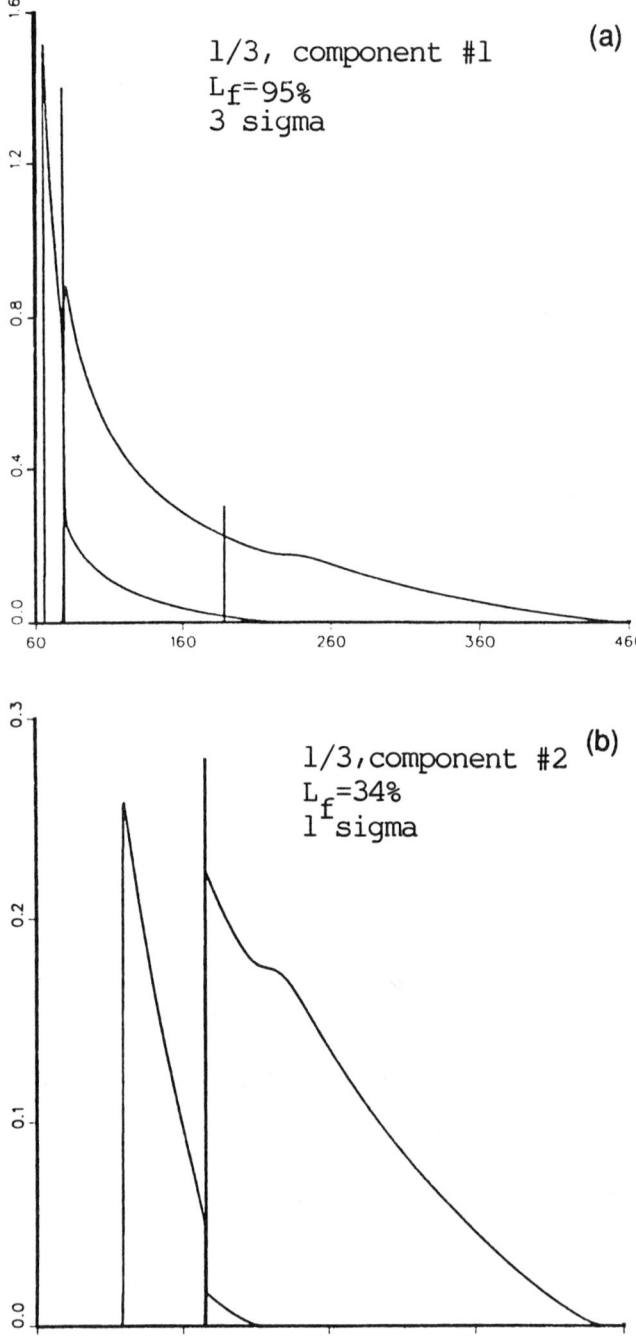

Fig. 36 Optimum sample size (as loading factor) for maximum production rate of either the first (left column) or the second (right column) component of a pair, using a high-efficiency column. Feed composition 1:3 (top) or 3:1 (bottom). The number of σ is the optimum feed

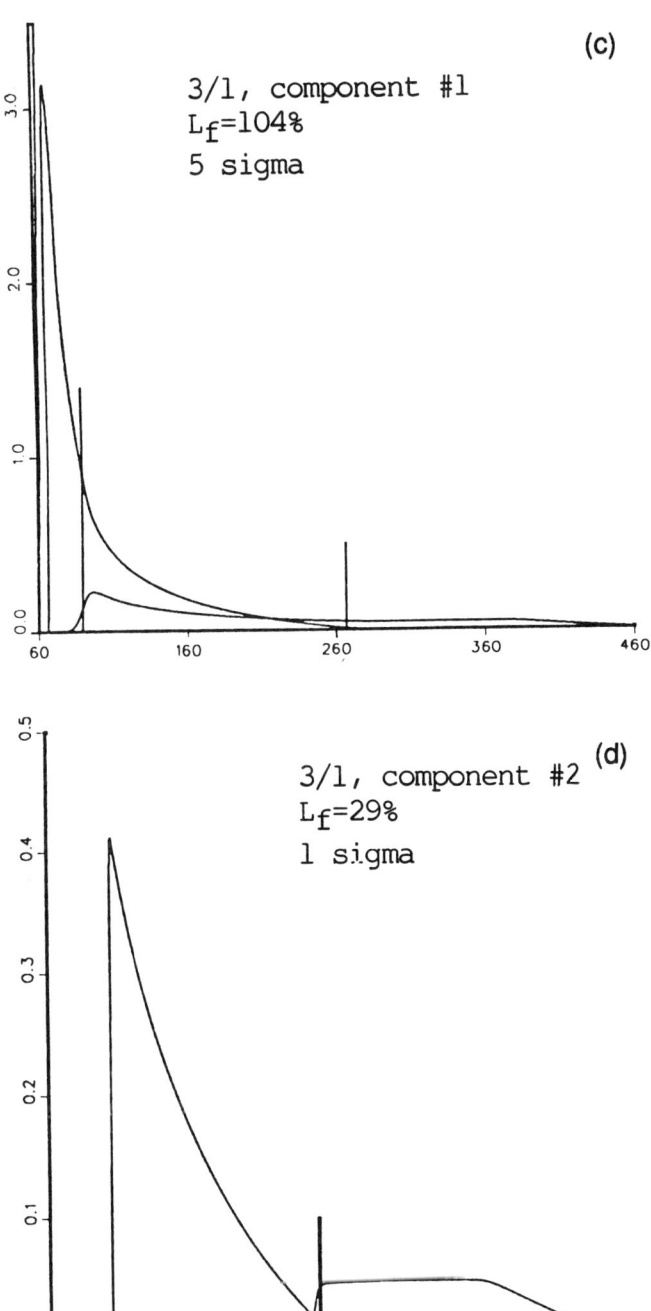

volume, given in volume unit of standard deviation of the band, under linear conditions. (Reproduced from Ref. 153 with permission of the American Chemical Society.)

phase and no overlap of the elution bands (i.e., 100% yield and 100% product purity). Their model includes a correction for the finite column efficiency which assumes shift invariant convolution of the thermodynamic effect (ideal model solution) by the mass transfer kinetics, i.e., additivity of band variance contributions. This simple model predicts an optimum throughput for the second component which is in agreement with the results of calculations using models which take into account the effect of competitive interaction and of axial dispersion, and which assume 99% purity [153-155]. For the first component, the throughput predicted is too low when the concentration of the second component is larger than that of the first one, since a strong displacement effect takes place in this case and band interference must be accounted for correctly.

The individual band profiles of a two-component mixture were calculated using finite difference techniques [46] in order to determine the effect of the individual parameters such as sample size [156], separation factor [157], flow velocity [158], and feed composition [159] on the production rate [160]. Optimization of the production rate of the second component with high-efficiency columns as a function of the sample size for a binary mixture shows that one should overload the column until the amount of overlap meets the purity requirements [153]. Optimization of the production rate of the first component as a function of the sample size for the same binary mixture shows that one should overload the column considerably, much beyond the degree of overlap for which the resolution between both bands has vanished [153]. Figure 36 shows the chromatograms corresponding to the maximum production rate on a given column, for the first and second components for two mixtures of different composition at a product purity of 99%. These results are in agreement with those of the simplex optimization made by Ghodbane and Guiochon [160].

The properties of the analytical solution of the ideal model in the case of Langmuir competitive isotherms permits the detailed investigation of the effect of sample size, separation factor, feed sample composition, limiting pressure drop, particle size, column nominal efficiency (i.e., maximum efficiency under linear conditions), and the required purity of the products on the production rate of the second component of a binary mixture and of the sample amount [154,155,161].

Several strategies have been considered by Golshan-Shirazi and Guiochon [154,155,161-163], such as finding the maximum production rate without or with recovery yield constraints. The optimum value of the sample size in each case is remarkably independent of the column efficiency (but the recovery yield and the production rate do depend on the nominal efficiency). There are no separate optima of the column length (L) and the particle size (d_p), or these optima are shallow (Figs. 37a and 38) [154]. Figure 37a (curves 2 and 3) show that columns with different lengths and particle sizes but the same ratio d_p^2/L give nearly the same production rate for the second component. On the other hand, there is an optimum value for the ratio d_p^2/L for a given pressure drop (Fig. 37b). Note in this figure that a given curve corresponds to columns of different lengths and particle sizes, operated at different flow velocities. Lastly, the maximum production rate which can be obtained with a given chromatographic system (i.e., constant isotherms) increases with increasing inlet pressure at which the column can be operated (Fig. 37c) [154]. Note in Fig. 37c that each point corresponds to a different value of the ratio d_p^2/L (i.e., a different column).

The production rate increases rapidly with ($\alpha - 1$), with the square of ($\alpha - 1$)/α for a given column (see Fig. 38b), and with the cube of ($\alpha - 1$)/α if the optimum column is used [164]. The production rate also increases rapidly with decreasing product purity. Finally, the optimum mobile phase flow velocity is high, corresponding to values of the operating efficiency (i.e., under linear conditions, at the optimum flow velocity) for which the resolution between the two components is below 1 (easy separations, $\alpha \geq 1.3$) or around 1 ($\alpha \leq 1.1$) (Fig. 38). In this figure, d_p^2/L is the same (i.e., 5×10^{-8} cm) for the four columns considered; the maximum production rate for a given pressure is nearly the same but the optimum velocity increases very fast with increasing particle size.

Preparative chromatography can be conducted in the displacement mode (see section on "Theory of Separations by Displacement Chromatography" in III.C). The potential advantages over the elution mode are the recovery of more concentrated fractions and the injection of larger sample amounts. The inconvenience is the addition of one experimental parameter which must be optimized: the displacer concentration. A suitable displacer has to be found, making the method development longer and more tedious. From the detector response, it is not easily possible to determine the individual component profiles. Thus, fraction collection or on-line analysis must be performed. After a feed batch has been separated, the column must be regenerated and the displacer eluted from the column before another cycle can be started. Therefore, the amount of purified material per cycle must be larger than in elution to compensate for these additional costs. Finally, the recovery yield can never achieve 100% because the adjacent bands in an isotachic train overlap.

Several studies have investigated the effect of the variation of a single parameter (i.e., sample size, flow velocity, displacer concentration, column length), holding all other parameters constant, on the production rate in displacement chromatography [26,177,181,182]. These studies show that there is an optimum velocity, an optimum

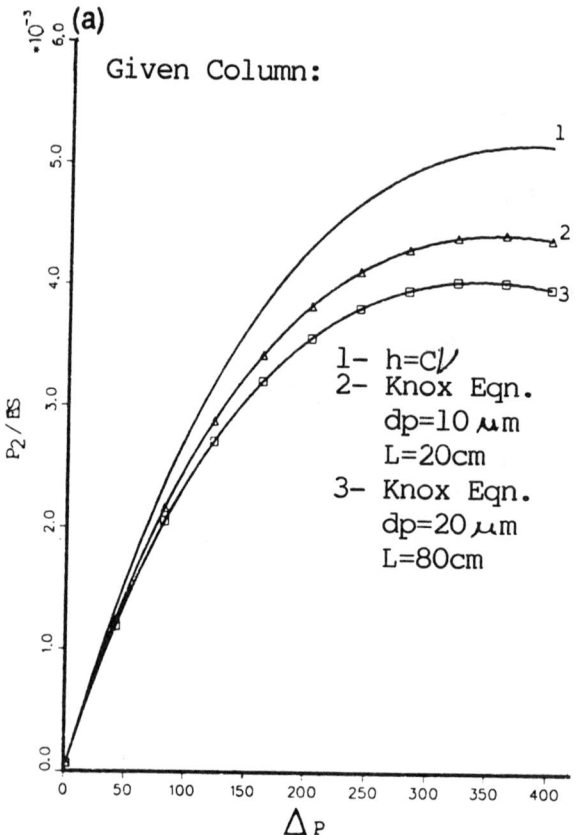

Fig. 37 Optimization of the experimental conditions in preparative chromatography. (a) Influence of the inlet pressure on the production rate of a given column. (b) Influence of the ratio d_p^2/L (d_p = particle size; L = column length) for three different pressures. (c) Influence of the available pressure on the production rate. (Reproduced from Ref. 154 with permission of the American Chemical Society.)

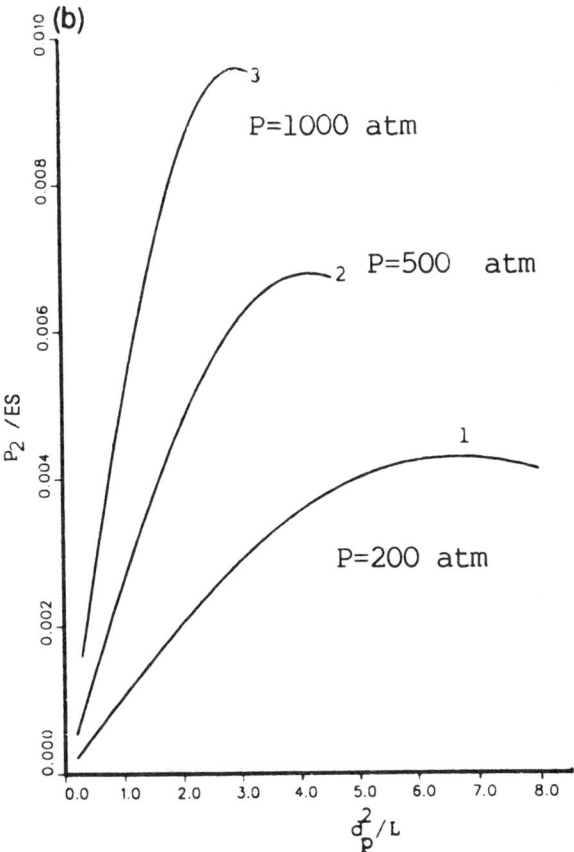

Fig. 37 (Continued)

displacer concentration, and an optimum column length. However, these parameters are coupled; thus the scope of the conclusions is limited.

Recent results based on multidimensional optimization of the sample size, displacer concentration, and flow velocity using a simplex algorithm show that the maximum production rate is achieved when the displacement train is nonisotachic (Fig. 39a) [165]. Moreover, this study shows that displacement chromatography should be carried out at high flow velocities (Fig. 40). These results are in agreement with the observation by Cramer and Subramanian that the individual band profiles in displacement chromatography become more diffuse with increasing mobile phase velocity [167].

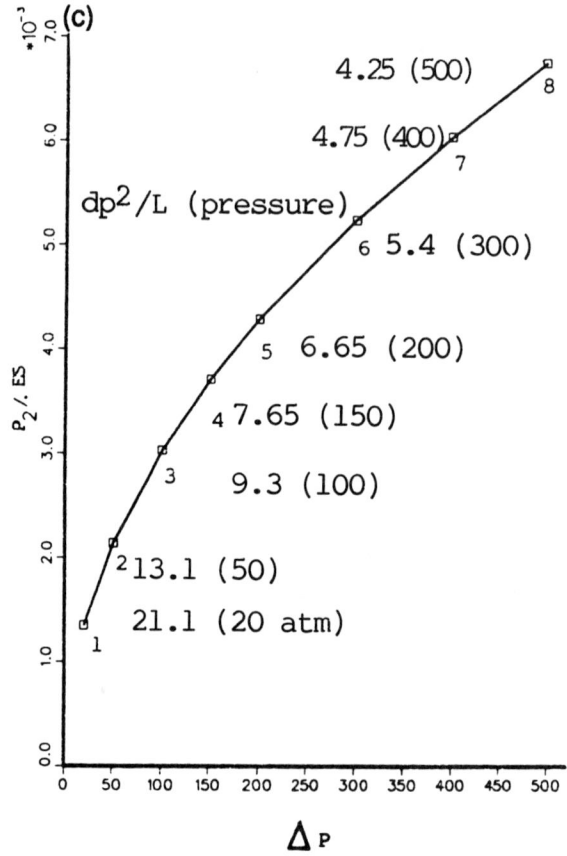

Fig. 37 (Continued)

Preliminary results in the comparison of overloaded elution and displacement chromatography have been presented [26,207]. They appear to be biased in favor of the latter method. A systematic multidimensional optimization shows that displacement chromatography has the advantage of giving high concentrations of the collected fractions but the disadvantage of low yields (Fig. 40) [165]. On the other hand, overloaded elution gives high yields but, since isocratic elution is a dilution process, the concentrations of the collected fraction are low. Figure 39 illustrates the profiles at optimum conditions for a binary mixture, in displacement and overloaded elution [165].

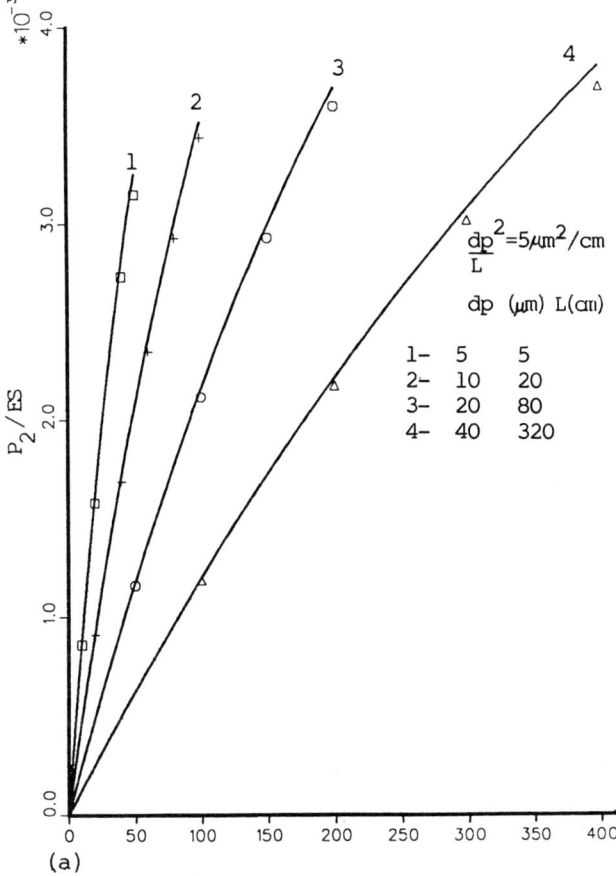

Fig. 38 Optimization of the experimental conditions in preparative chromatography. (a) Variation of the production rate for the second component with the mobile phase velocity. Influence of the ratio d_p^2/L on the maximum production rate. (b) Variation of the maximum production rate of a given column for the second component with the ratio $[(1 - \alpha)/\alpha]^2$ for different conditions. (Reproduced from Refs. 154 and 161 with permission of the American Chemical Society.)

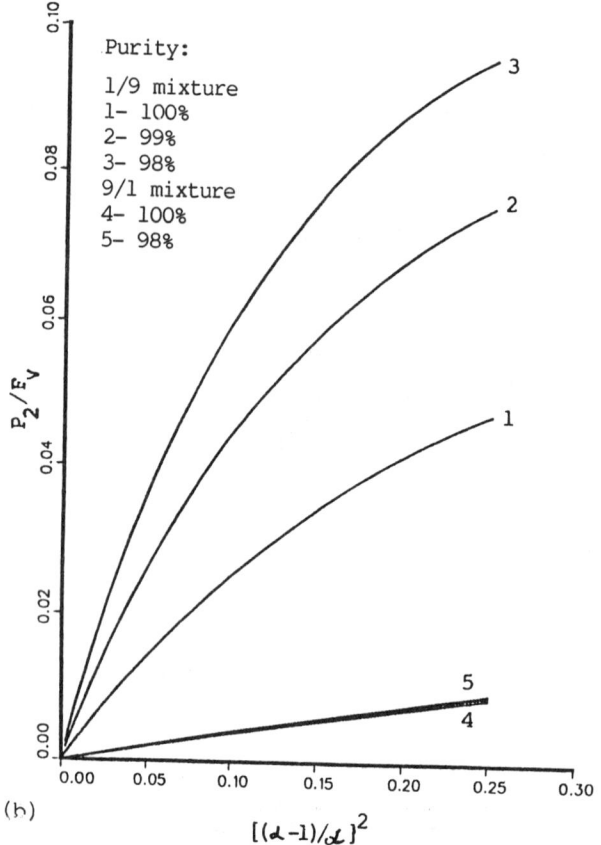

Fig. 38 (Continued)

Other Applications

The enrichment of a trace component in a mixture can be carried out using the properties of nonlinear chromatography [168-170]. This is done by increasing the loading factor to maximize the displacement effect. Calculations show that a fraction containing ~90% of the trace component and some amounts of the compounds eluted just after and/or before it, but at a much enhanced relative concentration of the trace component can be isolated [168]. Detection limits of trace components can be decreased by one to two orders of magnitude (Fig. 41) [169]. Extremely small amounts of thymine-thymine dimer have been extracted from thymine solutions by shaving a narrow fraction at the front of the overloaded, later eluting

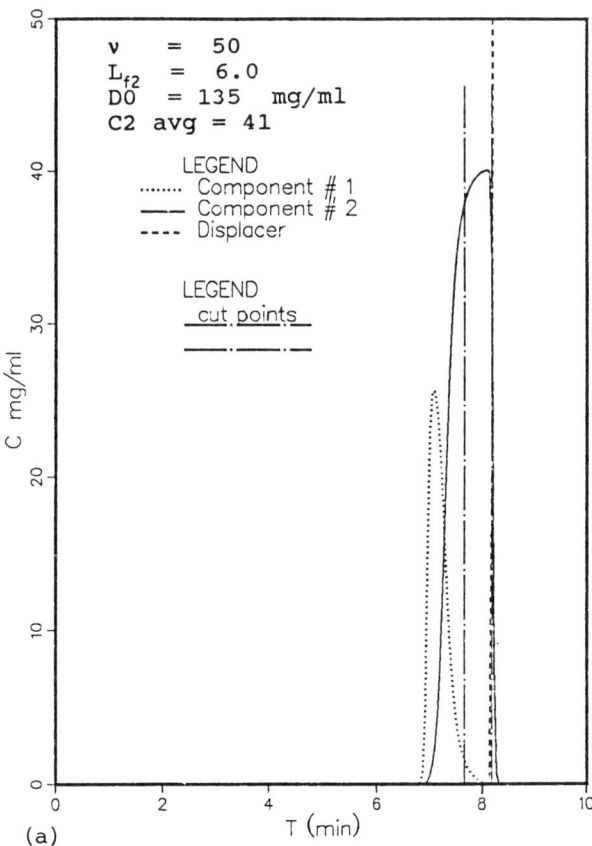

Fig. 39 Comparison between the optimum chromatograms for maximum production rate obtained in overloaded elution and in displacement chromatography for a given column and a given feed composition. 25-cm-long column, packed with 20-μm particles. 1:3 mixture, $\alpha = 1.20$. Constraints: recovery yield, 60%; fraction purity, 98%. (a) Displacement chromatography. (b) Overloaded elution. (Reproduced from Ref. 166 with permission of Elsevier.)

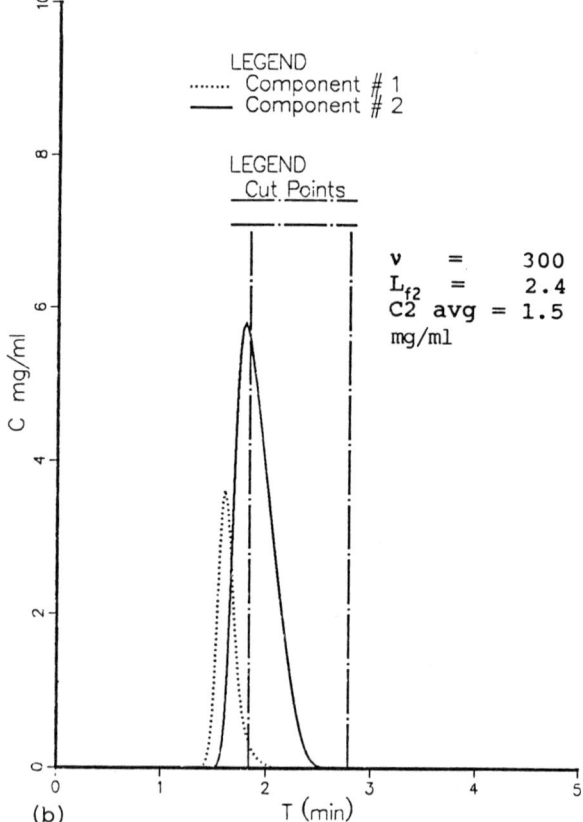

Fig. 39 (Continued)

band of thymine. A multistep enrichment procedure permitted a demonstration of the validity of the approach [169].

El Fallah and Guiochon studied the interaction between two compound bands which are injected separately [171]. The band of the less retained compound is injected first. The second one follows after an adjustable delay. A converse displacement effect was identified. In this case, the band of the more retained compound is displaced by the less retained band. Experimental evidence of this new effect was demonstrated using two different chromatographic systems. Excellent agreement between calculated and recorded elution band profiles was demonstrated with 2,4-dimethylphenol and 2-methyl-2-phenylethanol, while only fair agreement was achieved with 2,4-dimethylphenol and 2-phenylethanol. In the latter case,

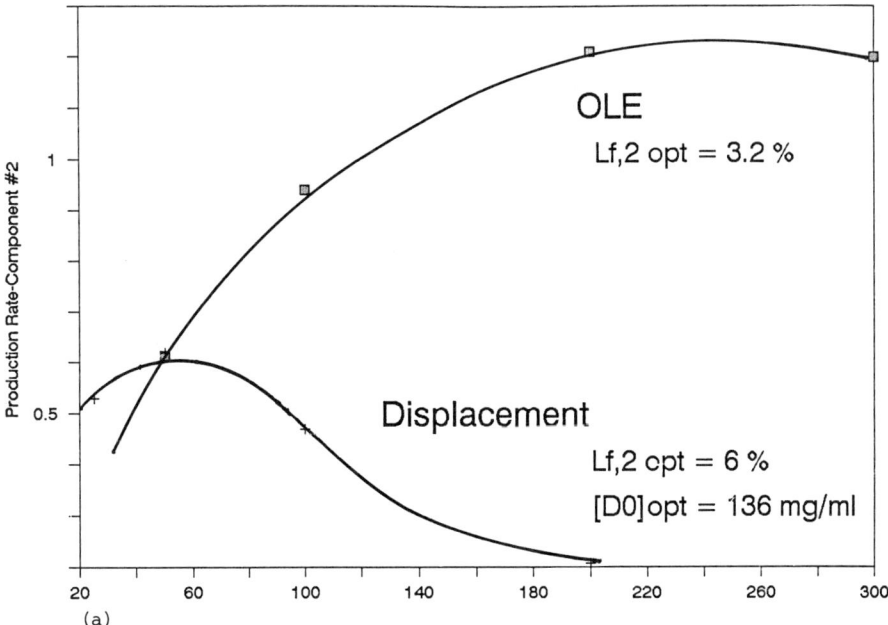
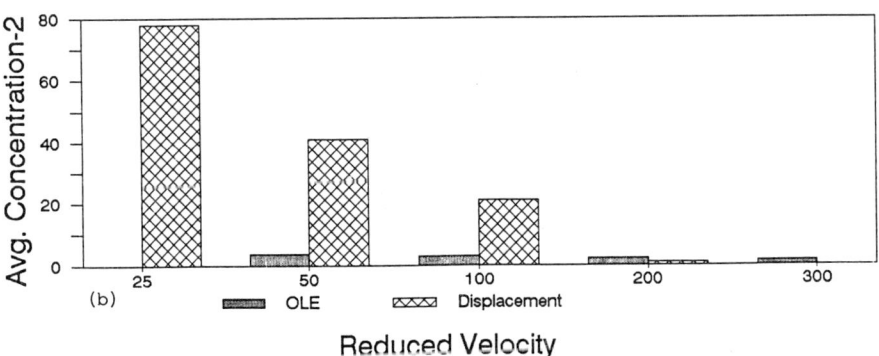

Fig. 40 Comparison between the maximum production rate and the fraction purities obtained in overloaded elution and in displacement chromatography, with a given column and a given feed composition. Top, variation of the maximum production rate for the second component with increasing flow velocity. Bottom, concentration of the recovered fraction. (Reproduced from Ref. 166 with permission of Elsevier.)

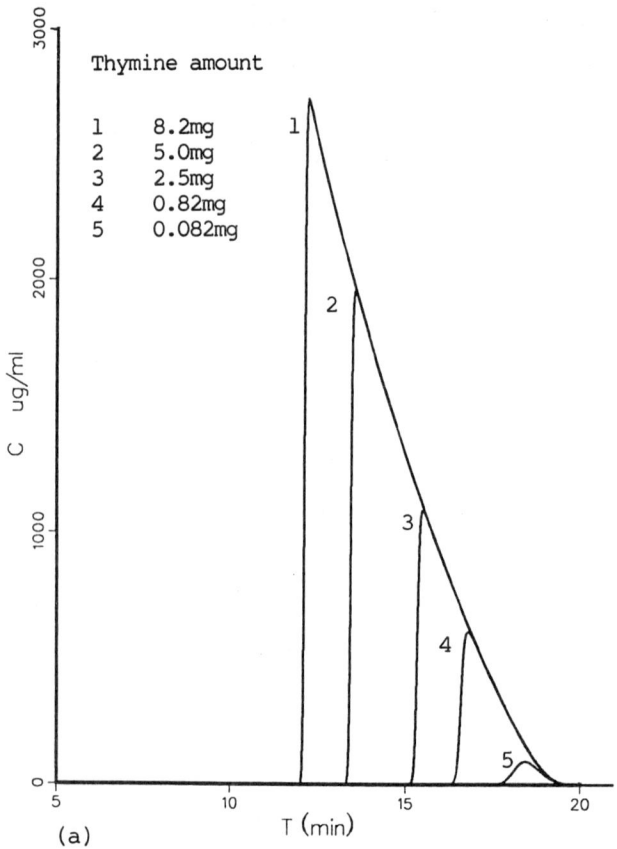

Fig. 41 Elution profile of the bands of a trace component and of the major component. Influence of sample size. Column 25 × 0.46 cm, packed with C18 HS silica, 7 μm. Mobile phase, water, 1 ml/min. (a) Predicted band profiles for thymine samples of increasing amount. (b) Predicted band profiles for thymine dimer samples diluted with increasing amount of thymine, as indicated in (a). Constant amount of thymine dimer: 82 ng. (Reproduced from Ref. 168 with permission of Elsevier.)

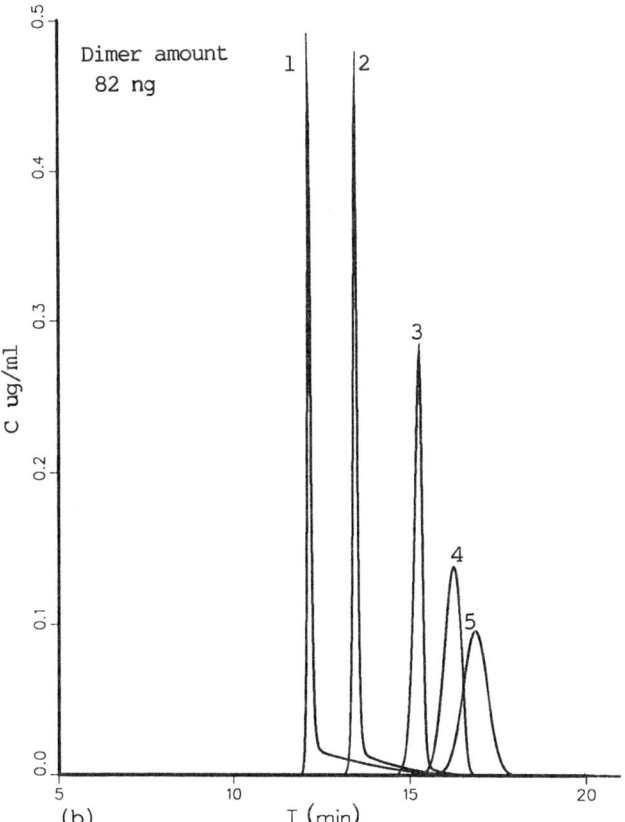

Fig. 41 (Continued)

the deviation is due to the failure of the competitive Langmuir isotherm to account correctly for the competitive adsorption behavior of the two compounds for which the difference in column saturation capacity is more important than for the former couple.

C. Three-Component and Multicomponent Problems

With three components, the number of possible initial and boundary conditions increases considerably. However, the really new cases, those which cannot be considered as the mere superimposition of two separate two-component problems, are still few. As we will show, most three-component separation problems in elution can be understood simply in terms of the proper combination of the results obtained in the two-component cases. Three-component systems

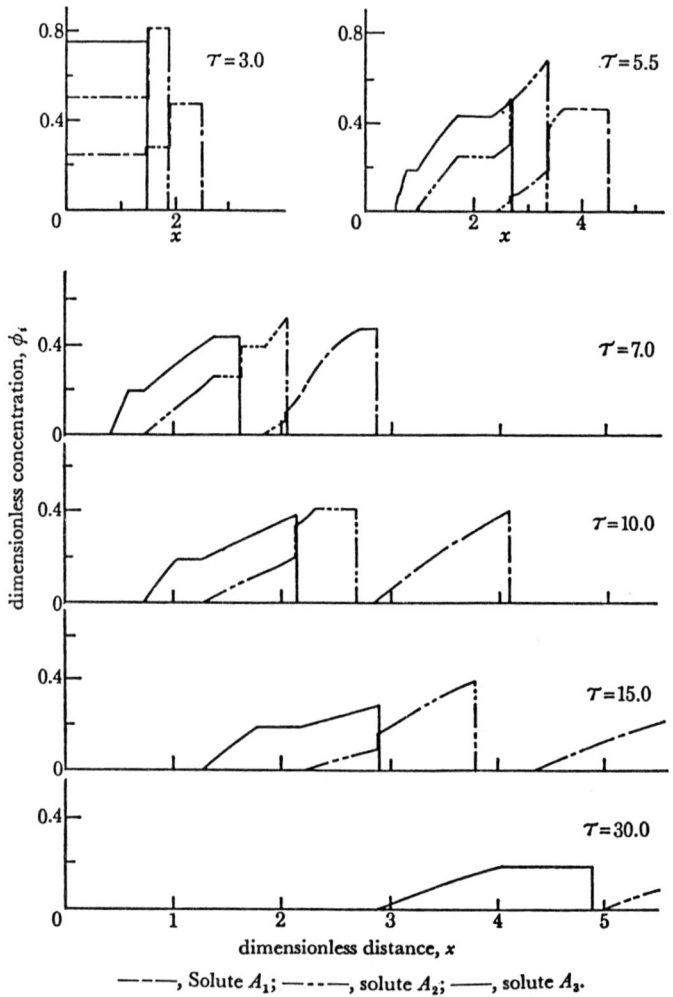

Fig. 42 Separation of the components of a ternary mixture as predicted by the ideal model. (Reproduced from Ref. 20 with permission of the Royal Society, London.)

nonetheless permit the study of binary separation problems in displacement chromatography and when a strong additive is introduced in the mobile phase (gradient elution, system peaks). The calculation of the band profiles in these cases requires the use of ternary competitive isotherms for which the Langmuir model is still less satisfactory than in the case of binary mixtures.

We consider in this section only the following cases: elution of a three-component sample (using the semiequilibrium model), separation of a binary mixture in displacement mode, separation of a binary mixture with a mobile phase containing a strong additive, and separation of a binary mixture in gradient elution. Frontal analysis loses its interest for multicomponent mixtures, unless one is interested in the preparation of pure fractions of the first eluted component of the mixture and recovery yield is not a problem [143]. The determination of ternary competitive equilibrium isotherms is still in its infancy, the problem being as yet unsolved for binary isotherms.

Elution Mode

Rhee et al. [20] calculated the individual band profiles at a given time for the three components of a ternary mixture during its progressive separation (Fig. 42). The solution of the ideal model are calculated by using a variable transformation. The tag-along effect is seen clearly on the last two components at $\tau = 7.0$. As the separation progresses, the tag-along effect is seen only on the third component at $\tau = 10$ and $\tau = 15$. Complete separation has occurred at $\tau = 30$.

Using the semi-ideal model, calculated profiles have been generated for a series of mixtures of three components of variable relative compositions using different sample sizes [172]. In most instances, e.g., in Fig. 43a, the patterns obtained could be explained by a combination of a displacement effect (the compression of an early eluting band by a later eluting one) and a tag-along effect (the spreading of a late eluting band by the band which is eluted just before it). When the concentration of the second component of a ternary mixture is low and the concentrations of the other two components are high, however, the second-component band is squeezed between the other two and its profile is unusual (Fig. 43b). Schematically, this profile looks like a half-Gaussian with a rear shock followed by a long tail of low concentration. This profile results from the combination of intense displacement and tag-along effects. This phenomenon can be used for trace enrichment [168].

Displacement Chromatography

In displacement chromatography, a sample is injected in a column saturated by a weak solvent for which its components have a high retention. Immediately after the feed is introduced onto the column, a displacer solution is pumped into the column. The displacer is more strongly adsorbed by the stationary phase than any sample component. The feed components are displaced from the stationary phase and form a series of bands before the displacer breakthrough front. After steady state is achieved, a series of bands, forming

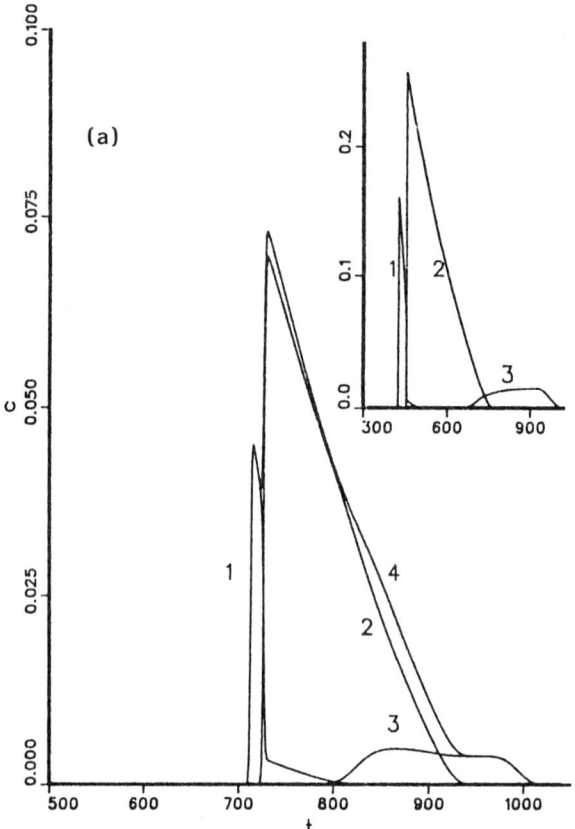

Fig. 43 Separation of the components of a ternary mixture as predicted by the semi-ideal model. Both displacement and tag-along effects are apparent. (Reproduced from Ref. 172 with permission of Elsevier.)

an "isotachic train," moves along the column at the rate of the displacer front. The concentration of each component band depends on the displacer concentration and can be determined from the intersection of the operating line and the adsorption isotherms of each solute [32,173]. The length of each band depends on the amount of the corresponding component injected with the feed.

The formation of the isotachic train requires a certain time and is a complex phenomenon [165]. Because of the need of a displacer which has to be more strongly retained than any feed component (acetonitrile cannot be used as a displacer in reversed phase

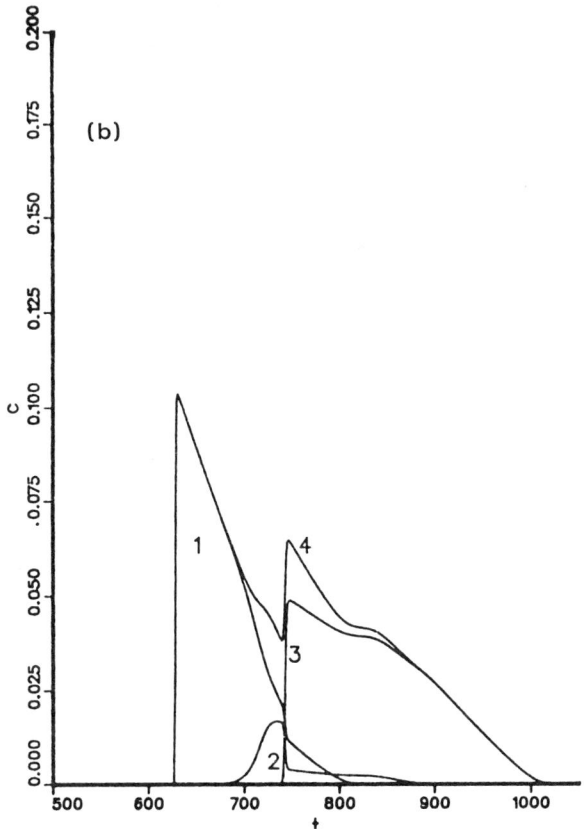

Fig. 43 (Continued)

chromatography, while it can be used as a strong solvent in gradient elution), the separation of a binary mixture by displacement is a three-component problem.

The difference between elution and displacement chromatography was recognized by Tswett in the early 1900s [174]. However, it was Tiselius [175,176] who first clearly defined and discussed the various possible modes of chromatography (elution, frontal analysis displacement). Glueckauf [3,5] gave the first theoretical analysis of displacement development. He also showed the effect of the mobile phase flow rate on displacement development [32].

In the last 40 years, the enormous growth of linear elution chromatography for analytical separations has overshadowed the use of displacement chromatography, which remained essentially ignored

from the chromatographic community until the early 1980s, in spite of several important industrial applications. In the early 1980s Horvath and coworkers [25,173,177,178] reintroduced this technique. They capitalized on various technological advances, high-efficiency columns, and fast high-performance liquid chromatography [179].
A comprehensive review of the history and development of displacement chromatography and of its applications was published by Frenz and Horvath [25]. Only the major advances and recent applications will be reviewed in this section.

Theory of Separations by Displacement Chromatography

On the theoretical front, Helfferich and Klein presented a general theory that could be applied to multicomponent displacement separations by a transformation of variable called the h transform [21]. This theory assumes ideal chromatography with equilibrium competitive Langmuir isotherms but can be extended to introduce column efficiency [108,180]. A clear summary account of this theory and of its application to the determination of the boundaries of the isotachic train for an infinitely efficient column was written by Frenz, Van der Schrieck, and Horvath [181] who reported good agreement in their comparison between the results of the calculation and those of experimental determination of band profiles.

The validity of the theory developed by Helfferich and Klein is limited, however, to the cases when the competitive Langmuir isotherm model accounts correctly for the equilibrium behavior of the components involved *and* when the column efficiency is very high (more than ~50,000 theoretical plates). Experimental results suggest that the former condition is not generally verified [50,71,72,141] but could rather happen only when the column saturation capacities of the two components are identical (e.g., isotopic compounds, enantiomers [70].

Rhee and Amundson [93] applied the theory of multicomponent chromatography to elucidate the characteristics of displacement systems for Langmuir competitive isotherms. They calculated the band profiles at different column lengths using the ideal model to show displacement development. Figure 44 shows the individual band profiles obtained for a three-component sample. During the development, strange band shapes appear. After enough time elapses, however, a train of adjacent "box cars" forms. The real importance of this result has been blown up out of proportion in many textbooks, which explains how the solution of the ideal model has led people to associate displacement chromatography and vertical zone boundaries. As we show in the next paragraphs, the size of the boundary regions where bands interfere can be very significant.

Later, numerical methods were developed to study displacement separations when the column has a finite efficiency [165]. Individual

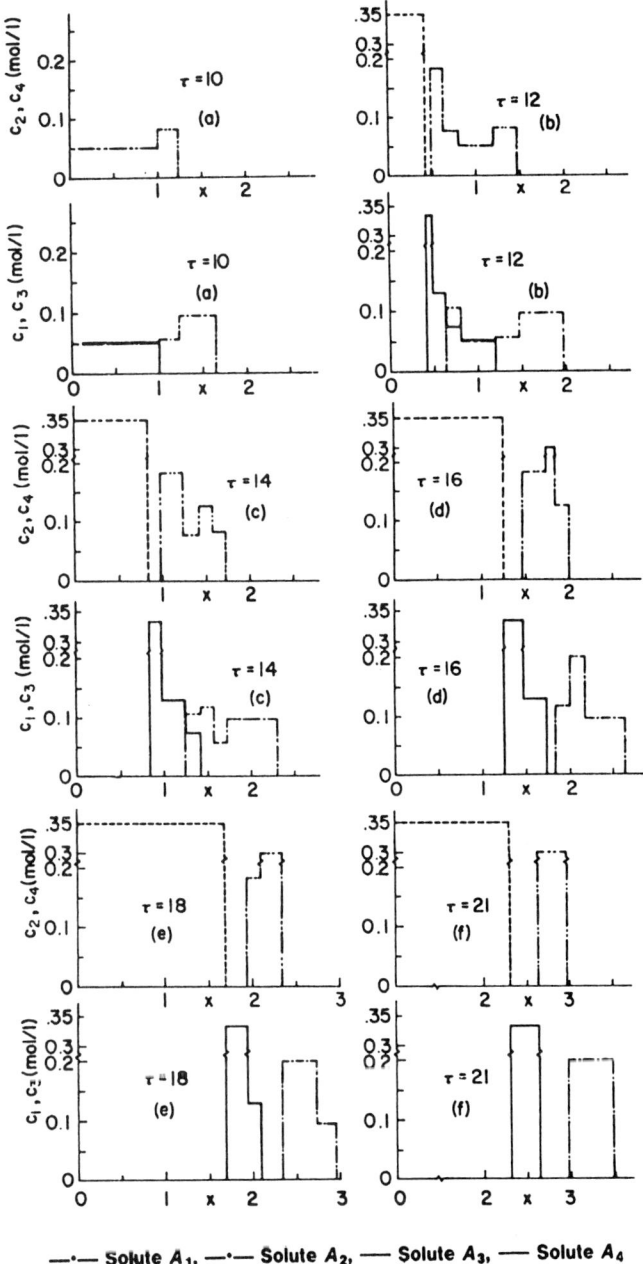

—·— Solute A_1, —·— Solute A_2, —— Solute A_3, —— Solute A_4

Fig. 44 Prediction of the intermediate band profiles of a multicomponent mixture during the formation of an isotachic train in displacement chromatography. Calculation made with the ideal model. (Reproduced from Ref. 93 with permission of the American Institute of Chemical Engineers.)

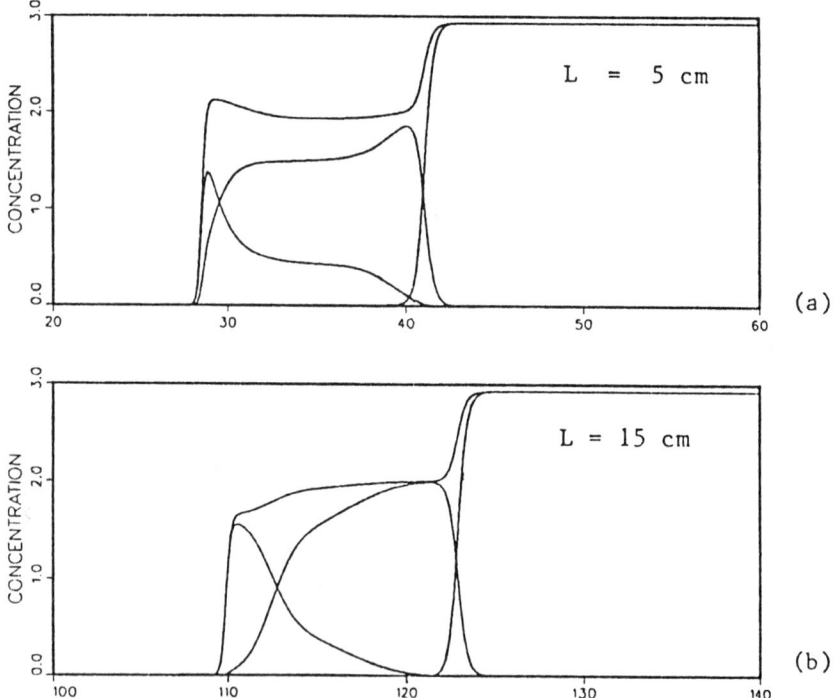

Fig. 45 Prediction of the intermediate band profiles of a multicomponent mixture during the formation of an isotachic train in displacement chromatography. Calculation made with the ideal model. (Reproduced from Ref. 165 with permission of Elsevier.)

band profiles were calculated using the semi-ideal model. The evolution of the profiles of the various components during the formation of the isotachic train was calculated for a binary separation (Fig. 45). Comparing the band profiles in the isotachic trains in Figs. 44f (ideal model) and 45d (semi-ideal model), we see the importance of the mixed zones at each band boundary. The finite column efficiency (1000 theoretical plates in Fig. 45d) reduces significantly the yield of each component at 99% purity. This can reduce many of the perceived advantages of displacement chromatography. Comparison of theory and experiments is shown at two different column lengths in Figs. 46a–46d. Good agreement is also observed between the experiments and the profiles calculated using the semi-ideal model [24,72].

The influence of the displacer concentration on the band profiles was also studied [165]. Very low displacer concentrations lead to overloaded elution, not displacement, because the displacer front

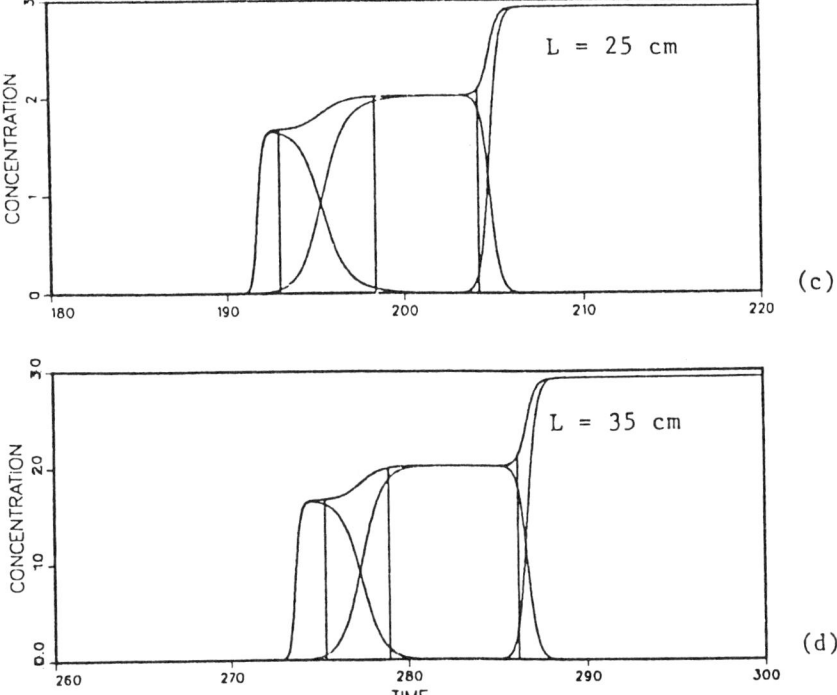

Fig. 45 (Continued)

cannot overtake the tail of the second-component band. Extremely large displacer concentrations result in the lack of a separation (overdisplacement).

Phillips et al. [182] used a kinetic and dispersive model to determine the independent effect of the mobile phase velocity, feed volume, and particle size on the throughput. He also shows the smoothening of the band boundaries with decreasing kinetic rate constant.

A detailed study of the effect of kinetic parameters has been made. With a high mass transfer coefficient, the band profiles obtained are the same as that predicted by the semi-ideal model (Fig. 45). As the mass transfer coefficient decreases, the width of the mixed zones in the isotachic profile increases (Fig. 47a) and eventually the plateaus disappear (Fig. 47b). Furthermore, by keeping the sample amount constant while increasing the column length (i.e., decreasing the loading factor), the constant pattern behavior seen in Fig. 47b is observed. By keeping the loading factor constant (i.e., increasing the sample amount in proportion to the column

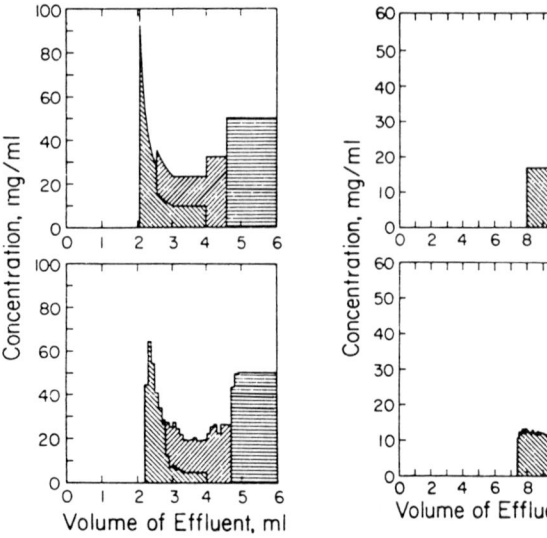

Fig. 46 Comparison of theoretical (ideal model) and experimental results in displacement chromatography. (Reproduced from Ref. 220 with permission of the American Institute of Chemical Engineers.)

length), we return to the profiles in Fig. 47a. Therefore, at moderate mass transfer rates the smoothing of the sharp front and rear boundaries is observed and an isotachic train cannot be observed if the rate of mass transfer is sufficiently slow, unless the column length is very long [183].

Preparative Applications of Displacement Chromatography

In the 1940s, as a part of the Manhattan Project, rare earth oxides were separated in the displacement mode and large amounts of the pure metals were produced [184,185]. Displacement chromatography was employed in the mid-1940s at the American Petroleum Institute to fractionate petroleum distillates in 52.4 ft × 3/4 in. i.d. columns [186]. In the late 1960s, Sorbex and several other similar processes were developed at UOP using various novel technologies to achieve what is essentially displacement chromatography on a moving bed column without physically moving the bed [187]. The technology was recently reviewed by Wankat [188].

Carrier displacement chromatography has been employed for the displacement of proteins using carboxymethyldextrans as spacers as well as a displacer [189,190]. Other workers have investigated the separation of lithium isotopes for either the enrichment or the isolation of pure isotopes [191-194] by ion exchange displacement

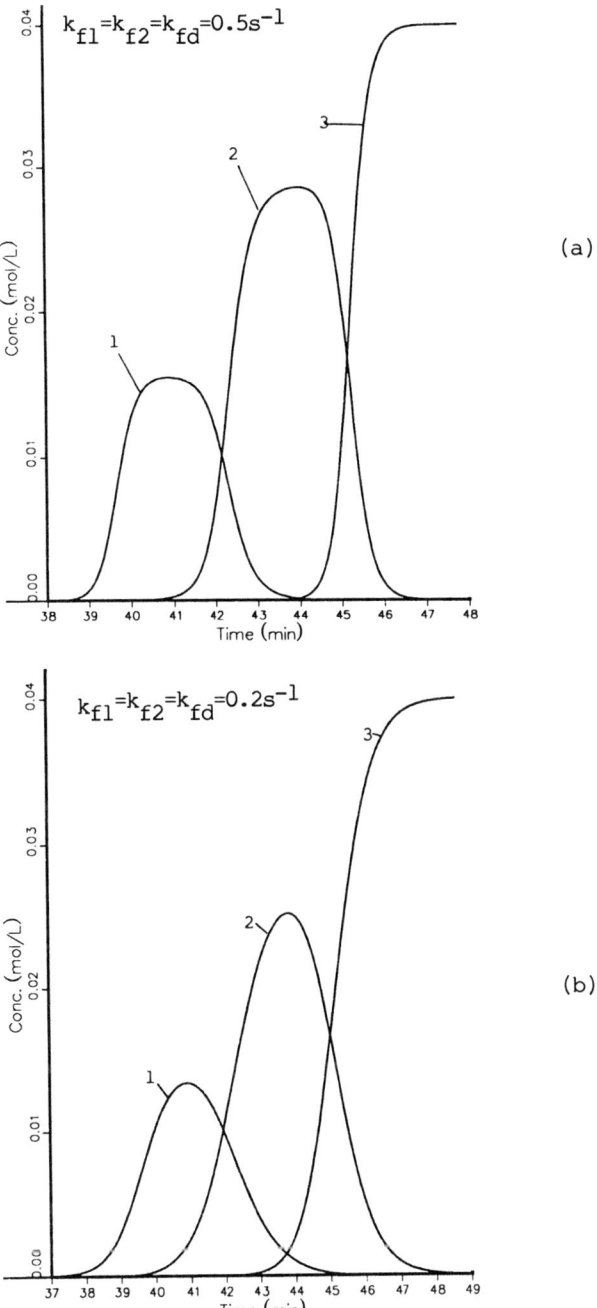

Fig. 47 Kinetic effect in displacement chromatography. (Reproduced from Ref. 183 with permission of the American Chemical Society.)

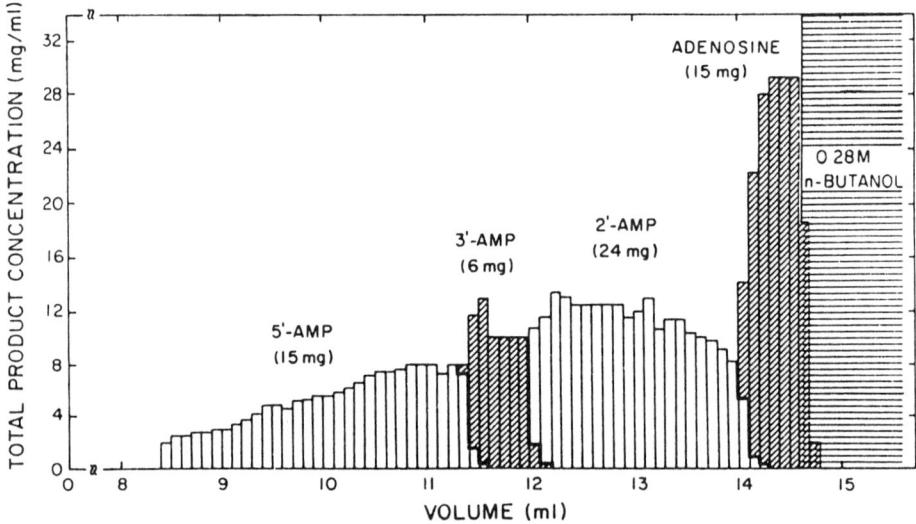

Fig. 48 Separation of nucleotides by displacement chromatography. Column: 40 × 0.46 cm, packed with μm Supelcosil LC-18. Mobile phase 10 mM acetate buffer, pH = 5, in water. Flow rate: 0.1 ml/min. (Reproduced from Ref. 177 with permission of Elsevier.)

chromatography. The literature presented thus far has utilized chromatography in the batch mode.

Methods for separating small organic molecules [173], polymyxin antibiotics [181], oligomycin antibiotics [195], steroids [196], enantiomers on chiral columns [197,198], amino acids [199], short peptides [177,200,201], polypeptides [202-204], insulin [205], nucleotides [206], and proteins [207-209] were developed. Two examples are shown in Figs. 48 and 49. Figure 48 shows the separation of nucleotides in reversed phase chromatography using n-butanol as a displacer [206]. Figure 49 shows isotachic and nonisotachic trains for the separation of β-lactoglobulin A and B in anion exchange chromatography [209]. An extensive review of experimental results in displacement chromatography was published by Frenz and Horvath [25].

These studies demonstrate the potential utility of displacement chromatography for the separation of various mixtures and its capacity to achieve high yields and high purities. A theoretical investigation by Katti and Guiochon [166] shows, however, that the main advantage of the displacement mode over elution is in the much higher (around one order of magnitude) concentration of the purified

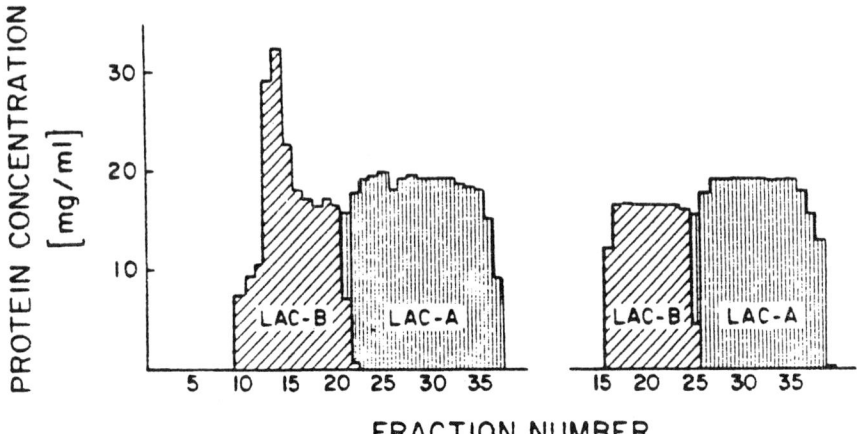

Fig. 49 Separation of proteins by displacement chromatography. (Reproduced from Ref. 207 with permission of Pergamon Press.)

fractions recovered. However, the recovery yield achieved in displacement rarely exceeds 70% and the production rate obtained under optimum conditions remains quite comparable for both modes. A limited advantage to one or the other technique in terms of production rate depends on the nature of the separation problem and the composition of the feed. The special nature of displacement forces one to consider separately the issue of column regeneration, which is not negligible and in almost all cases will nullify the production rate advantage theoretically possible in the displacement mode.

Preparative chromatography does not need to be a batch operation. Recently, the continuous annular chromatograph [210-212], originally developed in the elution mode, was extended to the displacement mode. By rotating an annular bed of sorbent past a stationary feed entry section and a stationary displacer entry section, the components separate along helical trajectories and exit at different radii of the column annulus (Fig. 50). A typical continuous annular chromatograph is 0.6 m long, has an annulus 12.7 mm wide and a diameter of 279 mm, and is packed with 50-μm-diameter particles. This separation process has been modeled and used to purify amino acids [213,214] (Fig. 51). Reasonable agreement is shown.

Analytical Applications of Displacement Chromatography

The application of displacement chromatography to the enrichment of trace components in a ternary mixture was studied theoretically by Katti and Guiochon [169], theoretically and experimentally by Ramsey et al. [170]. Experimental results agree well with those of Frenz et

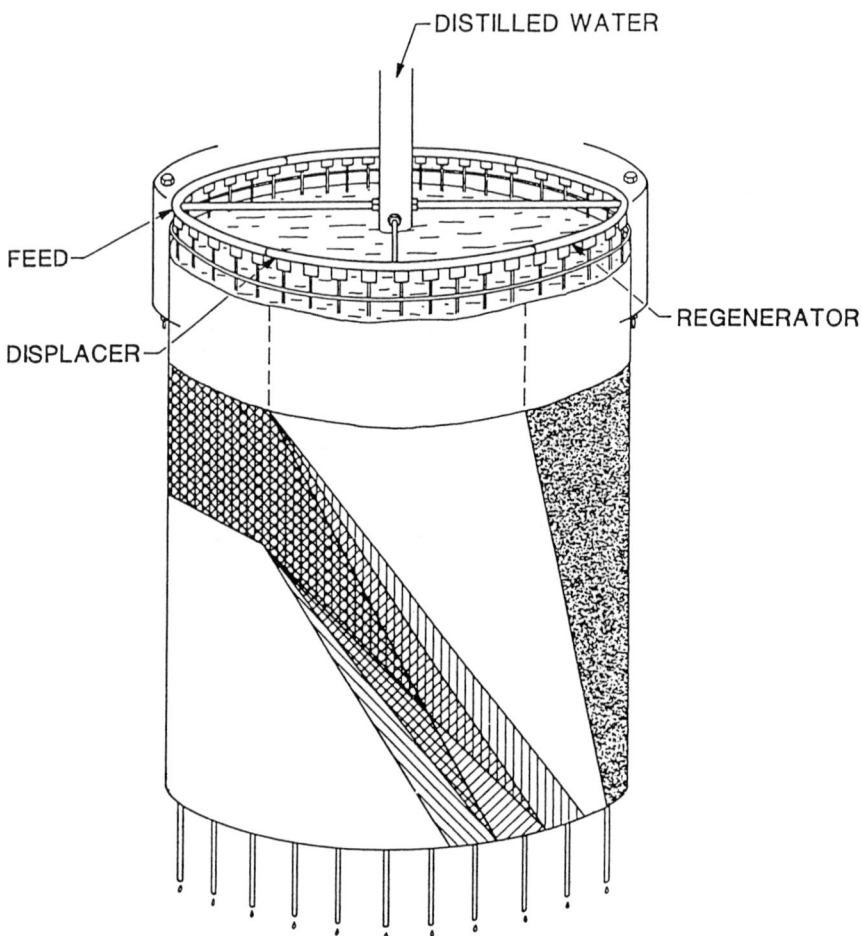

Fig. 50 Design of a continuous annular chromatograph and application to displacement chromatography. (Reproduced from Ref. 214 with permission of the American Institute of Chemical Engineers.)

al. [215]. In ideal displacement chromatography and when the isotachic train is fully developed, rectangular bands of each component are eluted in the order of increasing slope of the single-component isotherms. The height of these bands depends only on the experimental conditions and on the single-component isotherm of the corresponding compound, not on its concentration in the feed. When a component is in trace concentrations, the ideal model predicts that the height of its band remains the same as its plateau concentration

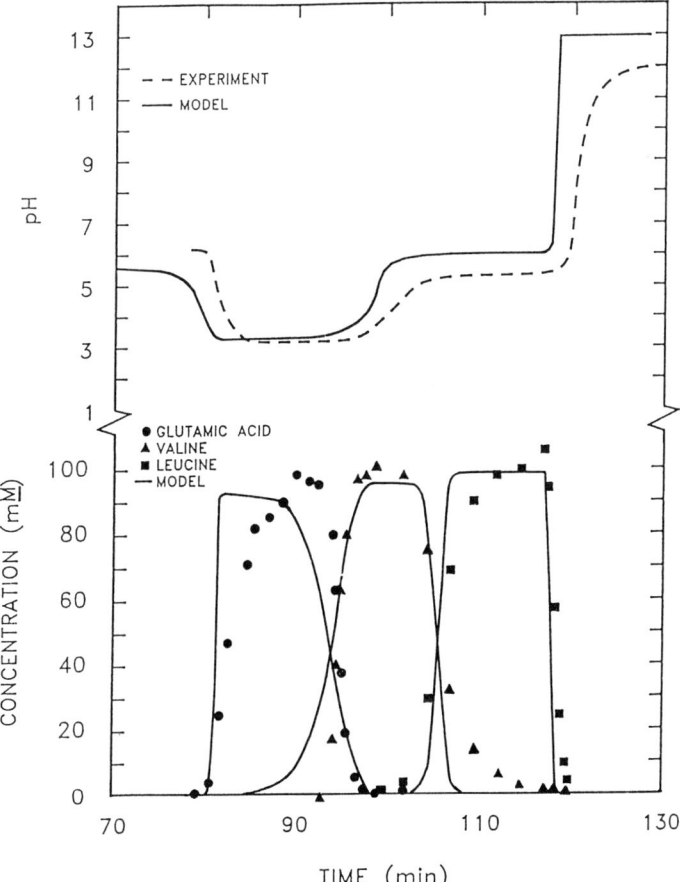

Fig. 51 Separation of peptides by continuous annular displacement chromatography. (Reproduced from Ref. 214 with permission of the American Institute of Chemical Engineers.)

determined from the equilibrium isotherms (see section on "Displacement Chromatography" in III.C) and that the width of its band is proportional to the amount of the component in the sample. The width would be very small for a trace component.

In practice, this is impossible. The finite column efficiency prevents the formation of rectangular band profiles, as it does prevent the formation of a true concentration discontinuity in overloaded elution. Shock layers appear at each zone boundary [77]. The thickness of these shock layers is a function of the column

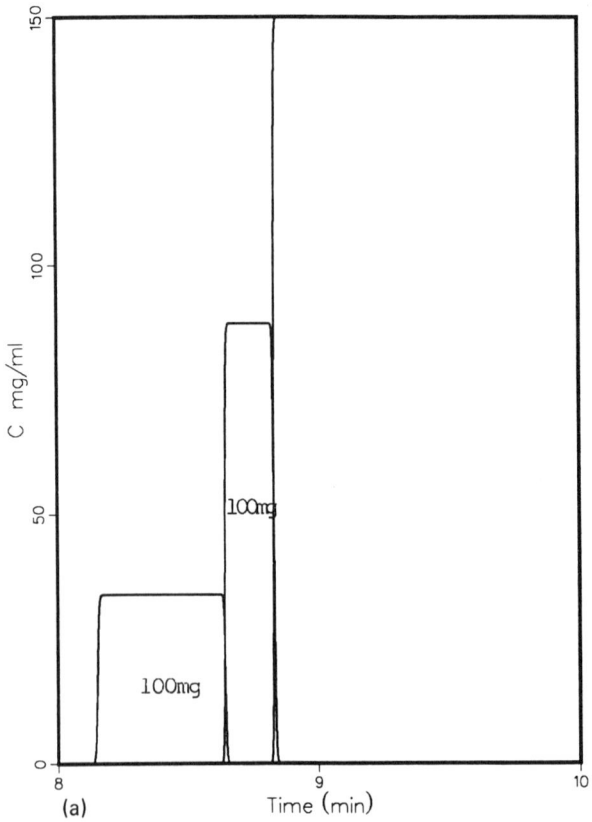

Fig. 52 Band profile of a trace component in displacement chromatography. (a) Isotachic displacement train calculated for a binary mixture. Column efficiency, 16,700 theoretical plates. (b) Profile of a trace component of the mixture in Fig. 52a (5 ppm), if eluted first (a), second (b), third (c), or under linear conditions (d). (Reproduced from Ref. 169 with permission of the French Academy of Sciences, Paris.)

efficiency and also of the intensity of the nonlinear effects, i.e., of the height of the ideal shock which tends to be formed. The presence of the two main component zones, one on each side of the trace component band, tends to narrow the width of this band while the apparent dispersion tends to spread it and relax the concentration gradients. The end result for low concentration traces is a nearly Gaussian band profile, whose height is proportional to the

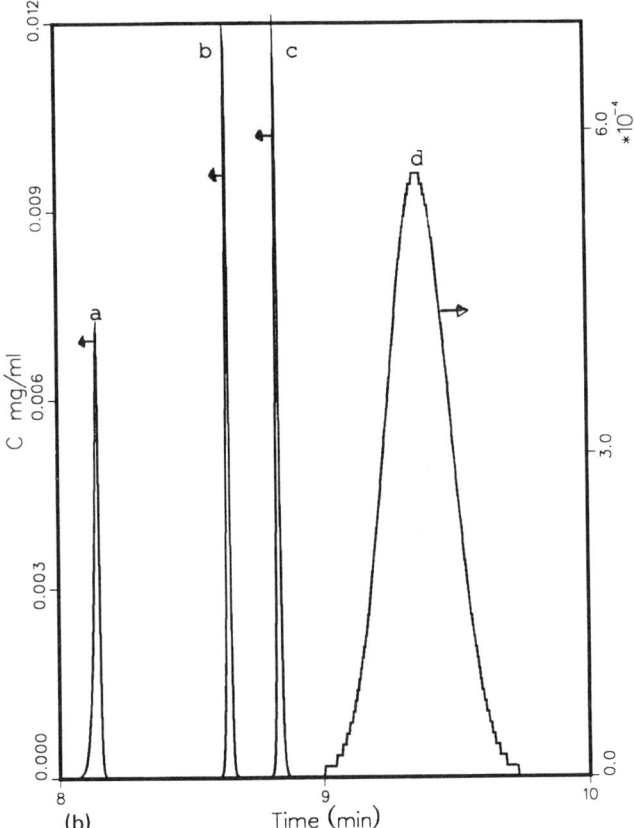

Fig. 52 (Continued)

amount of the trace component in the sample but which is much narrower than the band obtained for the same amount of the trace component if injected under linear chromatographic conditions. Typically, it is possible to generate a band having an apparent plate number of a few million using a column with a nominal efficiency of 10,000-15,000 plates.

Figure 52a illustrates the isotachic profile of two major components and Fig. 52b the profile of the impurity as if it were eluted first (a), second (b), and third (c). Profile d shows the linear elution peak of the impurity. In this example, the nominal efficiency of the column, given by peak d, is 16,700 plates while the apparent efficiency of peak c is 7.5 million plates.

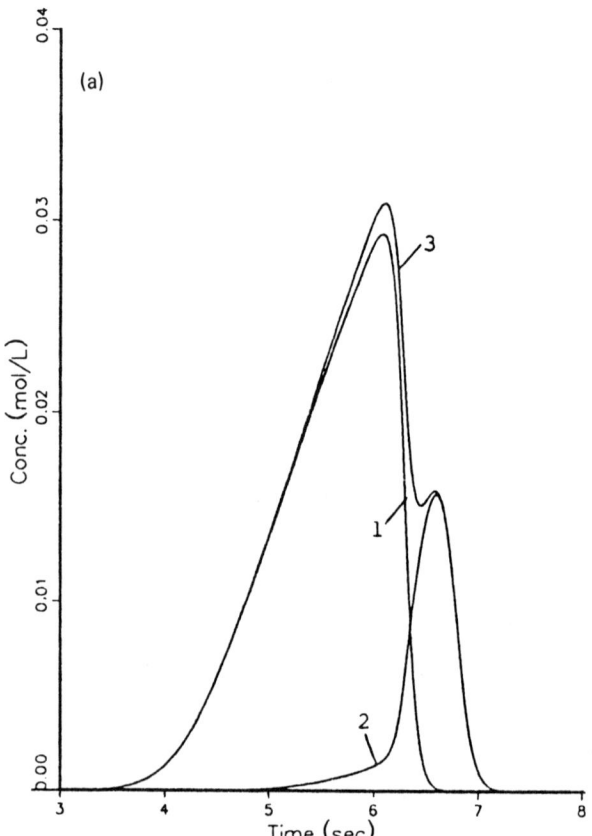

Fig. 53 Individual band profiles of the components of a binary mixture eluted with a strongly retained additive in the mobile phase. Comparison between calculated profiles (a) and experimental profile (b). Column: 25 × 0.46 cm, packed with Nucleosil silica, 15-25 μm. Mobile phase: dichloromethane with 0.133 M 2-propanol; flow rate, 2 ml/min. Sample: 4.3:1 mixture of 2-phenylethanol (74.3 mmol) and 3-phenylpropanol (17.2 mmol). (Reproduced from Ref. 217 with permission of the American Chemical Society.)

This effect predicted by the semi-ideal model has been demonstrated empirically, using a trace component detectable by spectrofluorescence [170]. In this preliminary study, the determination of the exact band profile of the trace component was complicated by the quenching of its fluorescence emission by the displacer (diethylphthalate), an effect which could also be simulated. The efficiency

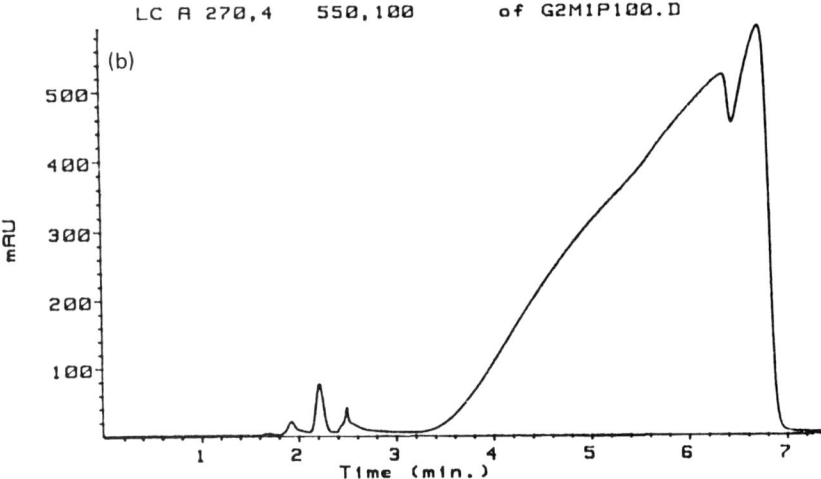

Fig. 53 (Continued)

of the column used was below 4000 plates. Furthermore, the curvature of the equilibrium isotherm of diethylphthalate under the experimental conditions is low and its retention relative to the probe moderate. Thus, the experimental conditions were not particularly favorable. Nevertheless, the efficiency enhancement exceeded one order of magnitude.

Similarly, Frenz et al. [215] reported that very narrow peaks of minor or trace peptides are observed between the bands of the main peptides formed in the trypsin digest of proteins in displacement chromatography. Recording the chromatogram in LC/MS, they reported the bandwidths of these trace components to be less than the period of spectrum acquisition.

System Peaks

The injection of a feed sample creates a perturbation of the equilibrium of the strong additive between the two phases of the system. The perturbation will strongly affect the individual elution profiles of the feed components if they are nearly as strongly adsorbed as or more strongly retained than the additive [216,217]. A theoretical investigation has shown that the band profiles are very similar to those observed with a pure mobile phase when the additive is weak compared to the feed component. As the strength of the additive retention increases, the band profiles change. If the additive is as retained as one of the components of a binary mixture, its band profile remains nearly Gaussian at all sample sizes while the profile of the other band becomes highly unsymmetrical and a shock layer appears on the side of the band directed toward the nearly symmetrical band.

At high additive strength, the band profiles are inverted and appear as anti-Langmuir profiles. Their front is diffuse while their rear is sharp. When the two-component bands interfere, new effects appear. These effects are analogous to the displacement and the tag-along effects, respectively, but act in the inverse direction. The chromatograms exhibit either a retainment effect (when the first component has a higher loading factor, its band tends to delay the elution of the second component band) or a pull-back effect (when the loading factor for the second component is higher, the first band tends to tag along the second one and to be eluted later and wider). At very high additive strength, abnormal chromatograms are observed, with three maxima for a binary mixture.

Experimental results have been published which qualitatively agree with the theoretical predictions [217] (Fig. 53). Figure 53a shows the theoretical elution profiles and 53b the detector response profiles for a binary mixture of 2-phenylethanol and 3-phenylpropanol on a silica column, using dichloromethane with 2-propanol as strong solvent.

Similar to the one-component system peak problem, at low propanol concentrations, the band profiles are comparable to those observed with the same components on a C18 column. These are the classical profiles obtained with Langmuir-type equilibrium isotherms, with very steep fronts and displacement or tag-along effect, depending on the ratio of the loading factors for the two components of the feed. At very high concentrations of propanol, the chromatograms exhibit the conventional anti-Langmuir band profile. For intermediate concentrations, the band profiles are most unusual and the predicted three maxima are observed for the binary misture in a range of propanol concentrations and sample sizes.

Separation of a Binary Mixture in Gradient Elution

Recently, there has been a surge of interest in overloaded gradient elution chromatography as applied to preparative scale purifications. Gradient elution is important in many of the applications of preparative chromatography, especially for the separation/purification of proteins where the retention factor is a strong function of the concentration of the mobile phase modifier [218]. In order to model gradient chromatography, a mass balance is required for the mobile phase modifier and for each solute. Individual band profiles have been modeled for the separation of a binary mixture using orthogonal collocation on finite elements [132]. Figure 54 illustrates the effect of load on the band shapes. Due to their interaction with the mobile phase modifier, band profiles in gradient elution may have complex

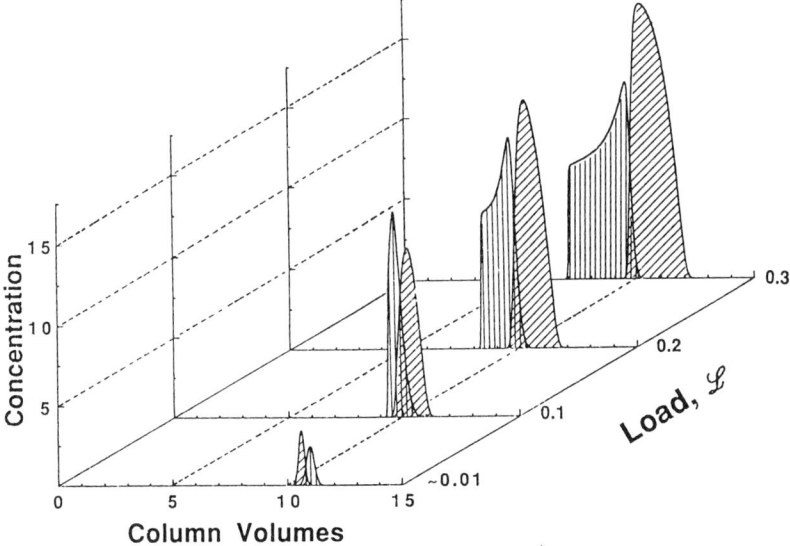

Fig. 54 Individual band profiles of the components of a binary mixture in gradient elution. (Reproduced from Ref. 132 with permission of Elsevier.)

shapes. In the example shown in Fig. 54d, the first component exhibits sharp front and rear boundaries and the second component has a concave down rear boundary. The shark fin shape of the rear boundaries of bands in gradient elution was also observed experimentally [120]. Optimization of the gradient slope and the sample size is critical to obtain high yields and highly concentrated collected fractions. In an industrial setting, however, step gradient methods may be preferred due to the simplicity of the pumping system.

Mant et al. showed that the displacement effect can be observed in gradient elution [219]. Extremely sharp boundaries have been observed between overlapping bands. This permits the recovery of nearly pure fractions with a good yield from the effluent of a column under experimental conditions when little resolution is observed between these bands.

IV. CONCLUSION

The theory of nonlinear chromatography has progressed to the point that any concentration signal obtained from a high-performance column can be accurately predicted provided the necessary competitive isotherms have been determined previously and can be accounted for properly by an algebraic equation. Alternately, the use of data tables to represent the isotherm is possible, but difficulties arise in finding proper mathematical methods for interpolation of these data. Accordingly, the current problem has become the investigation of the competitive isotherms.

There is another area of nonlinear chromatography where our understanding is still insufficient, when the kinetics of mass transfer controls the band profiles. A number of mechanisms are involved: diffusion through the mobile phase flowing around the particles, diffusion through the boundary layer at the particle surface, diffusion through the mobile phase stagnant in the particles and through the particle pores, kinetics of adsorption/desorption or of other retention mechanisms. Detailed analysis of the kinetics of these various mechanisms which are obviously coupled have been published in the case of pure components [22,23,29,30,90,108,126,135,137,138,213,214]. The influence of the presence of one component on the mass transfer coefficients of the other ones is still poorly known and there is a dramatic paucity of experimental data in this area.

Chromatography being a multistage separation method based on phase equilibria, it is not surprising that its understanding relies on the detailed knowledge of the thermodynamics and the kinetics of competitive equilibria. It offers new challenges but also new methodology for studies in these areas.

ACKNOWLEDGMENTS

This work was supported in part by grant CHE-8901382 of the National Science Foundation and by the cooperative agreement between the University of Tennessee and the Oak Ridge National Laboratory. We acknowledge support of our computational effort by the University of Tennessee Computing Center. We thank the various authors and the publishing companies for permission to reproduce their figures.

REFERENCES

1. J. N. Wilson, J. Am. Chem. Soc. 62, 1583 (1940).
2. D. DeVault, J. Am. Chem. Soc. 65, 532 (1943).
3. E. Glueckauf, Proc. Roy. Soc. (London), A186, 35 (1946).
4. E. Glueckauf, J. Chem. Soc., 1321 (1947).
5. E. Glueckauf, Disc. Faraday Soc. 7, 12 (1949).
6. S. Goldstein, Proc. Roy. Soc. (London) A219, 151 (1953).
7. H. C. Thomas, J. Am. Chem. Soc. 66, 1664 (1944).
8. J. C. Giddings, *Dynamics of Chromatography*, Marcel Dekker, New York, 1965.

9. G. J. Houghton, J. Phys. Chem. 67, 84 (1963).
10. P. C. Haarhoff and H. J. Van der Linde, Anal. Chem. 38, 573 (1966).
11. L. Lapidus and N. R. Amundson, J. Phys. Chem. 56, 984 (1952).
12. J. J. Van Deemter, F. J. Zuiderweg, and A. Klinkenberg, Chem. Eng. Sci. 5, 271 (1956).
13. R. Aris and N. R. Amundson, *Mathematical Methods in Chemical Engineering*, Prentice-Hall, Englewood Cliffs, NJ, 1973.
14. G. Guiochon and L. Jacob, Chromatogr. Rev. 14, 77 (1971).
15. G. Guiochon, L. Jacob, and P. Valentin, Chromatographia 4, 6 (1971).
16. P. Valentin and G. Guiochon, Sep. Sci. 10, 245 (1975).
17. P. Valentin and G. Guiochon, Sep. Sci. 10, 271 (1975).
18. J. R. Conder and J. H. Purnell, Chem. Eng. Sci. 25, 353 (1970).
19. H.-K. Rhee and N. R. Amundson, Chem. Eng. Sci. 29, 2049 (1974).
20. H.-K. Rhee, R. Aris, and N. R. Amundson, Trans. Roy. Soc. A267, 419 (1970).
21. H. Helfferich and G. Klein, *Multi-Component Chromatography. A Theory of Interference*, Marcel Dekker, New York, 1970.
22. A. J. Howard, G. Carta, C. H. Byers, Ind. Eng. Chem. Res. 27, 1873 (1988).
23. M. S. Saunders, J. B. Vierow, and G. Carta, AICHE J. 35, 53 (1989).
24. G. Guiochon, S. Ghodbane, S. Golshan-Shirazi, J. X. Huang, A. Katti, B. C. Lin, and Z. Ma, Talanta 36, 19 (1989).
25. J. Frenz and C. Horvath, in *High-Performance Liquid Chromatography: Advances and Perspectives*, Vol. 5 (C. Horvath, ed.), Academic Press, New York, 1988.
26. A. L. Lee, A. Velayudhan, and C. Horvath, in *Proceedings of 8th Int. Biotechnol. Symp.* (G. Durand, L. Bobichon, and J. Florent, eds.), Societe Francaise de Microbiologie, Paris, Paris, 1989, p. 593.
27. G. Viscomi, S. Lande, and C. Horvath, J. Chromatogr. 440, 157 (1988).
28. J. H. Knox and H. M. Pyper, J. Chromatogr. 363, 1 (1986).
29. C. K. Lee, Q. Yu, S. U. Kim, and N.-H. L. Wang, J. Chromatogr. 484, 29 (1989).
30. Q. Yu and N.-H. L. Wang, Computers Chem. Eng. 13, 915 (1989).
31. J. L. Wade, A. F. Bergold, and P. Carr, Anal. Chem. 59, 1286 (1987).
32. E. Glueckauf and J. I. Coates, J. Chem. Soc. (London), 1315 (1947).
33. N. K. Hiester and T. Vermeulen, Chem. Eng. Prog. 48, 509 (1952).

34. B. C. Lin, S. Golshan-Shirazi, and G. Guiochon, J. Phys. Chem. 93, 3363 (1989).
35. S. Golshan-Shirazi, B. C. Lin, and G. Guiochon, J. Phys. Chem. 93, 6871 (1989).
36. I. Langmuir, J. Amer. Chem. Soc. 40, 1361 (1918).
37. G. M. Schwab, Ergebnisse der Exacten Naturwissenschaften, Vol. 7, Julius Springer, Berlin, 1928, p. 276.
38. A. L. Myers, and J. M. Prausnitz, AIChE J. 11, 121 (1965).
39. M. D. LeVan and T. Vermeulen, J. Phys. Chem. 85, 3247 (1981).
40. B. C. Lin, S. Golshan-Shirazi, Z. Ma, and G. Guiochon, J. Chromatogr. 475, 1 (1989).
41. S. Golshan-Shirazi and G. Guiochon, Anal. Chem. 60, 2364 (1988).
42. S. Golshan-Shirazi and G. Guiochon, J. Phys. Chem. 94, 495 (1990).
43. S. Golshan-Shirazi and G. Guiochon, J. Phys. Chem. 93, 4341 (1989).
44. S. Golshan-Shirazi and G. Guiochon, J. Chromatogr. 506, 495 (1990).
45. G. Guiochon, S. Golshan-Shirazi, and A. Jaulmes, Anal. Chem. 60, 1856 (1988).
46. G. Guiochon, and S. Ghodbane, J. Phys. Chem. 92, 3682 (1988).
47. S. Golshan-Shirazi, S. Ghodbane, and G. Guiochon, Anal. Chem. 60, 2630 (1988).
48. S. Golshan-Shirazi and G. Guiochon, Anal. Chem. 60, 2634 (1988).
49. S. Golshan-Shirazi and G. Guiochon, Anal. Chem. 61, 462 (1989).
50. A. M. Katti and G. Guiochon, J. Chromatogr. 499, 21 (1990).
51. J. Weiss, J. Chem. Soc. (London) 297 (1943).
52. A. J. P. Martin and R. L. M. Synge, Biochem. J. 35, 1358 (1941).
53. L. C. Craig, J. Biol. Chem. 155, 519 (1944).
54. D. Graham, J. Phys. Chem. 57, 665 (1953).
55. R. J. Laub, ACS Symp. Series 297, 1 (1986).
56. A. I. M. Keulemans, Gas Chromatography, (C. G. Verver, ed.), Reinhold, New York, 1957, pp. 107-115.
57. A. Klinkenberg and F. Sjenitzer, Chem. Eng. Sci. 5, 258 (1956).
58. J. Villermaux, in Percolation Processes (A. E. Rodrigues and D. Tondeur, eds.), Sijthoff and Noordhoff, Rockville, MD, 1981, p. 83.
59. S. Seshadri and S. N. Deming, Anal. Chem. 56, 1567 (1984).
60. J. E. Eble, R. L. Grob, P. E. Antle, and R. L. Snyder, J. Chromatogr. 384, 25 (1987).
61. J. J. Van Deemter, F. J. Zuiderweg, and A. Klinkenberg, Chem. Eng. Sci. 5, 271 (1956).
62. C. Horvath and H. J. Lin, J. Chromatogr. 149, 43 (1978).
63. J. F. K. Huber, Ber. Bunsen-Ges. Phys. Chem. 77, 179 (1973).

64. E. V. Dose and G. Guiochon, Anal. Chem. *61*, 1723 (1990).
65. B. C. Lin, Z. Ma, S. Golshan-Shirazi, and G. Guiochon, J. Chromatogr. *500*, 185 (1990).
66. A. W. J. De Jong, H. Poppe, and J. C. Kraak, J. Chromatogr. *148*, 127 (1978).
67. J. E. Eble, R. L. Grob, P. E. Antle, and L. R. Snyder, J. Chromatogr. *384*, 45 (1987).
68. J. E. Eble, R. L. Grob, P. E. Antle, and R. L. Snyder, J. Chromatogr. *405*, 31 (1987).
69. S. Golshan-Shirazi, M. Z. El Fallah, and G. Guiochon, J. Chromatogr. *541*, 195 (1990).
70. S. Jacobson, S. Golshan-Shirazi, and G. Guiochon, J. Am. Chem. Soc. *112*, 6492 (1990).
71. A. M. Katti, Z. Ma, and G. Guiochon, AIChE J. *36*, 1722 (1990).
72. A. M. Katti, Ph.D. thesis, University of Tennessee, 1990.
73. A. M. Katti, J.-X. Huang, and G. Guiochon, Biotechnol. Bioeng. *36*, 288 (1990).
74. S. Golshan-Shirazi and G. Guiochon, J. Chromatogr. *461*, 1 (1989).
75. S. Golshan-Shirazi and G. Guiochon, J. Chromatogr. *461*, 19, (1989).
76. P. Rouchon, M. Schonauer, P. Valentin, and G. Guiochon, Sep. Sci. Technol. *22*, 1793 (1987).
77. B. Lin, S. Golshan-Shirazi, Z. Ma, and G. Guiochon, Anal. Chem. *61*, 2647 (1988).
78. F. Helfferich and J. Peterson, J. Chem. Educ. *41*, 410 (1964).
79. B. C. Lin, Z. Ma, and G. Guiochon, J. Chromatogr. *484*, 83 (1989).
80. M. Czok and G. Guiochon, Anal. Chem. *62*, 189 (1990).
81. D. D. Do and R. G. Rice, Chem. Eng. Sci. *42*, 2269 (1987).
82. S. Golshan-Shirazi and G. Guiochon, J. Phys. Chem. (in press).
83. J. C. Giddings, Anal. Chem. *35*, 1999 (1963).
84. J. C. Giddings and P. D. Schettler, Anal. Chem. *36*, 1483 (1964).
85. Villermaux, J., in *Column Chromatography* (E. sz Kovats, ed.), Sauerlander, Aarau, Switzerland, 1970, p. 66.
86. J. Villermaux, Chem. Eng. Sci. *27*, 1231 (1972).
87. J. Villermaux, J. Chromatogr. Sci. *12*, 822 (1974).
88. S. Goldstein, Proc. Roy. Soc. (London) *A219*, 171 (1953).
89. C. A. Lucy, J. L. Wade, and P. J. Carr, J. Chromatogr. *484*, 61 (1989).
90. G. Carta, Chem. Eng. Sci. *43*, 2877 (1988).
91. D. O. Cooney and E. N. Lightfoot, Ind. Eng. Chem. Fund. *5*, 212 (1966).
92. D. M. Ruthven, *Principles of Adsorption and Adsorption Processes*, John Wiley and Sons, New York, 1984.

93. H.-K. Rhee and N. R. Amundson, AIChE J. *28*, 423 (1982).
94. G. Schay and G. Szekely, Acta Chim. Hung. *5*, 167 (1954).
95. D. H. James and C. S. G. Phillips, J. Chem. Soc. (London) (*1954*), 1066.
96. E. Glueckauf, Trans. Faraday Soc. *51*, 1540 (1955).
97. E. Cremer and J. F. K. Huber, Angew. Chem. *73*, 461 (1961).
98. Y. A. Eltekov, Y. V. Kazakevich, A. V. Kiselev, and A. A. Zhuchkov, Chromatographia *20*, 525 (1985).
99. Y. A. Eltekov and Y. V. Kazakevitch, J. Chromatogr. *365*, 213 (1986).
100. J. Jacobson, J. Frenz, and C. Horvath, J. Chromatogr. *316*, 53 (1984).
101. J.-X. Huang and C. Horvath, J. Chromatogr. *406*, 275 (1987).
102. J.-X. Huang and C. Horvath, J. Chromatogr. *406*, 285 (1987).
103. E. V. Dose and G. Guiochon, Anal. Chem. *62*, 816 (1990).
104. E. V. Dose and G. Guiochon, Anal. Chem. (in press).
105. J. A. Jonsson and P. Lovkvist, J. Chromatogr. *408*, 1 (1987).
106. J. I. Coates and E. Glueckauf, J. Am. Chem. Soc. *69*, 1309 (1947).
107. Z. Ma and G. Guiochon, Comput. Eng. Chem. (in press).
108. R. W. Geldart, Y. Qiming, P. C. Wankat, and L. N.-H. Wang, Sep. Sci. Technol. *21*, 873 (1986).
109. S. Golshan-Shirazi and G. Guiochon, J. Chromatogr. *484*, 125 (1989).
110. E. Kvaalen, L. Neel, and D. Tondeur, Chem. Eng. Sci. *40*, 1191 (1985).
111. J. Newburger, L. Liebes, H. Colin, and G. Guiochon, Sep. Sci. Technol. *22*, 1933 (1987).
112. J. Newburger and G. Guiochon, J. Chromatogr. *484*, 153 (1989).
113. S. Golshan-Shirazi and G. Guiochon, Anal. Chem. *62*, 217 (1990).
114. M. Czok and G. Guiochon, Comput. Chem. Eng. *14*, 1435 (1990).
115. P. J. Karol, Anal. Chem. *61*, 1937 (1989).
116. B. Lin, Z. Ma, and G. Guiochon, Sep. Sci. Technol. *24*, 809 (1989).
117. L. R. Snyder, J. W. Dolan, and G. B. Cox, J. Chromatogr. *483*, 63 (1989).
118. M. Czok and G. Guiochon, J. Chromatogr. *537*, 497 (1991).
119. J. E. Eble, R. L. Grob, P. E. Antle, and L. R. Snyder, J. Chromatogr. *405*, 1 (1987).
120. J. E. Eble, R. L. Grob, P. E. Antle, and L. R. Snyder, J. Chromatogr. *405*, 51 (1987).
121. L. R. Snyder, G. B. Cox, and P. E. Antle, Chromatographia *24*, 82 (1987).

122. L. R. Snyder and G. B. Cox, J. Chromatogr. *483*, 85 (1989).
123. A. M. Katti and G. Guiochon, Am. Lab. *21*(10), 17 (1989).
124. G. B. Cox and L. R. Snyder, J. Chromatogr. *483*, 95 (1989).
125. M. Z. El Fallah, S. Golshan-Shirazi, and G. Guiochon, J. Chromatogr. *511*, 1 (1990).
126. Q. Yu, J. Yang, and N.-H. Wang, Reactive Polymers *6*, 33 (1987).
127. E. J. Wilson and J. Geankoplis, Ind. Eng. Chem. Fund. *5*, 9 (1966).
128. S. Yamamoto, M. Nomura, and Y. Sano, AIChE J. *33*, 1426 (1987).
129. L. R. Snyder, G. B. Cox, and P. E. Antle, J. Chromatogr. *444*, 303 (1988).
130. L. R. Snyder, J. W. Dolan, D. C. Lommen, and G. B. Cox, J. Chromatogr. *484*, 425 (1989).
131. L. R. Snyder and J. W. Dolan, J. Chromatogr. *484*, 437 (1989).
132. F. Antia and C. Horvath, J. Chromatogr. *484*, 1 (1989).
133. A. L. Lee, A. W. Liao, and C. Horvath, J. Chromatogr. *443*, 31 (1988).
134. A. Velayudhan and C. Horvath, J. Chromatogr. *443*, 13 (1988).
135. G. Carta, J. P. DeCarli, C. H. Byers, and W. G. Sisson, Chem. Eng. Commun. *79*, 207 (1989).
136. S. Clacsson, Disc. Faraday Soc. *7*, 34 (1949).
137. M. S. Saunders, F. Mawengkaug, and G. Carta, AIChE annual meeting, Washington, DC, November 1988, paper 7a.
138. M. S. Saunders, Ph.D. thesis, University of Virginia, 1989.
139. H. A. Chase, G. L. Skidmore, and B. J. Horstmann, Communication to PREP '89, Washington, DC, May 1989.
140. J. M. Jacobson, J. H. Frenz, and C. Horvath, Ind. Eng. Chem. Res. *26*, 43 (1987).
141. J. Zhu, A. M. Katti, and G. Guiochon, J. Chromatogr. (in press).
142. B. H. Arve and A. I. Liapis, Biotechnol. Bioeng. *30*, 638 (1987).
143. D. A. Hill, P. Mace, and D. Moore, Communication to PREP '90, Gent, Belgium, April 1990.
144. W. G. Rixey and C. J. King, AICHE J. *35*, 69 (1989).
145. D. Basmadjian and P. Coroyannakis, Chem. Eng. Sci. *42*, 1723 (1987).
146. D. J. Solms, T. W. Smuts, and V. J. Pretorius, J. Chromatogr. Sci. *9*, 600 (1971).
147. M. Denkert, L. Hackzell, G. Schill, and J. Sjogren, J. Chromatogr. *218*, 31 (1981).
148. J. J. Kirkland, J. Chromatogr. *83*, 149 (1973).
149. J. Puncocharova, J. Kriz, L. Vodicka, and D. Prusova, J. Chromatogr. *191*, 81 (1980).

150. Z. Ma, A. M. Katti, and G. Guiochon, J. Phys. Chem. 94, 6911 (1990).
151. J.-X. Huang and G. Guiochon, J. Coll. Interf. Sci. 128, 577 (1989).
152. G. Guiochon and A. M. Katti, Chromatographia 24, 165 (1987).
153. A. Katti and G. Guiochon, Anal. Chem. 61, 982 (1989).
154. S. Golshan-Shirazi and G. Guiochon, Anal. Chem. 61, 1368, 2464 (1989).
155. S. Golshan-Shirazi and G. Guiochon, J. Chromatogr. 536, 57 (1991).
156. S. Ghodbane and G. Guiochon, J. Chromatogr. 444, 275 (1988).
157. S. Ghodbane and G. Guiochon, J. Chromatogr. 450, 27 (1988).
158. S. Ghodbane and G. Guiochon, J. Chromatogr. 452, 209 (1988).
159. S. Ghodbane and G. Guiochon, J. Chromatogr. 440, 9 (1988).
160. S. Ghodbane and G. Guiochon, Chromatographia 26, 53 (1989).
161. S. Golshan-Shirazi and G. Guiochon, Anal. Chem. 61, 1276 (1989).
162. S. Golshan-Shirazi and G. Guiochon, J. Chromatogr. 517, 229 (1990).
163. S. Golshan-Shirazi and G. Guiochon, Am. Biotechnol. Lab 8(6), 26 (1990).
164. S. Golshan-Shirazi and G. Guiochon, J. Chromatogr. 523, 1 (1990).
165. A. M. Katti and G. Guiochon, J. Chromatogr. 449, 25 (1988).
166. A. M. Katti, E. V. Dose, and G. Guiochon, J. Chromatogr. 540, 1 (1991).
167. S. M. Cramer and G. Subramanian, Sep. Purif. Meth. 19, 31 (1990).
168. A. M. Katti, R. Ramsey, and G. Guiochon, J. Chromatogr. 477, 119 (1989).
169. A. M. Katti and G. Guiochon, C. R. Acad. Sci. (Paris) 309(II), 1557 (1989).
170. R. Ramsey, A. M. Katti, and G. Guiochon, Anal. Chem. 62, 2557 (1990).
171. Z. M. El Fallah and G. Guiochon, J. Chromatogr. 522, 1 (1990).
172. S. Jacobson, S. Golshan-Shirazi, A. M. Katti, M. Czok, Z. Ma, and G. Guiochon, J. Chromatogr. 484, 103 (1989).
173. C. Horvath, A. Nahum, and J. H. Frenz, J. Chromatogr. 218, 365 (1981).
174. M. Tswett, Ber. Deutsch. Botan. Gesell. 24, 316 (1906).
175. A. Tiselius, Arkiv. Kemi, Mineral. Geolog. 16A, 1 (1943).
176. A. Tiselius, Disc. Faraday Soc. 7, 7 (1949).
177. C. Horvath, J. Frenz, and Z. El Rassi, J. Chromatogr. 255, 273 (1983).
178. C. Horvath, in *The Science of Chromatography* (F. Bruner, ed.), Elsevier, Amsterdam, 1985, p. 179.

179. Y.-F. Man and C. Horvath, J. Chromatogr. *445*, 71 (1988).
180. J. Frenz, Ph.D. thesis, Yale University, 1983.
181. J. Frenz, P. Van der Schrieck, and C. Horvath, J. Chromatogr. *330*, 1 (1985).
182. M. W. Phillips, G. Subramanian, and S. M. Cramer, J. Chromatogr. *454*, 1 (1988).
183. S. Golshan-Shirazi, B. C. Lin, and G. Guiochon, Anal. Chem. *61*, 1960 (1989).
184. F. H. Spedding, E. I. Fulmer, T. A. Butler, and J. E. Powell, J. Am. Chem. Soc. *72*, 2349 (1950).
185. F. H. Spedding, Disc. Faraday Soc. *7*, 38 (1949).
186. B. D. Mair, A. L. Gaboriault, and F. D. Rossini, Ind. Eng. Chem. *39*, 1072 (1947).
187. D. B. Broughton, Sep. Sci. Technol. *19*, 723 (1984).
188. P. C. Wankat, *Large-Scale Adsorption and Chromatography*, Vols. 1 and 2, CRC Press, Boca Raton, 1986.
189. A. R. Torres, S. C. Edberg, and E. A. Peterson, J. Chromatogr. *389*, 177 (1987).
190. E. A. Peterson, Anal. Biochem. *90*, 767 (1987).
191. R. Jacques, Sep. Sci. Technol. *15*, 533 (1980).
192. Y. Fujii, M. Aida, M. Okamoto, and T. Oi, Sep. Sci. Technol. *20*, 377 (1985).
193. S. Fujine, Sep. Sci. Technol. *17*, 1049 (1982).
194. S. Fujine, K. Saito, and K. Shiba, Sep. Sci. Technol. *18*, 15 (1983).
195. K. Valko, P. Slegel, and J. Bati, J. Chromatogr. *386*, 345 (1987).
196. H. Kalasz and C. Horvath, J. Chromatogr. *215*, 295 (1981).
197. G. Vigh, G. Quintero, and G. Farkas, J. Chromatogr. *484*, 237 (1989).
198. G. Vigh, G. Farkas, and G. Quintero, J. Chromatogr. *484*, 251 (1989).
199. H. Kalasz and C. Horvath, J. Chromatogr. *239*, 423 (1982).
200. S. M. Cramer and C. Horvath, J. Prep. Chromatogr. *1*(1), 29 (1988).
201. S. M. Cramer, Z. El Rassi, and C. Horvath, J. Chromatogr. *394*, 305 (1987).
202. G. C. Viscomi, S. Lande, and C. Horvath, J. Chromatogr. *440*, 157 (1988).
203. G. C. Viscomi, A. Ziggiotti, and A. S. Verdini, J. Chromatogr. *482*, 99 (1989).
204. G. C. Viscomi, J. Chromatogr. *499*, 37 (1990).
205. G. Vigh, Z. Varga-Puchony, G. Szepesi, and M. Gazdag, J. Chromatogr. *386*, 353 (1987).
206. Z. El Rassi and C. Horvath, J. Chromatogr. *266*, 319 (1983).

207. A. W. Liao, Z. El Rassi, D. M. LeMaster, and C. Horvath, Chromatographia 24, 881 (1987).
208. G. Subramanian, M. W. Phillips, and S. M. Cramer, J. Chromatogr. 439, 341 (1988).
209. G. Subramanian, M. W. Phillips, G. Jayaraman, and S. M. Cramer, J. Chromatogr. 484, 225 (1989).
210. A. J. P. Martin, Disc. Faraday Soc. 7, 332 (1949).
211. C. D. Scott, R. D. Spence, and W. G. Sisson, J. Chromatogr. 126, 381 (1976).
212. C. D. Scott, Sep. Sci. Technol. 21, 905 (1986).
213. G. Carta, M. S. Saunders, J. P. DeCarli, and J. B. Vierow, AIChE Symp. Ser. 84, 54 (1988).
214. J. P. DeCarli, G. Carta, and C. H. Byers, AIChE J. 36, 1220 (1990).
215. J. Frenz, J. Bourell, and W. S. Hancock, J. Chromatogr. 512, 299 (1990).
216. S. Golshan-Shirazi and G. Guiochon, Anal. Chem. 61, 2373 (1989).
217. S. Golshan-Shirazi and G. Guiochon, Anal. Chem. 61, 2380 (1989).
218. K. Nakamura and Y. Kato, J. Chromatogr. 333, 29 (1985).
219. C. T. Mant, T. W. Lorne Burke, and R. S. Hodges, Chromatographia 24, 565 (1987).
220. J. Frenz and C. Horvath, AIChE J. 31, 400 (1985).

2
Problems in Aqueous Size Exclusion Chromatography

Paul L. Dubin *Indiana-Purdue University, Indianapolis, Indiana*

I.	INTRODUCTION	119
II.	COMMERCIAL PACKINGS	121
III.	ELECTROSTATIC INTERACTIONS	126
IV.	HYDROPHOBIC INTERACTIONS	134
V.	UNIVERSAL CALIBRATION	136
VI.	SEC OF PROTEINS	143
VII.	SEC OF MICELLES	145
	REFERENCES	147

I. INTRODUCTION

Twenty years ago, size exclusion chromatography (SEC) was practiced by two distinct and largely noncommunicating communities. Industrial chemists used porous, semirigid crosslinked organic resins, with organic solvents eluants, to determine the molecular weight (MW) and molecular weight distribution (MWD) of synthetic polymers by a technique referred to as "gel permeation chromatography" or GPC. Biochemists separated proteins on relatively compressible porous gels based on dextran or agarose, and called this process "gel filtration."

The two techniques possessed different terminology, literature, and commercial prospects. Lacking mechanically strong and small-particle gels for aqueous SEC, biochemists "poured" their own columns and were in the main satisfied with the resultant low resolution and speed. In contrast, commercial prepacked columns for GPC offered high speed and efficiency, and were therefore easily marketed—along with attendant instrumentation—to industrial polymer chemists. While porous glass packings occupied a sort of middle ground in this situation, being applicable to some biopolymers and most organic-soluble macromolecules, their unremarkable efficiency and highly adsorptive surface behavior limited their use in both camps.

The arbitrary distinction between gel filtration and GPC has faded during the last decade, largely because of the introduction of nonadsorptive, high-speed (high-pressure), packed columns for aqueous SEC. Protein chemists now take advantage of the same instrumental technology—e.g., automated sample injection, computerized data reduction, multiple detectors—introduced earlier for polymer analysis. This chapter intentionally excludes such technological aspects of SEC that are common to both organic and aqueous media. The omission of, for example, the optimization of column resolution and efficiency, the coupling of SEC with light-scattering or viscosity detectors, or computational procedures for dealing with broad-distribution calibration standards, is not a subordination of these topics, but rather a reflection of the satisfactory treatments to be found in various monographs and edited volumes on size exclusion chromatography in general [1-4]. Furthermore, this chapter does not attempt an exhaustive compilation of the literature on aqueous SEC; the interested reader is referred to a number of earlier comprehensive reviews on this topic [5-10].

This chapter deals with four topics that are unique to aqueous SEC. First, it is appropriate to provide a straightforward summary of the current commercially available packings; such tabulations appear regularly, but the frequency of introduction of new packings justifies an update. The second theme dealt with is solute-substrate interactions. "Nonideal" SEC may, of course, be observed in organic eluants; but here we deal specifically with electrostatic and hydrophobic interactions, which are characteristic of chromatography in aqueous media. Our third topic is the identification of the dimensional parameter that governs the chromatographic partition coefficient, i.e., the question of "universal calibration." This matter has certainly been addressed using data obtained in organic mobile phases, but it is becoming evident that a rigorous test of any proposed "universal" parameter requires a range of macromolecular shape factors that is difficult to come by in nonaqueous media. Lastly, attention is directed to special applications of aqueous SEC to two particular amphiphilic solute species: proteins and micelles.

II. COMMERCIAL PACKINGS

Stationary phases for aqueous SEC are conveniently divided into two classes: derivatized glass and silica, and crosslinked gels. The latter, in turn, may be categorized as either natural, i.e., based on bacterial or algal polysaccharides, or synthetic, i.e., based on hydrophilic monomers which undergo free radical polymerization.

Porous silica beads offer two main advantages: mechanical strength, and controllable and uniform particle size. These two properties lead, in principle, to extremely efficient packed columns, conducive to high-speed, high-resolution chromatography. The utilization of porous silica microspheres [11-13] for aqueous SEC was impeded at first by the highly adsorptive nature of the packing (these are similar to TLC substrates in this regard) and by the dissolution of silica in basic or neutral aqueous media. Derivatization of silica with silane reagents, to produce a hydrophilic but less acidic bonded phase [14-16], substantially diminished both of these unwanted effects. These bonded phases are most commonly glycols (in particular 3-glycidoxypropyl) but a number of other functional groups have been used.

Table 1 lists commercial packings for aqueous SEC based on porous silica. The nature of the bonded phase is noted (unless undisclosed by the manufacturer) along with other information. As noted in the table, packings based on porous glass or silica offer two key advantages: precise control of particle size and porosity, and mechanical strength. These features result in packings that offer very high efficiency, i.e., theoretical plates relative to elution volume, along with compatibility with high flow rates. The corresponding packed columns are ideally suited to rapid analyses. There are only two reasons that gel packings may be preferred to siliceous ones. First, slurry packing of "soft" or semirigid gel substrates is a well-established procedure by which the average practitioner may prepare his own inexpensive—albeit low-efficiency—columns. For many applications in the biochemistry research laboratory, such columns have been found adequate (or at least familiar). The second factor is the relative "inertness" of gel stationary phases, vis-a-vis siliceous packings. Despite intense research and development, no reagent for the surface derivatization of silica produces a stationary phase that is completely free of the adsorptive properties of the unreacted packing. This behavior (see below) presumably results from the incomplete reaction of SiOH groups, which, as pointed out by Iler [17], is unlikely to transform more than 50% of the starting silanol moieties, even for relatively nonbulky silanizing reagents. For these two reasons, the gel packings listed in Table 2 offer some advantages relative to silica-based packings, and are therefore commercially viable products.

Table 1 Commercially Available Siliceous Packings for Aqueous SEC (As of 1989)

Principal trade name (supplier) *commercial pseudonyms	Description, particle size	Typical efficiency (plates m^{-1})	Upper MW limit (standard)
A. *Underivatized Glass and Silica*			
CPG (Electronucleonics, USA)	Porous glass, 50 ± 20, 100 ± 25, 150 ± 25 μm	900 (2000)(a)	3000 Å
LiChrospher (E. Merck, FRG)	Porous silica, 10 μm (spherical)	20,000	5 × 10^6 (dextran)
Spherosil (Rhone Poulenc, France)	Porous silica, 7 μm (spherical)	1,000	2 × 10^6 (polystyrene)
Fractosil (E. Merck, FRG)	Porous silica, 50 ± 10, 90 ± 30 μm	—	—
Porasil (Waters Assoc., USA)	40–100 μm	1,000	> 2 × 10^6 (polystyrene)
PSM (DuPont, USA)	7 μm (spherical)	—	1 × 10^7 (polystyrene)
B. *Derivatized (Hydrophilic) Glass and Silica*			
Glycophase CPG (Pierce Chemical, USA)	Porous glass, 1-2-dihydroxylpropyl, 50 μm	800	3000 Å
LiChrosorb DIOL (E. Merck, FRG)	Porous silica, glycerolpropyl-bonded, 5, 7, 10 μm (irregular)	15,000	2 × 10^6 (dextran)

Column	Packing		Exclusion limit
SynChropak (Synchrom, USA) *Aquapore (Brownlee, USA) *Bio-Sil (Bio-Rad, USA)	Porous silica, glycerol-propyl-bonded, 10 μm (spherical)	10,000	1×10^7 (dextran)
TSK-SW (Toyo Soda, Japan)	Proprietary, 10 μm (spherical)	15,000	6×10^5 (dextran)
μBondagel (Waters Assoc., USA)	Proprietary, 10 μm (irregular)	10,000	1×10^7 (dextran)
Zorbax GF-250 (DuPont, USA)	Glycerolpropyl-bonded, 4 μm	—	—
Protein WS-300 (Showa Denko, Japan)	Proprietary, 9 μm	45,000	5×10^4 (pullulan)
Shim-Pak Diol (Shimadzu, Japan)	5 μm	—	1×10^6 (protein)
Shodex Ionpak (Showa Denko, Japan)	Sulfonated crosslinked polystyrene, 10 μm	15,000	5×10^6 (dextran)
Si Polyol, S? Polyol (Serva Feinbiochemica, FRG)	Proprietary, 60–150 μm	—	—
SynChropak CATSEC (Synchrom, USA)	"Polymerized polyamine," 10 μm	—	5×10^5 (polyvinyl-pyridine)

(a) Column packed with sieved glass.

Table 2 Commercially Available Polymeric Packings for Aqueous SEC (As of 1989)

Principal trade name (supplier); *other commercial names	Description, particle size	Typical efficiency (plates m^{-1})	Maximum flow rate
A. Rigid Hydrophilic Polymer Gels			
TSK-PW (Toyo Soda, Japan) *Spherogel PW (Altex/Beckman, USA) *Bio-Gel TSK (Bio-Rad, USA) *Micro-Pak TSK-PW (Varian, USA)	Acrylate polymer-based, 10 μm	12,000 (30,000)(a)	2 ml min
OHpak B-800 (Showa Denko, Japan)	Polyhydroxymethacrylate 12 μm	12,000 (30,000)(a)	1 ml min
OHpak Q-800 (Showa Denko, Japan)	Polyvinylalcohol gel, 9 μm	10,000	0.5-1 ml min
Asahipak GS (Asahi Kasei, Japan)	Polyvinylalcohol gel, 10 μm	30,000	2-3.5 ml min
PLaquagel (Polymer Laboratories, USA)	Polyacrylamide gel, 10 μm (spherical)	20,000	2 ml min (30 cm hr)
GELKO (Hitachi, Japan)	Polyacrylamide gel		1.5 ml min
Superose (Pharmacia, Sweden)	Crosslinked agarose 10-12 μm	12,000	0.1-1.5 ml min
Spheron (Lachema, Czechoslovakia)	Crosslinked poly(2-hydroxyethylmethacrylate), 20-40 μm	5,000	1 ml min (100 cm hr)(b)
Shodex Ionpak KS-800 (Showa Denko, Japan)	Sulfonated crosslinked polystyrene, 10 μm	40,000	1 ml min

B. *Compressible and Semicompressible Hydrophilic Polymer Gels*

Sephadex (Pharmacia, Sweden)	Crosslinked dextran 40–250 μm (dry)	600[c]	10 cm h[d]
Sephacryl (Pharmacia, Sweden)	Alkyl dextran crosslinked with N,N' methylene-bisacrylamide	1000[c]	40 cm hr
Sepharose (Pharmacia, Sweden)	Crosslinked agarose 100–150 (wet)		10 cm hr
Toyopearl (Toyo Soda, Japan)	Polyacrylate-based, 30 ± 10, 45 ± 15, 75 ± 25 μm	600–6000[c]	200 cm hr
Ultrogel AcA (LKB)	Composite polyacrylamide/agarose gel, 100 μm (wet)	2000	20 cm hr
Bio-Gel P (Bio-Rad, USA)	Polyacrylamide, 30–200 μm (wet)		
Bio-Gel A (Bio-Rad, USA)	Agarose, 50–20 μm (wet)	1500	50 cm hr
Enzacryl (Lachema, Czechoslovakia) (Koch-Light, UK)	Crosslinked poly(acryloyl-morpholine) 30–200 μm (dry)	7000	

[a] PW_{XL} (Toyo Soda) of KB Series (Showa Denko).
[b] Loss of column efficiency at higher flow rate.
[c] Depending on particle size and pore size.
[d] Gel compaction at higher flow rate.

III. ELECTROSTATIC INTERACTIONS

The early elution of inorganic salts [18-20] and ionic polymers [9, 21-24] from SEC packings has been noted for some time. Virtually all stationary phases for SEC bear some negative charge under typical use conditions. Underivatized glass and silica are intensely charged at all but very low pH due to the dissociation of silanol groups [17a]. Sephadex contains carboxylic acid groups [25], the concentration of which may become more abundant on storage under oxidizing conditions [7]; in addition, ionization of glucosidyl hydroxyl groups produces strong electrostatic effects at high pH [26,27], a condition sometimes encountered in the analysis of polysaccharides that are insoluble in neutral solutions [27]. The carboxylic acid content of PW gel, determined by pH titration, has been reported in detail [28]. The sulfate and carboxylate groups found in algal polysaccharides may be incorporated in agarose gels [7]; circumstantial evidence [29] suggests that Superose contains a modest surface concentration of such moieties.

It might at first blush appear that derivatization could nullify the ionic character of silica, by "steric protection" of the SiOH groups, if not by their chemical conversion. Such hypotheses fail to take into account the long-range nature of electrostatic interactions. Perhaps the most striking evidence for the effect of residual SiO^- groups is seen in the behavior of the partition coefficient K for the basic protein lysozyme on a variety of derivatized silica packings [30]. K is experimentally obtained as

$$K_{SEC} = \frac{V_e - V_0}{V_t - V_0} \quad (1)$$

where V_e is the measured peak elution volume, V_0 is the exclusion volume or interstitial volume of the column, and V_t is the total mobile phase volume, i.e., the sum of V_0 plus the column pore volume; since V_0 and V_t represent the range of normal elution volumes, the limits of K_{SEC} should be 0-1. As shown in Fig. 1, values of K_{SEC} in excess of unity are consistently seen for lysozyme on all packings at low ionic strength, revealing the adsorption of this positively charged protein (at pH below its isoelectric point) on these negatively charged packings.

Figure 1 also indicates the weakness of the commonplace assertion that all electrostatic effects in SEC may be dealt with by the addition of salt. While this statement is probably correct for hydrophilic synthetic polyelectrolytes, the situation is more complicated for amphiphilic and/or zwitterionic macromolecules such as proteins. The minima in the curves of Fig. 1 indicate that the suppression of charge effects on increase in ionic strength is followed by an

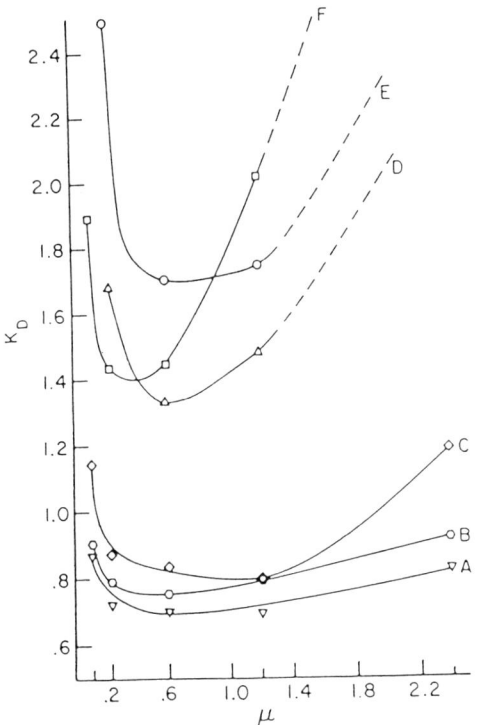

Fig. 1 Dependence of K_{SEC} of lysozyme on ionic strength of mobile phase (pH 7.05 phosphate buffer) for TSK SW 2000 (A); TSK SW 3000 (B); LiChrosorb Diol (C); Shodex OH pak B-804 (D); Waters I-125 (E); and SynChropak GPC 100 (F). (From Ref. 30.)

increase in adsorption at high salt concentration, possibly due to hydrophobic solute-substrate interactions, or simply salting out of the protein. It is clearly impossible to identify any particular ionic strength with "ideal" or purely "steric" chromatography. Indeed, only in the absence of ionic strength dependence of K_{SEC} over a reasonably broad range of salt concentration could one assert that charge interactions do not play a role in chromatography of ionic macromolecules.

While relatively subtle electrostatic effects may go unnoticed, the interaction of synthetic polycations with aqueous SEC packings often leads to complete retention. Presumably these adsorptive effects are highly cooperative, so that a sequence of cationic segments ionically bound to a surface of opposite charge is desorbed with difficulty. There are no reports of successful chromatography of

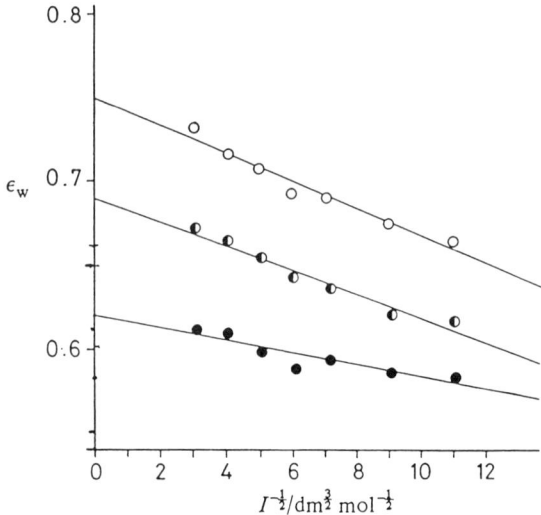

Fig. 2 Elution volumes of polystyrene latexes, relative to benzyl-alcohol, on mixed-bed CPG columns, in pH 6 phosphate buffer, as a function of ionic strength I. Latex diameters: 30 nm (○), 58 nm (◐), and 94 nm (●). (From Ref. 39.)

polycations on any conventional silica-based columns. Two approaches to SEC of such polyelectrolytes have been pursued. In one, cationic silanizing agents are used to create a positively charged stationary phase, so that ionic adsorption is replaced with ionic repulsion [31]. This procedure was first applied to porous glass by Talley and Bowman [31a], who found that low pH and moderate ionic strength were required to produce apparently normal elution behavior for poly-(4-vinylpyridine) on cationic CPG. Commercial packings that utilize this approach are Catsec [32] and DMAE-Fractosil [31c]. The principal disadvantage of such stationary phases may be the rather intense repulsions that arise between the polyions of interest and the packing. The high ionic strength needed to suppress this effect can limit the solubility of other polymers, so that, for example, the elution of a nonionic polymer, such as polyvinylpyrrolidone, may be abnormal [31b]. The second approach to this problem involves the use of hydrophilic gel packings. Sephadex [33], PW gel [34], and Superose [35] have all been applied to polycations. The concentration of supporting electrolyte required to suppress attractive interactions between polyion and substrate appears to be somewhat less than for siliceous packings. On the other hand, hydrophobic effects are observed for at least some of these gel phases [36], which become problematic for polymers with partially apolar moieties, such as polyvinylpyridine and polyvinylamine.

It is evidently easier to study repulsive charge effects in SEC than attractive ones, since the latter typically lead to irreversible adsorption. Many workers have observed the early elution of polyanions, such as lignosulfonates [37], carboxymethylcellulose [38], and sodium polystyrenesulfonate [23,24], on porous glass or silica, and have also noted the diminution of this effect with increasing ionic strength. There have been, however, very few attempts to account for this behavior in a quantitative model. Here, we shall discuss the semiempirical analyses of Booth and coworkers [39], Dubin and Tecklenburg [40,41], and Potschka [42], and also the theoretical approach of Smith and Deen [43].

Booth et al. [39] studied the ionic strength dependence of the retention of sulfonated polystyrene latexes on porous glass in dilute salt solution. The decrease in elution volume at low ionic strength was attributed to an increase in the effective size of the particle due to its electrical double layer (see Fig. 2).

We have used "universal calibration" plots (see below) to compare the retention times of sodium polystyrenesulfonate (NaPSS) with an uncharged polymer (dextran or pullulan), on CPG porous glass [40]. Plots of K_{SEC} vs. $\log[\eta]M$ (i.e., the log of the viscosity radius R_η) for the two polymers were congruent at high ionic strength (I = 0.5) but diverged progressively at lower I or higher pH. These divergences could be attributed to repulsion of the polyion from the walls of the CPG pores, since the use of the viscosity radius accounts for intrapolyion expansion effects. The diminution in K_{SEC} for the polyion relative to K_{SEC} for pullulan was then ascribed to a decrease in the effective pore size for the former, as shown in Fig. 3. Forced fit of universal calibration curves then led to the dimensions of the hypothetical restricted pore, and the difference between its radius and the geometrical pore radius was referred to as X_E, the repulsion distance [41]. Approximate calculations for the electrostatic potential at X_E gave values of ψ_E = 44 ± 0.5 mV, irrespective of ionic strength [40]. It was also noted that $X_E\kappa$, i.e., the repulsion distance in Debye lengths, was 2.6 ± 0.2, regardless of ionic strength. Subsequently, it was found that X_E was constant only over the ranges 20 Å < R_η < 60 Å (I = 0.01 M); 40 Å < R_η < 80 Å (I = 0.02 M); and 50 Å < R_η < 90 Å (I = 0.05 M). X_E decreased with solute size at lower R_η [40b]. This behavior was attributed to a transition from valence control, in which the polyion dimensions are small relative to κ and may therefore be approximated as a small ion whose repulsive free energy is proportional to its net charge; to macroion behavior, in which the solute-packing electrostatic interaction is controlled only by some portion of the polyion "sphere" which is able to approach within one Debye length of the wall.

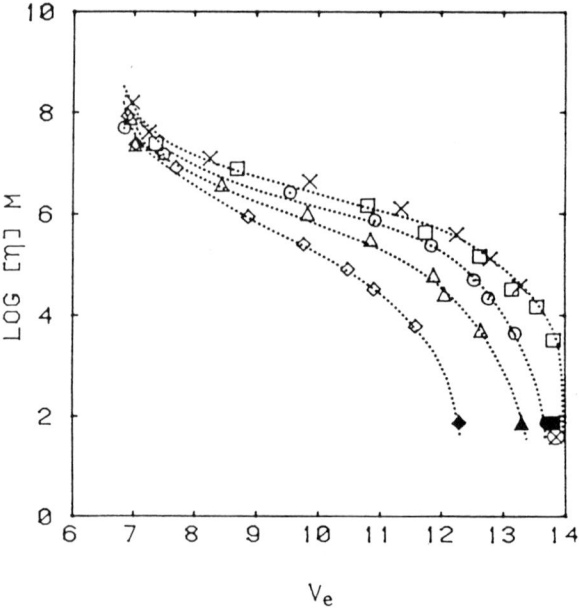

Fig. 3 "Universal calibration" plots on CPG glass (r_p = 500 Å) for pullulan in pure water (X), and sodium poly(styrenesulfonate) in pH = 5.00 eluants with ionic strengths I = 0.5 M (□), 0.05 M (○), 0.02 M (△), and 0.01 M (◊). Filled symbols are sodium citrate. (From Ref. 41.)

Potschka took a different approach to the question of electrostatic interactions in SEC [42], focusing on the electrical double layer around the solute. For several globular proteins, above and below their isoelectric points, it could be demonstrated that the true dimensions of the macromolecules, referred to by the author as their "bulk viscosity radii," were insensitive to either ionic strength or pH (in contrast to the behavior of synthetic polyelectrolytes). Consequently, the dependence of K_{SEC} on the ionic strength I can be ascribed to electrostatic substrate-solute interactions. Assuming that elution volumes V extrapolated to infinite I correspond to "ideal" or charge-unperturbed behavior, Potschka constructed plots of protein viscosity radius, R_η, vs. $V_{I\to\infty}$, from which one may obtain the *effective* R_η at any ionic strength. Proteins above their isoelectric points are negatively charged and therefore repelled by anionic groups on the gel packing, in this case carboxylic acid groups on the PW gels employed by Potschka. The consequent diminution of K_{SEC} with decreasing ionic strength

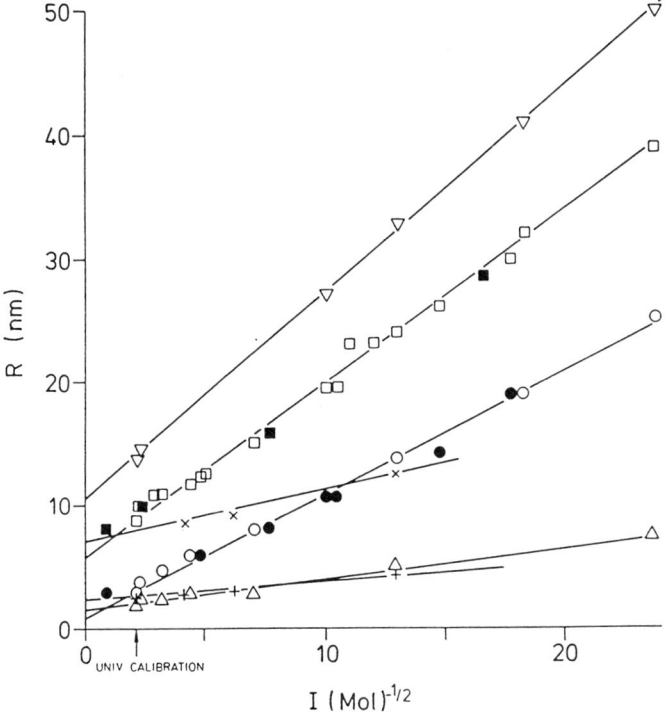

Fig. 4 Size of the effective charge repulsion layer surrounding the polyelectrolyte according to SEC analysis on a TSK 5000PW column; R is the effective radius, considered to be the total radius of the macromolecule plus the double layer. The viscosity radii of the proteins (thyroglobulin, ovalbumin, and horse myoglobin) were obtained by universal calibration at high ionic strength, and the observed deviation from that value was correlated with ionic strength. Measurements were done in a variety of buffers all containing 0.4 mM BME: MS2 at pH 8.0 (▽); thyroglobulin (pI = 4.5) at pH 8.0 (□), 6.6 (■), and 5.0 (x); ovalbumin (pI = 4.7-4.9) at pH 8.0 (○), 7.2 (●), and 5.0 (+); horse myoglobin (pI = 7.3) at pH 8.0 (△). (From Ref. 43.)

is then viewed as resulting from an increase in $R_{\parallel}^{eff}$ of the protein, as shown in Fig. 4, which in turn arises from its expanded electrical double layer. In this description, the gel "sees" the macromolecule as surrounded by ion atmosphere, the thickness of which depends primarily on the ionic strength as well as on the surface charge density and mean curvature of the macromolecule.

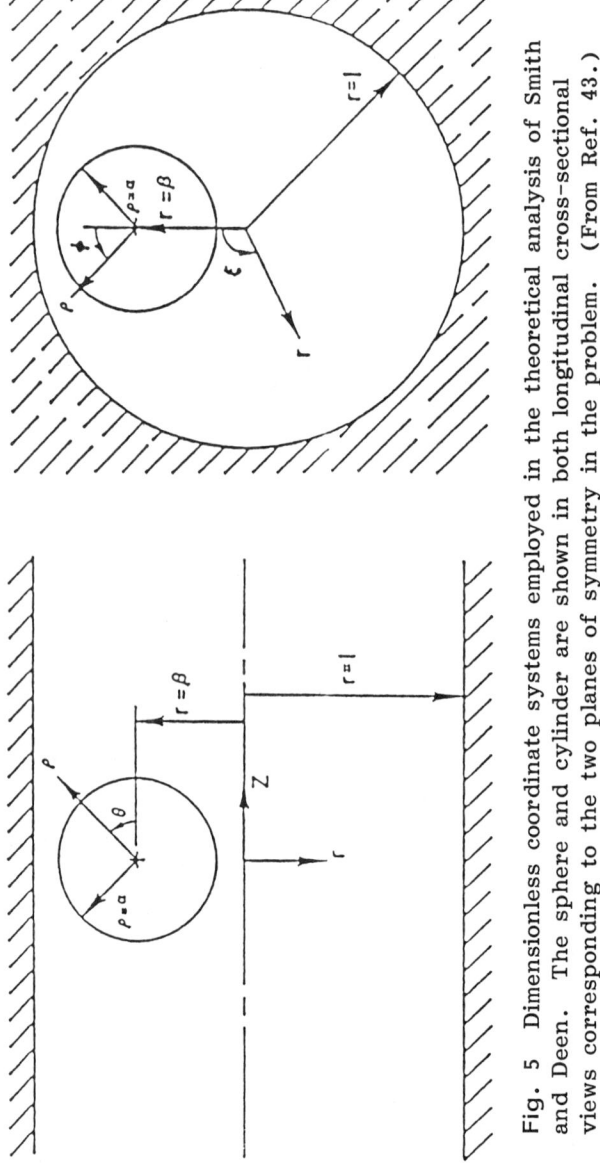

Fig. 5 Dimensionless coordinate systems employed in the theoretical analysis of Smith and Deen. The sphere and cylinder are shown in both longitudinal cross-sectional views corresponding to the two planes of symmetry in the problem. (From Ref. 43.)

The aforementioned models of Booth et al., Potschka, and Dubin are all simplifications of the problem of repulsive overlapping double layers, in which the former two workers ascribe the repulsive effect to the ion atmosphere around the solute, while the last author assigns the electrostatic component of K_{SEC} to the double layer on the surface of the stationary phase. The classical treatment of overlapping similarly charged double layers by Verwey and Overbeck [44] does not seem to have been applied to this problem, possibly because approximate analytical expressions are only obtained under conditions of low potential. Recently, however, a rigorous theoretical treatment of K for such charged systems was put forward by Smith and Deen [42]. Smith and Deen consider the electrostatic free energy for a spherical particle in a cylindrical cavity of identical charge sign (see Fig. 5). The key parameters are the particle dimension relative to cylinder size (α), surface charge density of the cylindrical cavity (σ_c), ionic strength (I), and charge of the sphere. This last factor is considered as either a (dimensionless) constant volume density arising from the fixed charges (Q_V) or a (dimensionless) constant surface density of fixed charges (Q_S). In addition, it is also possible to treat a sphere of constant surface potential. The subsequent procedure is conceptually straightforward but mathematically complicated. The sphere is allowed to assume any sterically acceptable position within the pore, and the distribution of small ions is then allowed to conform to a Poisson-Boltzmann distribution, essentially determined by the screened, simultaneous potentials of the spherical solute and the pore wall. The free energy is then calculated relative to the situation when the solute and cavity are well separated. This is the energy of transfer of solute from bulk medium to pore, and may be related through a Boltzmann expression to the relative mean solute concentration in the pore, i.e., to K:

$$K = \frac{C}{C_0} \tag{2}$$

While the foregoing treatment does not lead to explicit and simple algorithmic expressions for $K(\alpha, Q_c, Q_V, I)$, it is still possible to obtain theoretical curves of K vs. α, given values for the aforementioned variables. Smith and Deen make essentially no comparisons between their results and experimental data. Work in the author's laboratory [45] has been directed toward this end. Of particular interest is whether the free-energy function developed by Smith and Deen supports the perception of a well-defined electrostatic "barrier," i.e., "X_E," or produces a softer energy gradient when the solute approaches the pore wall. It is also interesting to consider whether any of the models proposed accounts well for the behavior of a random coil polyelectrolyte or, put differently, whether a real flexible polyion can be properly treated as some sort of "equivalent charged sphere."

IV. HYDROPHOBIC INTERACTIONS

Studies with low MW solutes provide the clearest evidence for nonsteric effects, inasmuch as observed values of K_{SEC} in excess of unity can be attributed to partitioning of the solute. Brown's distinction between adsorption and partitioning [46] is a useful one—the former refers to an interaction specifically involving a complementary pair of functional groups on the solute and packing surface, while the latter connotes interactions relatively independent of molecular orientation but highly sensitive to solvent polarity; consequently, the adsorption of oligomers and polymers—being cooperative—should increase with MW, while their partition—entailing intimate mixing of solute and gel—decreases with MW [47].

The literature on the partitioning of low MW amphiphiles in aqueous SEC gels is abundant, and deals with n-alcohols [36,48-50], monosubstituted aromatics [51], sodium dodecyl sulfate [52], and tetraalkylammonium chlorides [53]. Virtually all these studies were carried out with Sephadex packings. The mechanism of apolar partitioning on these crosslinked polysaccharide gels is complex, and three distinct explanations have been put forward. A central observation is that nonpolar affinity of Sephadex gels increases dramatically with the degree of crosslinking, and this effect (at least for aromatic solutes) has been attributed to interactions between the solute and the various units arising from the epichlorohydrin crosslinking reaction, possible involving the ether or hydroxyl units in these crosslinks [54,55]. On the other hand, crosslinking also increases the average matrix concentration, i.e., the fraction of water in the gel that is proximate to polysaccharide units. The aqueous solubility of aromatic hydrocarbons is increased in concentrated sugar solutions (e.g., from 10 to 50% for naphthalene in 2 M solutions of various pentoses and hexoses [56]) and effects of a similar magnitude are observed in dextran solution [57]. These workers therefore suggested that the polysaccharide network produces vicinal "structured water" (in the Nemethy and Scheraga sense) in which hydrocarbons may be accommodated at lower entropy cost, and they also present DSC results in support of the contention that water structure is greatly modified in crosslinked polysaccharides [58]. However, in view of the fact that most concentrated monosaccharide solutions do not exhibit hydrocarbon solubilization [56], one may argue that polysaccharides proffer hydrophobic "faces" to amphiphilic solutes, much as cyclodextrins are thought to interact with aliphatic alcohols [59]. Thus, the exact mechanism accounting for apolar partitioning in Sephadex is a matter of debate.

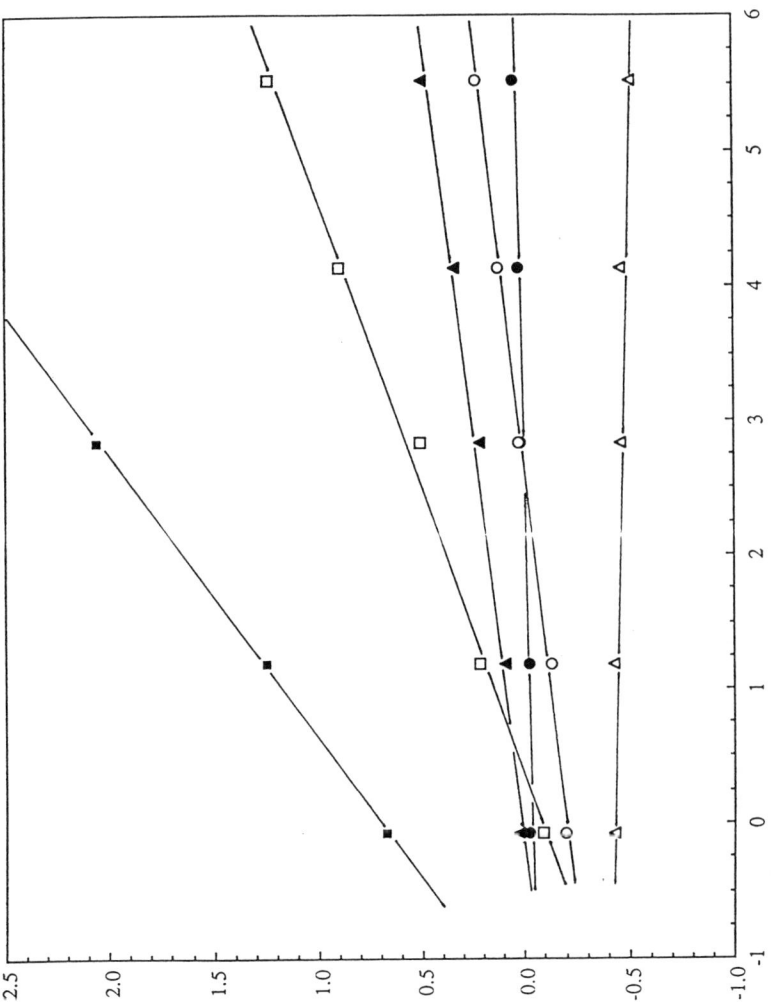

Fig. 6 Dependence of K_{SEC} on the molar solubility of n-alcohols (butanol-octanol) in pure water on PW gel (■); Superose 6 (▲); Sephadex G-100 (▴); Sephadex G-75 (○); Sephadex G-10 (□); and Bio-Gel A50 (●). (From Ref. 62.)

The relationship between hydrophobic partitioning on LC packings and the aqueous solubility of the solute has been pointed out by a number of authors [60]. It is understood that these correlations form a type of linear free-energy relation. Hafkenscheid and Tomlinson used the capacity factors of various solutes on reversed phase columns to assign hydrophobicities [61]. Marsden [7] pointed out that a contribution per methylene group to the partitioning component of K could be identified. Most recently, the author found [62] that plots of ln K_{SEC} vs. ln S, where S is the molar aqueous solubility, are linear for n-alcohols in SEC gels such as Sephadex, Superose, Biogel, and PW gel (see Fig. 6). It was furthermore demonstrated that the slope of such a plot may be identified as:

$$\gamma = -\frac{\Delta G^o_{SEC}(CH_2)}{\Delta G^o_s(CH_2)} \qquad (3)$$

where ΔG^o_{SEC} is the contribution of a methylene group to the free energy of solute transfer from water to gel, and $\Delta G^o_s(CH_2)$ is the analogous quantity for transfer from water to neat liquid (alcohol). γ thus provides a quantitative measure for gel hydrophobicity. Values of γ for, say, Biogel A50, Sephadex G75, Superose, Sephadex G10, and PW gel are 0.014, 0.080, 0.085, 0.23, and 0.56, respectively [62].

V. UNIVERSAL CALIBRATION

In 1967 Benoit and coworkers observed that a number of organophilic polymers with a variety of branching configurations (e.g., linear, comb-branched, star-branched) all conformed to a single plot of log[η]M vs. V_e [63]. This identification of [η]M as a "universal calibration" parameter has had considerable practical impact, inasmuch as it makes possible the transformation from one log M vs. K_{SEC} curve to another, if the necessary Mark-Houwink relations are known [64]. Perhaps more significantly, universal calibration guided theoretical treatments for the following decade. However, it should be noted that none of these treatments achieved much more than qualitative success; predictions of K_{SEC} from [η]M values and pore dimensions have not been quantitatively confirmed.

One must recognize two major obstacles to the modeling of SEC systems such as those studied by Benoit. First, the geometry of the pores is extremely irregular and heterogeneous. Second, the polymer dimension is not unambiguous: with statistical chain polymers, for example, a number of size parameters are all linearly related and therefore would be equally satisfactory as universal dimensions. Thus, the viscosity radius [65]

$$R_\eta = \left(\frac{3[\eta]M}{10\pi N_A}\right)^{1/3} \tag{4}$$

(where the units of $[\eta]$ are cm^3/g), the radius of gyration R_g, and the mean-square end-to-end distance $\langle r^2 \rangle^{1/2}$, will all correlate equally well with K.

Experimental conditions do exist which reduce these two types of uncertainties. The pore geometry is more uniform and easier to characterize for rigid stationary phases, such as porous glass. Second, one may envision polymer dimensions more easily for compact macromolecules such as globular proteins. Unfortunately, these two simplifications cannot be achieved simultaneously because globular proteins adsorb strongly to siliceous packings.

Despite the "nonideality" of solutes and substrates in general, it is interesting to test the validity of an "ideal" or "geometrical" model. It is readily shown that a spherical solute of radius r will display a diminished concentration within a cavity of radius a, according to the expression:

$$K = [1 - r/a]^\lambda \tag{5}$$

where λ is 2 or 3 for a cylindrical or spherical cavity, respectively, and K is the concentration of solute in the pore relative to the interstitial volume [66]. Waldman-Meyer [67] showed that Eq. 5 with $\lambda = 2$ agrees very well with data for dextrans on CPG glass, if one assumes $r = R_\eta$ (see Fig. 7). Furthermore, the slopes of plots of $K_{SEC}^{1/2}$ vs. R_η yield values of $1/a$ in very good agreement with CPG pore radii obtained by other methods.

The foregoing studies suggest that R_η is the dimensional parameter that determines K, hence also K_{SEC}. Such a conclusion would clearly be better supported with results for macromolecules with varying shape factors, and several studies have been directed toward this end. It should be evident that such divergent conformations coexist in aqueous media, in which many examples of random coil, ellipsoidal or spheroidal, and rodlike macromolecules may be identified. On the other hand, as already noted, electrostatic and hydrophobic polymer-stationary phase interactions are commonplace in aqueous SEC, and these effects must be avoided if the intention is to establish the relationship between elution and macromolecular dimensions.

Comparisons of the elution of globular proteins and flexible nonionic chain molecules (dextran) were made by Frigon and coworkers [68], who observed congruent behavior for the dependence of retention on the viscosity radius R_η but not for the Stokes radius R_S.

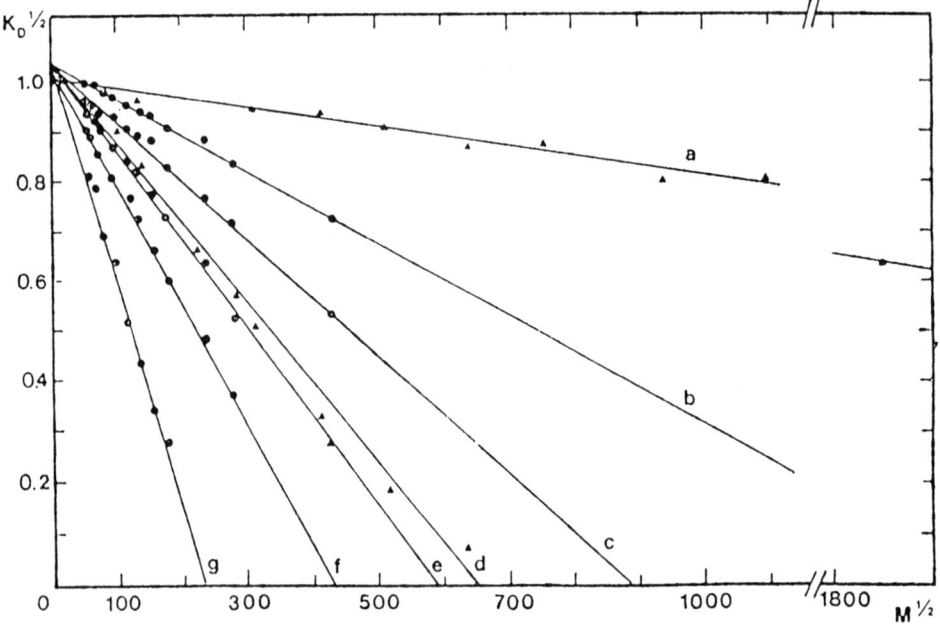

Fig. 7 Plots of $K_{SEC}^{1/2}$ vs. $MW^{1/2}$ for polystyrene (▲) on CPG $r_p = 905$ Å (a); 113 Å (b). Dextrans (●) on CPG $r_p = 276$ Å (c); 173 Å (d); 113 Å (e); 85 Å (f); and 4.5 Å (g). (From Ref. 67.)

It should be noted that the variation of the protein data around the dextran curve was ±5%, which the authors attributed to "nonsize exclusion, repulsion, or adsorptive interactions between the various proteins and the silica based column packing materials." Similar comparisons were made by the current author [69] who observed that the data for globular proteins and pullulan (a nonionic linear flexible-chain polysaccharide, available as narrow-MWD fractions) coincided in the form K_{SEC} vs. R_η, on either Superose or Biogel A. In a more comprehensive study [70], Potschka reported K for numerous globular proteins, rodlike proteins (tropomyosin, actinin), and selected samples of DNA. Potschka observed that all data collected on Superose conformed to a single plot of R_η vs. K_{SEC}, and reported that the results for rodlike macromolecules diverged from those for ellipsoidal biopolymers only when other size parameters, e.g., R_g, R_s, or the mean projection length ⟨L⟩, were used instead of the viscosity radius.

The preceding findings appear to run contrary to theoretical treatments. Giddings [71], and subsequently Cassasa [72], maintained on theoretical grounds that all macromolecules with the same ⟨L⟩ should coelute from a given column. This relationship would rule out the radius of gyration as the universal parameter. The viscosity radius also would be disqualified because, as pointed out by Cassasa, the dependence of R_η on ⟨L⟩ is different for different shape factors; more specifically, for random coils [72]:

$$[\eta]M = \Phi \left(\frac{\pi^{3/2}}{64}\right) \langle L \rangle^3 = 3.43 \times 10^{23} \langle L \rangle^3 \tag{6}$$

For spheres:

$$[\eta]M = \left(\frac{5\pi N_A}{12}\right) \langle L \rangle^3 = 7.88 \times 10^{23} \langle L \rangle^3 \tag{7}$$

And for rods:

$$[\eta]M = \left(\frac{2\pi N_A}{5}\right) \Omega \langle L \rangle^3 = 7.57 \times 10^{23} \langle L \rangle^3 \tag{8}$$

where Φ is Flory's universal constant, and Ω is a slowly decreasing function of the axial ration p (e.g., $\Omega = 0.27$ for p = 200), and all quantities are in cgs units. Thus, while experimental results appear to generate broad support for the argument that the viscosity radius determines retention, theoretical considerations appear to militate against this finding.

In view of this confrontation between theoretical models and experiment, we compared retention volumes of random coils (dextran and pullulan), ellipsoids (globular proteins), and rodlike macromolecules (DNA and schizophyllan) [73]. A number of precautions were taken to minimize or eradicate polymer-substrate interactions. First, the choice of schizophyllan as representative of rigid rods [74] was dictated by the fact that neutral polysaccharides virtually never adsorb or partition on aqueous SEC gels. The stationary phase chosen, Superose, has relatively little ionic character and shows only weak hydrophobic interactions with amphiphilic solutes [62]. A multivariant procedure was used to "optimize" the pH and ionic strength of the mobile phase; the criterion being the convergence of protein retention data to a smooth curve for K_{SEC} vs. R_S (regression coefficient larger than 0.99), regardless of isoelectric point, which was taken as strong evidence for separation based on size along. Lastly, the ionic strength, ~0.4 M, was large enough to suppress electrostatic solute-substrate interactions, even for the highly charged macroion DNA.

The results of this study are shown in Fig. 8. It is evident that rodlike macromolecules elute earlier than spherically symmetrical

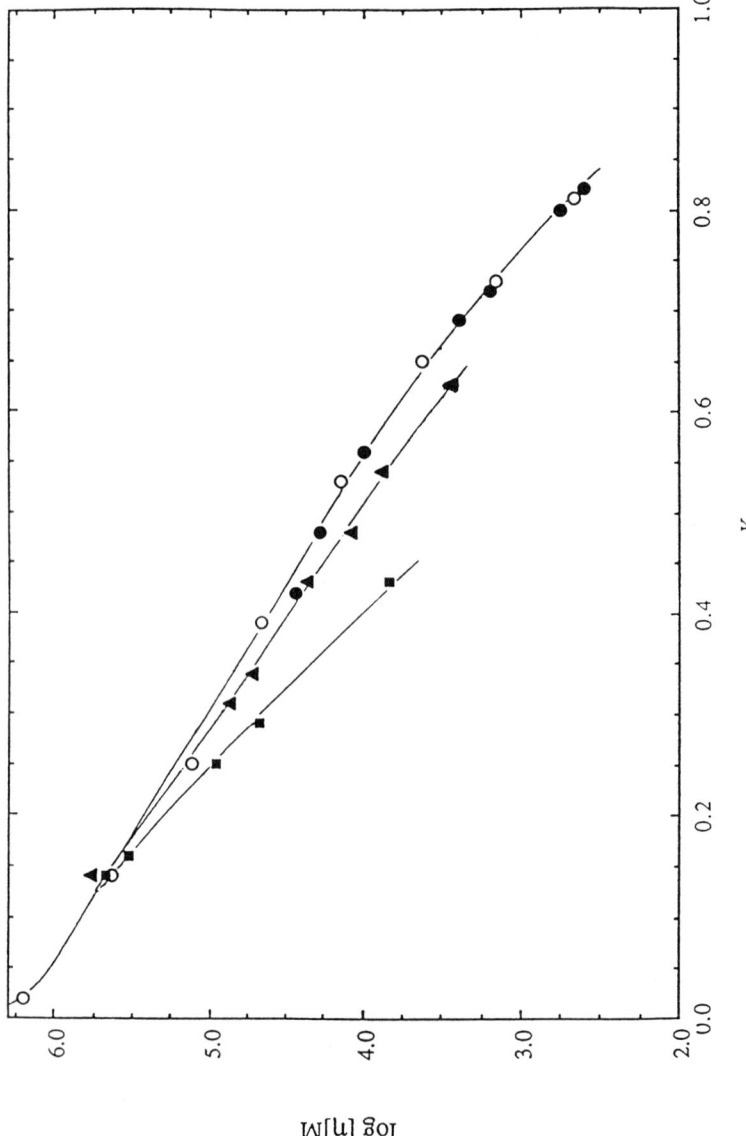

Fig. 8 "Universal calibration" plots on Superose 6, in pH 5.5, I = 0.38 M phosphate buffer, for pullulon (○), globular proteins (●), DNA (▲), and schizophyllan (■). (From Ref. 73.)

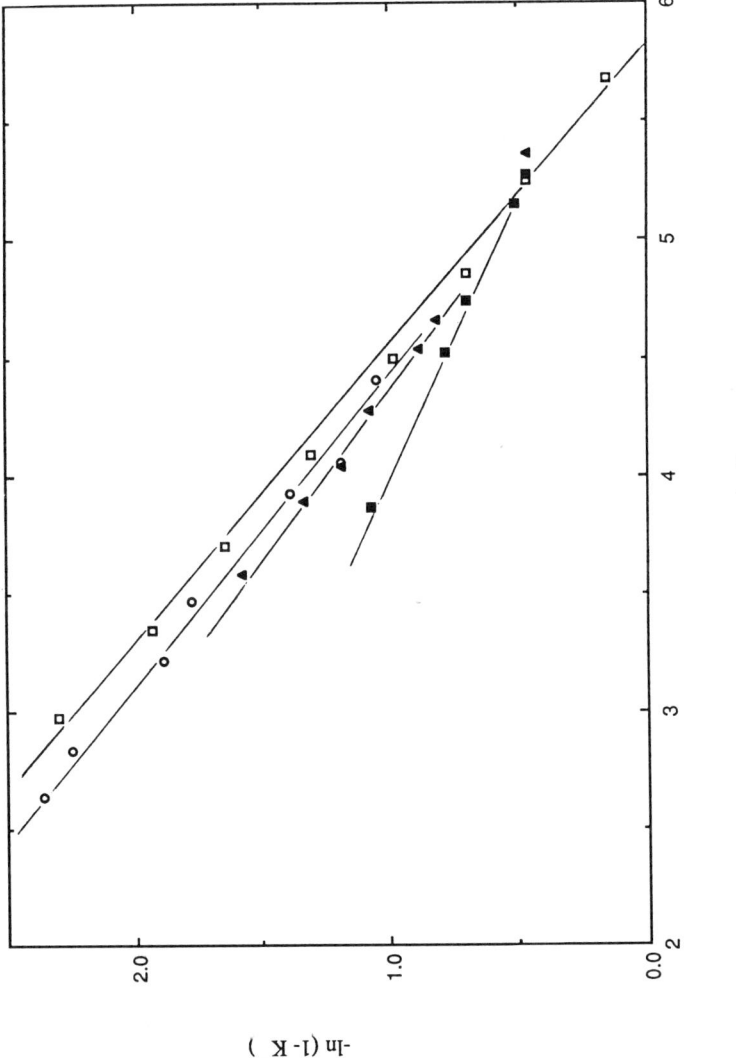

Fig. 9 Plots of $-\ln(1 - K^{1/2})$ vs. $\ln R_\eta$ for pullulan (□), globular proteins (○), DNA (▲), and shizophyllan (■).

polymers of equal R_η, although the plots converge at higher MW. Rods are clearly differentiated from the other species as long as they are short compared to their persistence lengths. At high MW their behavior is identical to that of flexible chains. The data for schizophyllan is most clearly separated from the other solutes because its persistence length is 1200 Å, which is about threefold larger than that of DNA. Since the DNA samples in [70] were all of high MW, the divergence observed in Fig. 8 were not seen.

The above-mentioned deviations from a universal plot of R_η vs. K_{SEC} are in agreement with theoretical predictions [72]. However, plots of the projection length $\bar{X}$ (calculated from Eqs. 6-8) on K_{SEC} reveal even less convergence with each macromolecular type displaying separate dependence. R_g may also be ruled out because the viscosity radius of rods is larger than the radius of gyration.

The failure of any previously defined dimensional parameter to provide universal calibration may indicate that the behavior of macromolecules in confined regions may not be directly comparable to dilute solution phenomena. The nature of R_{SEC} might be explored empirically by establishing by presupposing some general form:

$$R_{SEC} = \alpha R_\eta^\beta \tag{9}$$

and assuming the validity for a geometrical expression, for cylindrical cavities, namely:

$$K^{1/2} = 1 - \frac{R_{SEC}}{r_p} \tag{10}$$

Combination of Eqs. 9 and 10 lead to

$$\ln(1 - K^{1/2}) = \delta + \beta \ln R_\eta \tag{11}$$

where $\delta = \ln(\alpha/r_p)$. Equation 11 predicts that plots of $\ln(1 - K_{SEC}^{1/2})$ vs. $\ln R_\eta$ will be linear, with slopes corresponding to the scaling exponent in Eq. 9. Such plots are shown in Fig. 9, for the data of Fig. 8. Agreement with Eq. 11 is excellent for the rodlike macromolecules, corresponding to β values of 0.67 and 0.43, for DNA and schizophyllan, respectively. As expected, data for pullulan and proteins show somewhat poorer fit, with a value of $\beta \simeq 0.8$. The implications of these results with regard to the separation mechanism are the subject of continued investigation.

VI. SEC OF PROTEINS

The field of protein chromatography is characterized by intense activity in applied clinical and biotechnology research laboratories, in the development teams of instrument companies, and in academic laboratories. Some of the current progress is reviewed in recent monographs or edited volumes [75-80] and in proceedings of conferences, such as the International Symposium on HPLC of Proteins, Peptides and Polynucleotides. Currently, one of the primary goals of protein chromatography is the ability to accomplish separations of nearly identical proteins, i.e., variants differing only by virtue of several or even just one amino acid substitutions. To achieve this end, multiple chromatographic modes are employed, often involving gradients of pH, strength, and solvent.

Since proteins invariably possess a high surface density of charged groups, as well as hydrophobic regions of varying degrees of accessibility, electrostatic interactions and hydrophobic interactions are characteristic of protein chromatography. Indeed, protein migration governed by size alone is probably the exception. In practice, however, the overwhelming concern is separation efficiency, and mixed-mode chromatography is generally regarded as a benefit to be exploited. Generally, protein chromatography is categorized as ion exchange chromatography (IEC), reversed phase liquid chromatography (RPLC), hydrophobic interaction chromatography (HIC), and size exclusion chromatography (SEC). If numerous ionic substituents are intentionally introduced into the stationary phase (e.g., carboxylated Sepharose, sulfonated Styragel), separation is considered to involve an "ion exchange" mechanism. If the packing is an inorganic support with a bonded apolar silane phase such as C_8 or C_{18}, and elution involves a cosolvent (e.g., acetonitrile) gradient, the process is referred to as RPLC. If hydrophobic ligands are introduced onto a gel (e.g., phenyl Sepharose) and the protein is eluted with a salt gradient, the term HIC is applied. If geometrical factors predominate, migration is thought to correspond to SEC. In actuality, several mechanisms may be operating simultaneously in any separation. For example, RPLC packings contain in abundance unreacted silanol groups whose negative charge affects protein retention. Also, as noted above, many SEC packings display some hydrophobic character.

The foregoing comments indicate the complex state of protein chromatography. With regard to SEC, proteins offer one major simplification: their size and shape are sometimes better characterized and more easily described, than either statistical chain polymers, or semirigid asymmetrical macromolecules. Thus, for most globular proteins, the Stokes radius R_s or the viscosity radius R [81], which are nearly equal are appropriate dimensional parameters.

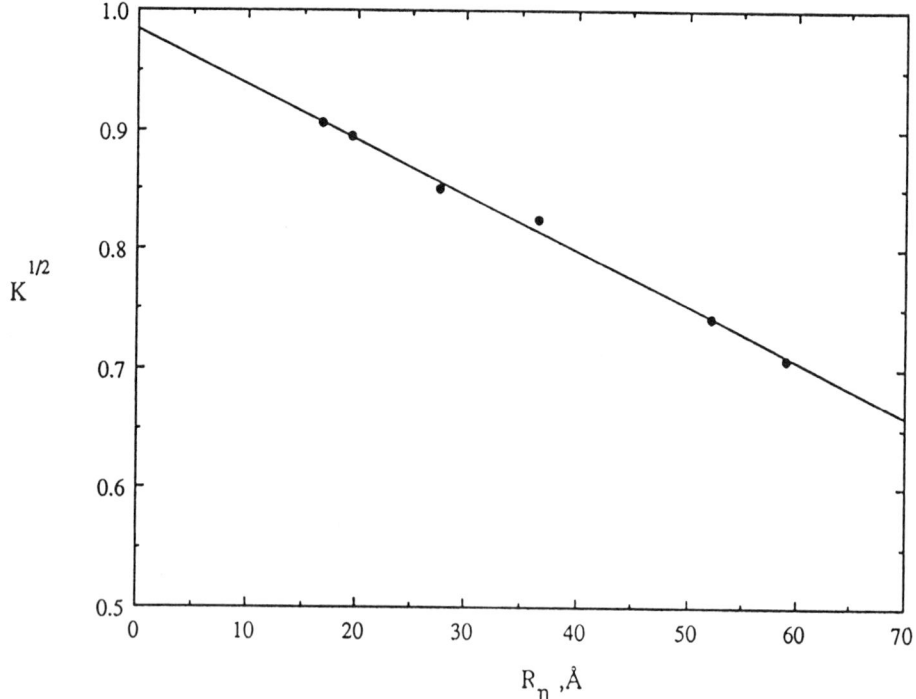

Fig. 10 $K^{1/2}$ vs. R_S for globular proteins on Superose 6 column in pH 5.5, I = 0.38 M phosphate buffer. (From Ref. 82.)

For this reason, proteins would seem to be the solutes of choice for the characterization of packing pore structure and the testing of SEC theories. However, these goals can only be achieved if nonideal effects are completely suppressed. Verification of such ideal behavior presents a dilemma in that one must assume a priori a theoretical form for $K(R_s, r_p)$, agreement with which demonstrates pure SEC.

We have attempted to establish such ideal behavior for globular proteins on Superose [82] by optimizing the mobile phase. The key variables are ionic strength (I), which influences both electrostatic and hydrophobic interactions, and pH, which affects the former. $K(I, pH)$ values were determined for eight globular proteins on a Superose 6 column. Since the proteins differed widely with respect to hydrophobicity and isoelectric point, one would expect that nonideal effects would obviate any simple dependence of K on R_s. The regression coefficient r of such dependence then becomes an inverse measure of adsorption or repulsion, and should approach unity for pure SEC. The search for a maximum in r(I,pH) involved a simplex optimization, which converged on the conditions pH = 5.5, I = 0.38. Under these conditions the data, shown in Fig. 10, conformed to a linear dependence of $K^{1/2}$ on R_s, in accordance with the geometrical expression, as in Eq. 9.

VII. SEC OF MICELLES

Size exclusion chromatography of surfactant micelles represents an interesting phenomenon in that these "macromolecular" species are in dynamic equilibrium with monomeric surfactant. Thus, the chromatogram of sodium dodecyl sulfate (SDS) in water or dilute salt displays a leading edge (micellar front) followed by micelles in equilibrium with surfactant monomer, and finally a peak corresponding to monomer alone [83]. Consequently, information about the monomer-micelle equilibrium as well as about micelle size can be obtained according to the way in which SEC is carried out. Most simply, micelle decomposition can be suppressed by incorporating surfactant at or above the CMC into the mobile phase. This technique has been used to obtain the size of micelles of Triton X-100 and octylphenylnonaethylene glycol [84]. Similarly, Tanford et al. [85] observed good agreement between the radius of $C_{12}E_6$ obtained by Sepharose SEC (calibrated with proteins) and the viscosity radius. We observed excellent correlation between the apparent micelle radius of Triton X-100/SDS mixed micelles by SEC with the value from dynamic light scattering, although the latter was consistently larger [69].

If a large volume of micellar solution is applied to the column and then eluted with mobile phase, the extrema of the elution profile contain information about the micelle aggregation number and micellization equilibrium constant [86], according to the asymptotic theory developed for the SEC of self-associating proteins [87]. Simulations of sec retention curves for micellar solutions based on such treatments have been carried out by Nakagawa and Jizomoto [88].

A related technique, applicable to UV-absorbing surfactants, is gel scanning chromatography. Direct scanning of the gel column as a function of time yields the apparent velocity of the micellar component. Comparison with the velocities of known solutes provides K_{SEC} for this species. Such data have been used, for example, to obtain the mean aggregation number and the micellization association constant of palmitoyl CoA [89].

SEC offers special insight on the behavior of mixed surfactant systems. In practice, mixed surfactants are found to be highly nonideal [90]. This nonideality, at first identified by measurement of mixed surfactant CMCs, has led in the case of highly positive interaction parameters to predictions of micelle demixing, i.e., the

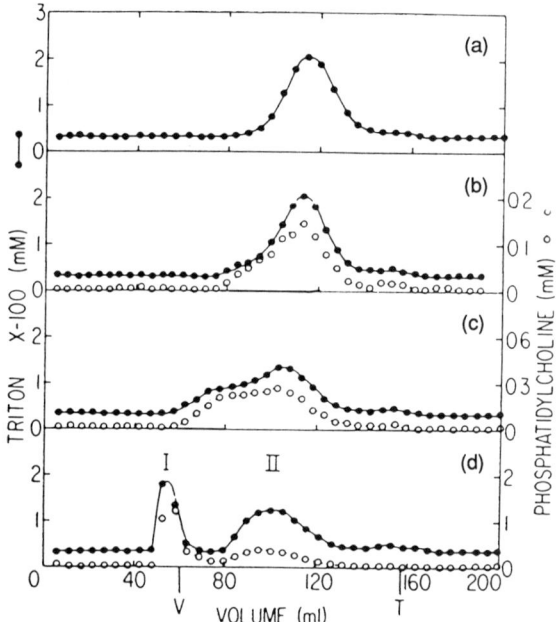

Fig. 11 Chromatograms of Triton X-100 (TX100)/phosphatidylcholine (PC) on 6% agarose in buffer containing 0.35 mM TX100. All injected samples 45 mM in TX100. Mole fraction of PC in sample: 0 (a); 0.091 (b); 0.22 (c); 0.40 (d). (From Ref. 94.)

coexistence of more than one type of mixed micelle [91]. Kinetic analyses of the fading rates of micelle-solubilized dyes offer rather indirect evidence of such heterodispersity [92] as does the surface tension behavior of mixtures of sodium 3,6,9-trioxaicosanoate and alkylpolyoxyethylene ethers [93]. More direct evidence arises from the appearance of multiple peaks in SEC. Thus, Dennis reported on the elution of mixtures of phosphatidylcholine (PC) with Triton X-100 on agarose gel [94]. Progressive addition of PC broadens and shifts the eluted peak to smaller retention times; at concentrations of 45 mM (Triton X-100) and 30 mM (PC), two well-resolved peaks appear, as shown in Fig. 11, with PC preferentially incorporated into the larger species. Asakawa et al. [95] studied mixtures of lithium perfluorooctanesulfonate (LiFOS) and lithium tetradecyl sulfate (LiTS) by SEC on Sephadex G-50. Over a wide range of compositions, 0.5-0.8 mol fraction LiFOS and above 30 mM total surfactant, two peaks are observed, corresponding to a LiTS-rich high MW micelle in equilibrium with a LiFOS-rich lower MW micelle.

It is interesting to note that the elution volumes were independent of the mixing ratio, indicating that the response to a change in surfactant composition in the demixing region was a change in the concentrations of the two (saturated) micelles without change in their respective composition or aggregation number.

ACKNOWLEDGEMENTS

Portions of the research described herein were supported by the Johnson's Wax Fund and by grant 21294-B7-C from the American Chemical Society. The author also thanks Prof. Phyllis Brown for her interest and encouragement.

REFERENCES

1. W. W. Yau, J. J. Kirkland, and D. D. Bly, Modern Size Exclusion Chromatography, John Wiley and Sons, New York, 1979.
2. T. Kemmer and L. Boross, Gel Chromatography, John Wiley and Sons, New York, 1979.
3. T. Provder, ed., Size Exclusion Chromatography (GPC), American Chemical Society Symposium Series, No. 138, ACS, Washington, D.C., 1980.
4. J. Janca, ed., Steric Exclusion Liquid Chromatography of Polymers, Chromatography Series Vol. 25, Marcel Dekker, New York, 1984.
5. H. G. Barth, J. Chromatogr. Sci. 18, 409 (1980).
6. P. L. Dubin, Sep. Purif. Meth. 10, 287 (1981).
7. R. P. Bywater and N. V. B. Marsden, in Chromatography Fundamentals and Applications of Chromatographic and Electrophoretic Methods, Journal of Chromatography Library, Vol. 25A (E. Heftmann, ed.), Elsevier, Amsterdam, 1983, Chap. 8.
8. H. G. Barth, Adv. Chem. Ser. No. 213, J. E. Glass, ed., ACS, Washington, D.C., 1986, Chap. 2.
9. A. R. Cooper and D. S. Van Derveer, J. Liquid Chromatogr. 1, 693 (1978).
10. P. L. Dubin, ed., Aqueous Size Exclusion Chromatography, Elsevier, Amsterdam, 1988.
11. J. J. Kirkland, J. Chromatogr. 125, 231 (1976).
12. F. A. Buytenhuys and F. P. van der Maeden, J. Chromatogr. 149, 489 (1978).
13. P. E. Barker, B. W. Hiatt, and S. R. Holding, J. Chromatogr. 174, 143 (1979).
14. F. E. Regnier and R. Noel, J. Chromatogr. 14, 316 (1976).
15. P. Roumeliotis and K. K. Unger, J. Chromatogr. 185, 445 (1979).
16. H. Engelhardt and D. Mathes, J. Chromatogr. 142, 320 (1977).

17. (a) R. K. Iler, The Chemistry of Silica, John Wiley and Sons, New York, 1976, Chap. 6.
    (b) R. K. Iler, J. Chromatogr. 209, 341 (1981).
18. A. Neddermeyer and L. B. Rogers, Anal. Chem. 41, 94 (1969).
19. B. Stenlund, Adv. Chromatogr. 14, 37 (1976).
20. N. Ohta and K. Kawasaki, in Ref. 10.
21. A. M. Posner, Nature 198, 1161 (1963).
22. B. Gelotte, J. Chromatogr. 3, 330 (1960).
23. A. L. Spatorico and G. Beyer, J. Appl. Polym. Sci. 19, 2933 (1975).
24. A. Dommard, M. Rinaudo, and C. Rochas, J. Polym. Sci. Polym. Phys. Ed. 17, 673 (1979).
25. Ref. 2, p. 258.
26. J. E. Rollings, A. Bose, J. M. Caruthers, M. R. Okos, and G. T. Tsao, Am. Chem. Soc. Div. Polym. Chem. Prepr. 22(1), 294 (1981).
27. S. Sarkarnen, D. C. Teller, E. Abramowski, and J. L. McCarthy, Macromolecules 15, 1098 (1982).
28. H. Sasaki, T. Matsuda, O. Ishikawa, T. Takamatsu, T. Tanaka, Y. Kato, and T. Hashimoto, Sci. Rep. Toyo Soda Mfg. Co. Ltd. 29(1), 37 (1985).
29. J. M. Principi, thesis, Purdue Univ., 1988.
30. E. Pfannkoch, K. C. Lu, F. E. Regnier, and H. G. Barth, J. Chromatogr. 18, 430 (1980).
31. (a) C. P. Talley and L. Bowman, Anal. Chem. 51, 2239 (1979);
    (b) A. Domard and M. Rinaudo, Polym. Commun. 25, 55 (1984);
    (c) M. Stickler and F. Eisenbess, Eur. Polym. J. 20, 849 (1984).
32. D. L. Gooding, M. N. Schmuck, and K. M. Gooding, J. Liquid Chromatogr. 5, 2259 (1982).
33. R. A. Stratton, Institute of Paper Chemistry, Appleton, WI, private communication, 1981.
34. P. L. Dubin and I. J. Levy, J. Chromatogr. 235, 377 (1982).
35. M. Strege and P. L. Dubin, J. Chromatogr. 463, 165 (1989).
36. P. L. Dubin, I. J. Levy, and R. Oteri, J. Chromatogr. Sci. 22, 432 (1984).
37. B. Stenlund, Adv. Chromatogr. 14, 37 (1976).
38. H. G. Barth and F. E. Regnier, J. Chromatogr. 192, 275 (1980).
39. M. G. Styring, C. J. Davison, C. Price, and C. Booth, J. Chem. Soc. Faraday Trans. 1 80, 3051 (1984).
40. P. L. Dubin and M. M. Tecklenburg, Anal. Chem. 57, 275 (1985).
41. P. L. Dubin, C. M. Speck, and J. I. Kaplan, Anal. Chem. 60, 875 (1988).
42. M. Potschka, J. Chromatogr. 441, 239 (1988).
43. F. G. Smith, III, and W. M. Deen, J. Colloid Interf. Sci. 91, 571 (1983).

44. E. J. W. Verwey and J. Th. G. Overbeek, *Theory of Stability of Lyophobic Colloids*, Elsevier, Amsterdam, 1948.
45. P. L. Dubin, R. M. Larter, C. Wu, and J. I. Kaplan, J. Phys. Chem. *94*, 7243 (1990).
46. W. Brown, J. Chromatogr. *59*, 335 (1971).
46. Å. Ch. Haglund and N. V. B. Marsden, J. Polym. Sci. Polym. Lett. Ed. *18*, 271 (1980).
47. P. L. Dubin, K. L. Wright, and S. W. Koontz, J. Polym. Sci. Polym. Chem. Ed. *15*, 2047 (1977).
48. K. Ujimoto and H. Kurihara, J. Chromatogr. *208*, 183 (1981).
49. Y. Yano and M. Janado, J. Chromatogr. *200*, 125 (1980).
50. N. V. B. Marsden, Ann. N.Y. Acad. Sci. *125*, 428 (1965).
51. (a) A. C. Haglund and N. V. B. Marsden, J. Polym. Sci. Polym. Lett. Ed. *18*, 271 (1980).
    (b) C. L. deLigny, J. Chromatogr. *172*, 397 (1979).
52. M. Janado, Y. Yano, H. Nakamori, and T. Nishida, J. Chromatogr. *193*, 345 (1980).
53. K. Ujimoto and H. Kurihara, J. Chromatogr. *208*, 183 (1981).
53. H. Determann and I. Walter, Nature *219*, 604 (1968).
54. A. J. W. Brook and K. C. Munday, J. Chromatogr. *51*, 307 (1970).
55. M. Janado and Y. Yano, Bull. Chem. Soc. Jpn. *52*, 1913 (1985).
56. M. Janado, in *Aqueous Size Exclusion Chromatography* (P. L. Dubin, ed.), Elsevier, Amsterdam, 1988, Chap. 2.
57. (a) M. Janado, K. Takenaka, H. Nakamori, and Y. Yano, J. Biochem. *87*, 57 (1980);
    (b) Y. Matsui and K. Mochida, Bull. Chem. Soc. Jpn. *52*, 2808 (1979).
58. Y. Ikada, M. Suzuki, and H. Iwata, in *Water in Polymers* (S. P. Rowland, ed.), ACS Symp. Ser., Washington, D.C., 1980, p. 287.
59. M. Janado and Y. Yano, J. Solution Chem. *14*, 891 (1985).
60. See Ref. 7, pp. 269-270, 296-297, 303-305.
61. T. L. Hafkenscheid and E. Tomlinson, J. Chromatogr. *218*, 409 (1981).
62. P. L. Dubin and J. M. Principi, Anal. Chem. *61*, 780 (1989).
63. Z. Grubisic, R. Rempp, and H. Benoit, J. Polym. Sci. Part B *5*, 753 (1967).
64. A. R. Weiss and E. Cohn-Ginsberg, J. Polym. Sci. A-2 *8*, 148 (1970).
65. P. J. Flory, *Principles of Polymer Chemistry*, Cornell University Press, Ithaca, NY, 1953, p. 606.
66. E. F. Cassassa, J. Phys. Chem. *75*, 3929 (1971).
67. H. Waldmann-Meyer, J. Chromatogr. *350*, 1 (1985).
68. R. P. Frigon, J. K. Leypoldt, S. Uyeji, and L. W. Henderson, Anal. Chem. *55*, 1349 (1983).

68. R. P. Frigon, J. K. Leypoldt, S. Uyeji, and L. W. Henderson, Anal. Chem. 55, 1349 (1983).
69. P. L. Dubin, B. A. Smith, J. M. Principi, and M. A. Fallon, J. Colloid Interf. Sci. 127, 558 (1989).
70. M. Potscha, Anal. Biochem. 162, 47 (1987).
71. J. C. Giddings, e. kucera, C. P. Russell, and M. N. Meyers, J. Phys. Chem. 78, 4397 (1968).
72. E. F. Casassa, Macromolecules, 9, 182
73. P. L. Dubin and J. M. Principi, Macromolecules 22, 1891 (1989).
74. (a) H. Enomoto, Y. Einaga, and A. Teramoto, Macromolecules 17, 1473 (1984);
    (b) T. Yanaki, T. Norisuye, and H. Fujita, Macromolecules 13, 1462 (1980).
75. M. T. W. Hearn, in *Advances in Chromatography*, Vol. 20 (J. C. Giddings, E. Grushka, J. Cazes, and P. H. Brown, eds.), Marcel Dekker, New York, 1982, p. 34.
76. F. E. Regnier, Anal. Biochem. 126, 1 (1982).
77. W. S. Hancock and J. T. Sparrow, HPLC 3, 49 (1983).
78. M. T. W. Hearn, F. E. Regnier, and C. T. Wehr, eds., *High Performance Liquid Chromatography of Proteins and Peptides; Proceedings of the First International Symposium*, Academic Press, Orlando, 1983.
79. K. M. Gooding, Biochromatography 1, 34 (1986).
80. S. Yamamoto, K. Nakanishi, and R. Masuno, *Ion-Exchange Chromatography of Proteins*, Marcel Dekker, New York, 1988.
81. C. R. Cantor and P. R. Schimmel, *Biophysical Chemistry, Part II. Techniques for the Study of Biological Structure and Function*, W. H. Freeman, New York, 1980, Chaps. 10, 12.
82. P. L. Dubin and J. M. Principi, J. Chromatogr. 479, 159 (1989).
83. H. Coll, in *Gel Permeation Chromatography* (K. Altgelt, ed.), Marcel Dekker, New York, 1971, p. 329.
84. R. J. Robson and E. A. Dennis, Biochem. Biophys. Acta 508, 513 (1978).
85. C. Tanford, Y. Nozaki, and M. F. Rohde, J. Phys. Chem. 81, 1555 (1977).
86. N. Funasaki, S. Hada, and S. Neya, J. Phys. Chem. 92, 7112 (1988), and references cited therein.
87. G. C. Ackers and T. E. Thompson, Proc. Natl. Acad. Sci. 53, 341 (1965).
88. T. Nakagawa and H. Jizomoto, J. Am. Oil Chem. Soc. 48, 571 (1971).
89. G. L. Powell, J. R. Grothusen, J. R. Zimmerman, C. A. Evans, and W. W. Fish, J. Biol. Chem. 256, 12740 (1981).
90. (a) P. M. Holland and D. N. Rubingh, J. Phys. Chem. 87, 1984 (1983);
    (b) R. Nagarajan, Langmuir 1, 331 (1985).

91 (a) K. L. Mysels, J. Colloid Interf. Sci. 66, 331 (1978);
    (b) R. F. Kamrath and E. I. Franses, J. Phys. Chem. 88, 1642 (1984);
    (c) T. Asakawa, S. Miyagishiand, and M. Nishida, J. Colloid Interf. Sci. 104, 279 (1985).
92. K. Ogino and M. Abe, in *Phenomena in Mixed Surfactant Systems* (J. F. Scamehorn, ed.), ACS Symposium Ser. 311, Washington, D.C., 1986, Chap. 5.
93. M. Abe, N. Tsubaki, and K. Ogino, J. Colloid Interf. Sci. 107, 503 (1985).
94. E. A. Dennis, Arch. Biochem. Biophys. 165, 764 (1974).
95. T. Asakawa, S. Miyagashi, and M. Nishida, Langmuir 3, 821 (1987).

3
Chromatography on Thin Layers Impregnated with Organic Stationary Phases

Jiří Gasparič *Charles University, Hradec Králové, Czechoslovakia*

I.	INTRODUCTION	154
II.	THE PARTITION SOLVENT SYSTEM	154
	A.	The Support	156
	B.	Choice of Stationary Phase	160
	C.	Impregnation Procedure	161
	D.	Amount of Stationary Phase	165
	E.	Mobile Phase	168
	F.	The Solute	176
III.	APPLICATIONS	179
	A.	Analytical Separations	179
	B.	Lipophilicity Determination and QSAR Studies	179
IV.	CONCLUSIONS	225
	REFERENCES	227

I. INTRODUCTION

The first thin-layer chromatographic procedures using layers impregnated with organic stationary phases appeared in the early 1960s to achieve the desired separations. Authors mostly took advantage of experience previously made in paper chromatography with nonaqueous polar (formamide, propylene glycol, methanol) or less polar (liquid paraffin, petroleum, olive oil) organic stationary phases [1-3]. Thus, many excellent separations were achieved based on partition mechanism in comparison with the original adsorption process in the following time period of 10-15 years, especially in the fields of steroids [4-6], alkaloids [7], and lipids [6,8]. Later on, the use of layers impregnated with organic liquids ceased and they seemed to have been definitely pushed out by chemically bonded phases [9]. At the same time, however, reversed phase partition chromatography has become a simple method for obtaining quantitative information about the hydrophobic character of biologically active compounds in QSAR (quantitative structure activity relationships) studies [10-13]. At present, chromatography on impregnated layers is routinely, and often mechanically, used by medicinal chemists without taking into consideration things discovered long before in paper and analytical applications of thin-layer chromatography. Authors of such papers often do not find it necessary to give any information on the experimental conditions under which their very important R_M measurements were carried out. On the other hand, further correlation of the data thus obtained with biological activity is given in full detail.

Therefore, the aim of this chapter is to summarize both the basic facts from literature and the author's own experience with chromatography on layers impregnated with organic stationary liquids.

II. THE PARTITION SOLVENT SYSTEM

The principal purpose of impregnating the layers is to establish conditions for a chromatographic process based on partition. Such a partition system requires a pair of phases: the stationary phase and the mobile one. When a nonaqueous polar stationary phase is used (formamide, dimethyl formamide, ethylene glycol), the mobile phase must be of low polarity (hexane, cyclohexane, carbon tetrachloride, benzene, etc.), and in the reversed case when a less polar stationary phase is used (liquid paraffin, silicone oil, ethyl oleate, 1-octanol) water in mixtures with organic solvents (methanol, acetone, acetonitrile, acetic acid, etc.) or an aqueous buffer solution is the mobile phase. The polarity of the solvent pair in the latter case is reversed and such a type of chromatography was denoted as reversed phase partition chromatography [14-17].

In partition chromatography separations are a function of the partition coefficient of the solutes between the mobile and stationary phases. If all the solutes in the mixture have the same partition coefficient, then the solute mixture elutes unresolved. However, if the partition coefficients of all components in the mixture are different, separations of all components results. Thus, if the solute mixture favors one phase over another and resides in that phase, poor or no separation is obtained. However, if the solutes reside in both phases and these retentions are different for each component of the mixture, resolution is obtained. In other words, to resolve the components of a mixture the chromatographer must create conditions under which the solutes are partitioned differently in both mobile and stationary phases. To achieve this aim either the stationary phase, the mobile phase, or both car be altered [18].

Equation 1 [19] expresses the relationship between the R_F value in partition chromatography and the partition coefficient, where P = partition coefficient (= the concentration in the mobile phase in

$$P = \frac{V_M}{V_S} \cdot \left[\frac{1}{R_F} - 1\right] \qquad (1)$$

mol/liters divided by the concentration in the stationary phase in mol/liters, V_M = volume of the mobile phase, and V_S = volume of the stationary phase. Upon substituting the symbol R_M, defined [20] as

$$R_M = \log\left[\frac{1}{R_F} - 1\right] \qquad (2)$$

into Eq. 1 and rearranging, we obtain

$$R_M = \log P + \log r \qquad (3)$$

where r is the phase/volume ratio V_S/V_M, which is a constant for a given chromatographic system provided the experimental conditions are constant.

Complications can be caused by dissociation of the solute. The pH of the phases must be adjusted in such cases so that the analytes are present unambiguously either as neutral molecules or as completely dissociated species.

Thus, the issues of a partition system in thin-layer chromatography involve the following partial problems: (a) choice of proper support, (b) choice of suitable stationary phase, (c) impregnation procedure, (d) amount of stationary phase, (e) choice of mobile phase, and (f) properties of the solutes (polarity, dissociation, etc.).

A. The Support

In thin-layer chromatography the sorbent plays the role of the support of the stationary phase. Whereas practically no interactions of cellulose paper with the analytes were observed under analogous conditions in paper chromatography (see, however, Refs. 21 and 22), the use of more active sorbents in thin-layer chromatography brought about the possibility of such interactions. It has been stated many times that the support should have as weakly as possible adsorbing properties. Many authors have recommended proceeding carefully when using active sorbents like silica gel or alumina [3], see also Refs. 23 and 24.

The problem should be considered from two points of view. Analytical applications do not require the partition mechanism to be strictly the sole process, provided the interactions of the analytes with the support, mostly due to adsorption, do not worsen the desired separation efficiency. QSAR studies, however, are based on partition coefficient measurements through R_M values determined in a particular partition system. Thus, the experiments arranged for such measurements should be freed from any interaction which would distort the results by structure-chromatographic behavior rules characteristic for any other mechanism besides partition [23]. The most simple experimental approach for confirming that partition is the sole process is based on the comparison of the chromatographic behavior of the solutes on the layer impregnated with the chosen stationary phase with the "blank," i.e., with the results obtained with the untreated layer using the same mobile phase [25]. Another possibility is to confirm that the R_F values of the solutes decrease with the increasing content of the stationary phase in the support, i.e., the increasing value of V_S (Eq. 1). It is also possible to use relationship 4 using the partition system oleyl alcohol/methanol-water, which is analogous to the reversed phase thin-layer chromatographic

$$\frac{1}{R_F} = 1 + {}_sP.k.C_{ol} \qquad (4)$$

system. $_sP$ is the partition coefficient in the system oleyl alcohol/methanol-water (the subscript s indicates that a mixed solvent, such as methanol-water, is involved), k is a constant, and C_{ol} is the oleyl alcohol concentration in the mixture used for impregnation of the plates. When no adsorption of the solutes onto the support phase takes place [26] during the migration, straight lines are obtained when $1/R_F$ is plotted against C_{ol}, with intercepts close to unity and slopes of $_sP.k$.

To understand the possible interactions of the sorbent, one must get rid of the notion that the sorbent plays only the role of an inert support. Silica gel, e.g., can be represented by a matrix

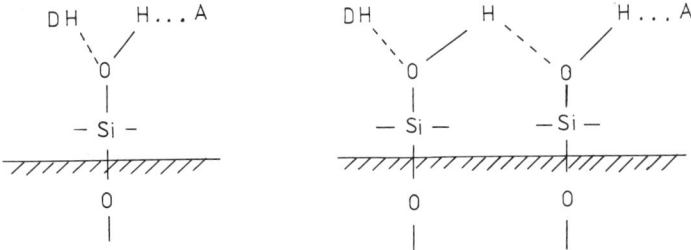

Fig. 1 Schematic representation of different types of silanol groups on silica gel surface: (left) single hydroxy group; (right) vicinal hydroxy groups; DH = hydrogen donor, A = hydrogen acceptor.

of siloxane bonds, on the surface of which different types of silanol groups are present [27,28] (Fig. 1). These silanol groups are assumed to be the active sites where interactions between the sorbent and both the solute and the solvent molecules can take place. The mechanism of these interactions in adsorption chromatography has been studied intensively (see, e.g., the system silica gel-solute-mobile phase in Refs. 29-38). In partition chromatography we cannot suppose that the particles of the sorbent are simply mechanically coated by the organic stationary phase. The operating forces will be quite different depending on whether the layer is impregnated with a polar or a less polar organic stationary phase.

The polar organic stationary phase, applied usually in a less polar organic solvent, can be expected to be immediately bound to the silanol groups. A decrease of the activity of the sorbent can be assumed to result when an unsufficient amount of the organic phase has been used. Higher amounts of the polar phase probably form a film bound to the adsorbed, molecules of the stationary phase via hydrogen bonds, e.g., as in the case of water molecules in partition chromatography with aqueous stationary liquid [39] (Fig. 2).

Whether all the silanol groups are saturated with the molecules of the polar organic stationary phase or whether some silanols are sterically unreachable is not clear. Furthermore, one must suppose that the solute can be (depending on its own physicochemical properties) bound to the polar stationary phase. However, one must also admit that some competition between the solute and the stationary phase for the active sites of the sorbent could take place. This competition between the solute and the mobile phase occurs in adsorption chromatography. Similar competitive interactions could occur here as well.

On the other hand, the coating of the silica gel particles by a nonpolar stationary phase (liquid paraffin) can be expected to cause

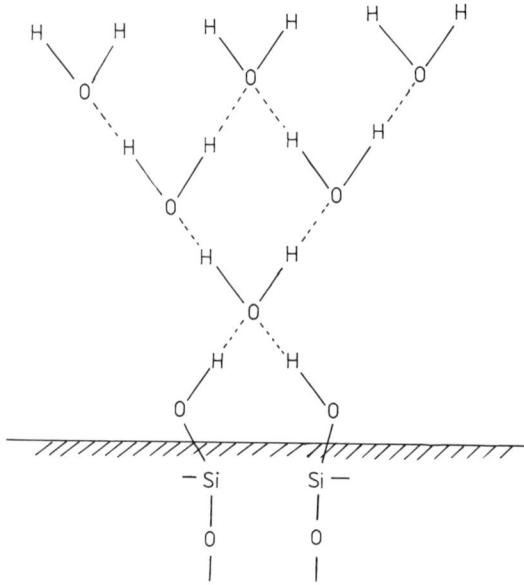

Fig. 2 Schematic representation of water molecules bound to silanol groups of silica gel via hydrogen bonds.

very weak (dispersive) interactions between the stationary phase and the sorbent. One would suppose that using an aqueous mobile phase the process of chromatographic separation (partition) will take place only between the two phases since the particles of the sorbent are protected from the mobile phase by the layer of the lipophilic stationary phase. However, it will be shown later that this cannot be true. Prior to any discussion the role of the stationary and the mobile phases, as well as the solute, must be summarized.

The above facts can explain why adsorption behavior of several supports in partition thin-layer chromatographic systems with impregnated layers has been observed [40]. More on this topic will be discussed in a later section. It should be remembered at this point that a similar situation has been observed in reversed phase chromatography using layers of alkyl-bonded silica gel. Many models of the so-called reversed phases and different mechanisms attempting to explain retention have been proposed (for references, see [41]). Originally it was suggested that retention is governed by only one mechanism. Later a dual (or mixed) retention mechanism with alkyl-bonded silica gel was reported [42-44] accounting for the presence of unreacted silanol groups on the surface of the derivatized silica gel matrix [27] (see Fig. 3). The residual silanol group content is

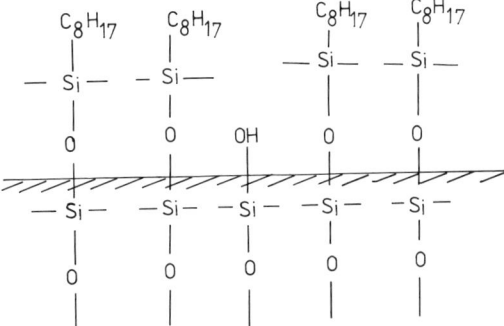

Fig. 3 Schematic representation of free silanol groups on the surface of C_8 chemically bonded silica gel.

an important factor influencing the retention behavior of solutes and their peak shape [45-47], especially if these are of polar character [36]. Under chromatographic conditions the silanol groups are considered to be hydrated, while the chemically bonded alkyl ligates can be partially or fully solvated by organic molecules of the eluent. Thus, the stationary phase of this type should be regarded as a heterogeneous surface composed of the alkyl chain, the layer of adsorbed liquid whose composition is different from the mobile phase, the surface of unreacted silanol groups and the solvation cover of the silanol groups consisting of water and organic modifier. With aqueous-organic eluents the solute retention is explained as being due to both silanophilic and hydrophobic-dispersive interactions.

Interaction with the free silanol groups is especially strong with basic compounds. It can be diminished by blocking the silanol sites with an amine [48,49] or by masking the basic site of the solute molecules by ion pair formation [50]. Another possibility is the capping of the silanol groups [51,52].

The presence of the residual silanol groups can be detected most readily by using methyl red indicator, which turns red in the presence of acidic silanol groups [53]. A more sensitive test is to chromatograph a polar solute using n-hexane as mobile phase [47,54].

Recently, it was shown that similar problems with a dual retention mechanism can be observed when commercial acetylcellulose materials are used as stationary phases in planar reversed phase chromatography using the same mobile phase but solutes of different polarity [55]. This is caused by free OH groups which are present in these materials due to incomplete acetylation.

Similarly, polyamide can operate either as a polar stationary phase, analogous to formamide, or as a reversed phase stationary phase depending on which mobile phase is used, i.e., a nonaqueous less polar (benzene-methanol) or an aqueous phase (methanol-water) [56,57].

Thus, it can be concluded that either the impregnation or the chemical reaction used for preparing chemically bonded phases cannot establish conditions under which the active sites of the support (silica gel, cellulose, etc.) would be completely blocked. The character of the mobile phase as well as the polarity of the solute will give rise to a straight or reversed phase mechanism generated by silanophilic or hydrophobic interactions.

Another factor which cannot be omitted is the presence of impurities in the support which could interact with the solutes and affect their solubilities. As example, the formation of chelates with metal impurities should be mentioned [58].

B. Choice of Stationary Phase

The most important requirement for a successfully operating partition process is the good solubility of the analyte in the stationary phase. The original paper chromatographic methods [19,59,60] using untreated paper containing water or even the complex cellulose-water as the stationary phase were suitable only for compounds of good solubility in aqueous media (amino acids, sugars, lower aliphatic acids). The application of paper chromatography to separation of compounds insoluble in water necessitated the use of organic stationary liquids [61-63]. Analogously, in thin-layer chromatography, polar stationary phases (formamide, dimethyl formamide, ethylene glycol, propylene glycol, phenoxyethanol, dimethyl sulfoxide, etc.) are suitable for compounds of middle polarity. Less polar stationary phases (liquid paraffin, silicone oil, 1-octanol, 1-butanol, etc.) are used for unpolar, lipophilic compounds. Thus, it is the polarity of the analyte which is decisive for the choice of the proper stationary phase.

The QSAR studies require that the properties of the stationary phase be very close to that of 1-octanol, since the partition coefficient of the solute in the system 1-octanol-water was recommended [64] as a measure of the ability of the solute to penetrate through biological membranes. It is the system which is most often used and for which the largest number of determinations containing the widest selection of functional groups have been made [65]. However, in view of what has been stated in the preceding paragraph, we can conclude that 1-octanol cannot always be a suitable stationary phase. The polarity of drugs covers a broad range: from very hydrophilic compounds through compounds of middle polarity up to lipophilic compounds. Many drugs are so lipophilic that when plates with 1-octanol as stationary phase are developed with aqueous solutions the analytes hardly move from the starting point. No retention, on the other hand, is observed of compounds which are hydrophilic because of low solubility in the stationary phase and very high solubility in the mobile phase. It has been shown that such polar compounds can be the source of apparent irregularities and problems (see later discussion).

The conclusions made in the preceding section should be remembered once more at this moment: the covering of the support with stationary phase does not necessarily destroy the original adsorption properties of the sorbent. It will be shown in Section II.D that the degree of impregnation will establish conditions favorable either for partition or adsorption processes. Thus, the property and suitability of the stationary liquid must be considered only in connection with the support used.

The purity and stability of the stationary liquid can, of course, affect the results of the chromatographic process. For example, in the case of formamide it has been shown that ammonium formiate, a possible decomposition product, can influence the quality of spots and therefore it is sometimes added to the stationary phase [66]. Free ammonia or formic acid as impurities can cause the problems met with acidic or basic analytes, respectively. On the other hand, free ammonia [67,68] or formic acid [69-71] is added to formamide to achieve buffering and variation of the solvent power of formamide.

It should be mentioned that there are several other reasons for thin layers to be impregnated in addition to establishing conditions for the partition system. The support can be impregnated with agents forming charge transfer complexes [72] or inclusion compounds (clathratography [73-74]) with the solute enabling specific separations, selected counterions [75], anionic or cationic detergents (soap thin-layer chromatography [76]), chiral stationary phases [77], liquid ion exchangers [78-83], chelating agents [84], etc. The presence of such agents creates deviations in the behavior of the analytes from that expected in the partition system.

C. Impregnation Procedure

There are two principal ways to introduce the organic liquid into the support material.

(1) The "direct method," i.e., impregnation "before coating": the organic liquid is added to the slurry prepared for coating the glass plates. An auxiliary organic solvent must be used instead of water in such cases. Examples of the composition of particular slurries are given in Table 1.

(2) The "indirect method," i.e., the ready, dry layer is impregnated with the stationary phase in one of the following ways [104].

Table 1 Examples of the Slurry Compositions Used for Preparing Impregnated Layers

Stationary phase	Support	Composition	Ref.
Formamide	Silica gel	5 g of silica gel G + 4 ml of formamide in 7.5 ml of ethanol	85
		20 g of silica gel HF + 50 ml of a 5-30% solution of formamide in acetone	86
		20 g of silica gel G + 18 ml of formamide + 4 ml of 5% ammonia + 45 ml of ethanol	87
	Cellulose	13.5 g of cellulose MN 300 HR + 65 ml of a 6.0 M solution of formamide in acetone	88
	Talc	4 g of talc + 7 ml of ethanol + 1 ml of formamide-acetone (3:7)	89
Polyamide	Cellulose	13.5 g of cellulose MN 300 HR + 11 ml of a solution of polyamide (Nylon 66) in formic acid (0.13 g/1 ml) + 74 ml of formic acid	90
	Kieselguhr	5 g of polyamide (Nylon 66) are left overnight in 60 ml of the mixture benzene-methanol (1:1) and the homogeneous solution obtained after short heating on a water bath is slurried with 30 g of Kieselguhr G	91
β,β'-Oxydipropionitrile	Silica gel	20 g of silica gel + 100 ml of acetone + 4 ml of β,β'-oxydipropionitrile + 2 g of starch + 60 ml of petrol ether	92
Polyglycols	Silica gel	40 g of silica gel G + 10 ml of polyethylene glycol 200 or 400 + 80 ml of water	93
	Kieselguhr	40 g of kieselguhr G + 10 g of Carbowax 400 + 80 ml of water	94
		30 g of kieselguhr G + 15 g of PEG 600 + 40 ml of water	95

Polyesters	Kieselguhr	30 g of kieselguhr G + 12 g of the 80% polyester solution in methyl glycol + 40 ml of acetone + 20 ml of water + 0.05 g of sodium diethyldithiocarbamate	96
Ethyl oleate	Cellulose	15 g of cellulose MN 300 HR + 70 ml of a 0.75% solution of ethyl oleate in diethyl ether	97
Oleylalcohol	Kieselguhr	24 g of kieselguhr G + 1.25% oleylalcohol + 7 ml of acetone + dioxane to 60 ml	98
Plant oils (triglycerides)	Kieselguhr	10 g of kieselguhr G + 20 ml of dioxane + 3 ml of acetone + 2 ml of plant oil	99
Liquid paraffin	Silica gel	20 g of silica gel G + 3 g of liquid paraffin + 30 ml of benzene + 30 ml of acetone	100
	Cellulose	15 g of cellulose MN 300 + 100 ml of diethyl ether + 3 ml of liquid paraffin	101
	Talc	30 g of talc + 45 ml of diethyl ether + 2 ml of liquid paraffin	101
	Starch	To a 10% solution of liquid paraffin in petroleum ether about 10 g of starch is added to obtain a fairly spreadable slurry	102
Silicone oil	Silica gel	25 g of silica gel GF 254 + 5% of silicone oil + 6 ml of acetone + dioxane to 50 ml	103

(i) Dipping technique: the plate with the layer or the ready-made sheet with the layer is carefully immersed in the solution of the stationary phase in a volatile solvent for several seconds (some authors recommend several hours). Next the plate is left to stand in a vertical position on absorbing tissue (a pack of filter paper) so that any excess of the impregnating solution would be sucked off and the volatile solvent evaporated. Usually, 5-15 min is sufficient. Care should be taken in the case of rather volatile phases (dimethyl formamide, ethylene glycol) to control this time period in order to avoid the loss of the stationary phase due to volatility [105-107]. The concentration of the organic liquid in the impregnating solution, and the auxiliary solvent as recommended by different authors, can be taken from the tables in the review of applications in Section III.

(ii) Development technique: This procedure is a modification of the preceding one. It is based on ascending flow of the impregnating solution through the layer in a chromatographic chamber, similar to ascending development of a chromatogram, until the solvent front has reached the upper end of the plate. Some authors recommend prolonging this process for several hours or overnight, probably to avoid formation of a concentration gradient. The volatile solvent is then evaporated as in (i). The concentration of the solutions used can be taken from the tables in Section III.

(iii) The spraying technique: Another possibility is to apply the impregnating solution by spraying uniformly on the plate and evaporating the solvent as in the preceding cases. Uneven impregnation might be the disadvantage of this procedure.

(iv) Impregnation in the gas phase: The ready-made or homemade silica gel layer is activated for 2 hr at 110°C. Then it is placed, in a vertical position (!), in a 30-cm exsicator containing a dish with ~100 ml of dimethyl formamide. Evenly impregnated layers, with a defined impregnation rate, can be obtained according to the corresponding adsorption and desorption curves [108].

The treatment of silica gel layers by organic solvent vapors ("vapor preadsorption chromatography") must be interpreted in this connection very cautiously. The preloading with acetonitrile [109, 110], nitromethane [109,110], nitrobenzene [111-113], chlorobenzene [114], methanol [115-117], acetone [115,116], or formic acid [118] brings about the deactivation of the active sites of silica gel probably due to formation of a loose complex between adsorbent and substrate. Mobile phases used in such cases are usually miscible with the preadsorbed liquids and one can suppose that they are washed out. The preloading of the covalently bonded reversed phases by gas molecules of hydrocarbons (hexane, cyclohexane) leads to a change in selectivity when aqueous mobile phases are used and the results on untreated and preadsorbed layers are similar. The reason for this behavior may be that preloading with hexane or cyclohexane builds up a new matrix on the original reversed phase matrix. Partition of the analytes now takes place between the preloaded phase and the mobile phase. It is a prerequisite that this new stationary phase be maintained throughout the chromatographic run, i.e., it should not be soluble in the mobile phase. The hydrocarbon phase is less lipophilic than the original chemically bonded phase, thus permitting chromatography with mobile phases having a higher water content [119]. Polar compounds can of course interact with free silanol groups.

The opinion and recommendations of different authors concerning the individual impregnation techniques differ considerably. Probably it will be the quality and properties of the type of the layer used which will determine the degree of handling required by the impregnation technique in order not to disturb the integrity of the layer (crumbling of the layers).

The technique used for impregnation can also influence the content of the stationary phase in individual zones of the layer from the origin to the solvent front (see, e.g., Refs. 105 and 106). The requirement of a constant stationary phase content is not so severe in the case of analytical applications so far as it does not negatively influence the analytical results. The QSAR studies, however, require more exact measurements and thus care should be taken to establish constant conditions over the whole range of the R_F values.

The importance of the degree of loading of the support will be discussed in the next part.

D. Amount of Stationary Phase

Equation 1 was derived for a common partition process in paper chromatography. Our knowledge of the stationary phase in paper chromatography with untreated paper is not good enough to enable us the calculation of the ratio V_M/V_S. Usually, this ratio is supposed to be constant for a particular solvent system. The same situation seems to be in partition thin-layer chromatography using untreated layers and water-containing mobile phases (1-propanol-ammonia, 1-butanol-acetic acid-water, etc.). The use of organic stationary phases in combination with immiscible mobile phases offers the possibility of changing purposefully the loading of the support and thus the ratio V_M/V_S. It is well known from paper chromatography that the R_F values decrease with increasing impregnation grade of the paper [105,106,120]. The results obtained with layers impregnated with both polar and unpolar stationary phases have also shown that the R_F values decrease with an increase in the impregnation coefficient of the stationary phase [121-124]. This is a logical consequence of the increase of the V_S term in Eq. 1 and of Eq. 4.

Equations 1 and 2 can be combined and written as:

$$R_M = \log P - \log V_M + \log V_S \qquad (5)$$

Provided that V_M remains constant, P being constant by definition, then plots of R_M vs. log V_S (or, more specifically, the log of the concentration of the stationary phase in the solvent used for slurrying the support) should indicate the constancy of the V_M term. This has been shown to be approximately so in the case of cellulose impregnated with dimethyl formamide, monomethyl formamide, and formamide over a certain concentration range. Deviations from

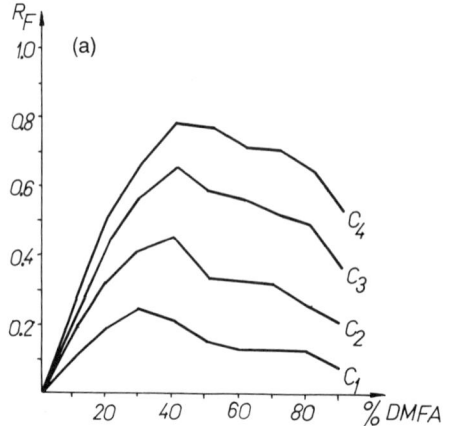

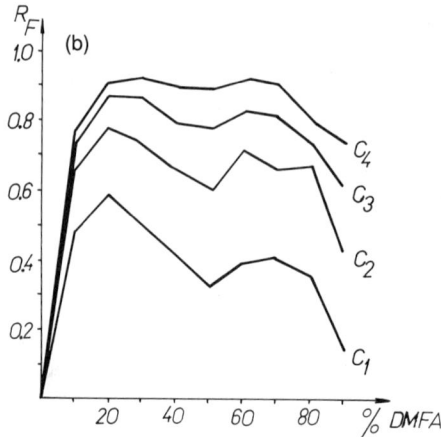

Fig. 4 Influence of the impregnation degree of thin layers with dimethyl formamide (DMFA) on R_F values of 3,5-dinitrobenzoates of C_1-C_4 alcohols. (a) Silica gel; (b) aluminum oxide; (c) kieselguhr. (From Ref. 129.)

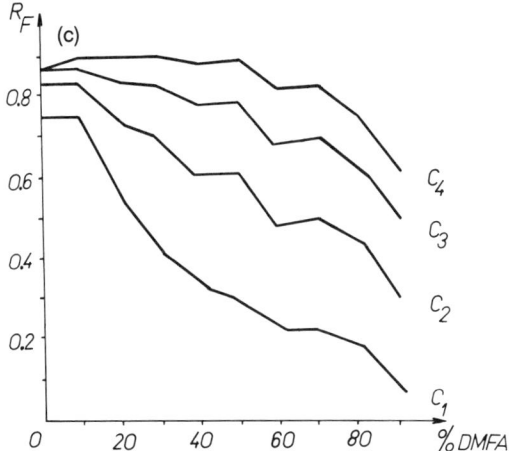

Fig. 4 (Continued)

linearity, however, were observed at very low or very high loadings. These deviations at low concentrations seem to be attributed to incomplete coverage of the support by the stationary phase, while those at high concentrations were considered to be a consequence of excess stationary phase being washed off the layer and pushed ahead of the mobile phase [88,125,126].

Quite different results are obtained when silica gel or alumina layers are impregnated with different amounts of polar stationary phases, e.g., starting from 1% of dimethyl formamide or formamide up to higher concentrations. At low concentrations the increasing impregnation degree brings about only a successive deactivation of the adsorbent resulting in an increase of R_F values. Then, after the full saturation of the sorbent has been reached, at a certain loading degree, the conditions for a partition system have been created and a decrease of R_F values is observed with increasing polar phase content [127-129]. This fact is illustrated in Fig. 4 showing chromatograms of alkyl-3,5-dinitrobenzoates [127]. Full saturation of silica gel establishing conditions for a partition mechanism with formamide as stationary phase needs a 57% content of formamide [86]. The comparison of reaching a separation process based on partition on cellulose and silica gel layers is illustrated in Fig. 5. 1-Methoxyanthraquinone was chromatographed on cellulose and silica gel layers impregnated with 1-40% dimethyl formamide solutions in ethanol. A decrease in R_F values was observed with increasing dimethyl formamide content on cellulose, whereas on silica gel layers R_F values increased under the same conditions [130]. At the highest impregnation degree the results were identical.

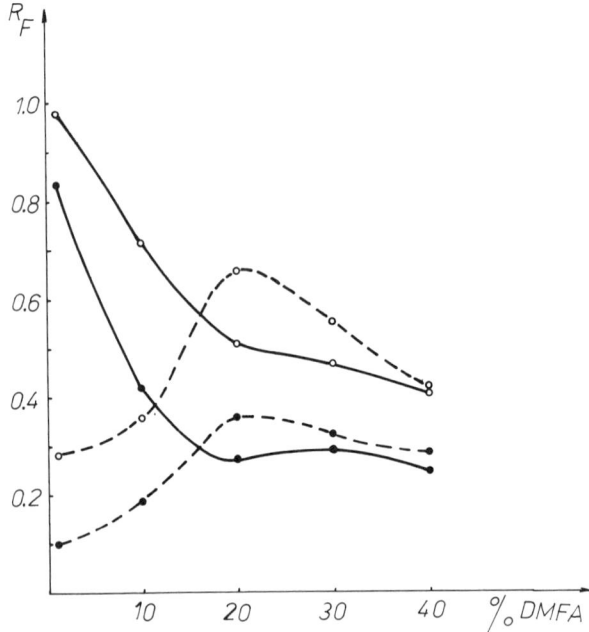

Fig. 5 Influence of the impregnation degree with dimethyl formamide (DMFA) on R_F values of 1-hydroxyanthraquinone (○) and 2-methoxyanthraquinone (●). ———, Cellulose; -----, silica gel. (From Ref. 130.)

The mode of drying the layers after the impregnation procedure in the vertical position (the starting line of the chromatogram at the top or at the bottom) can cause some irregularities in the impregnation degree (see Refs. 105-107) along the whole migration distance. A carefully selected series of homologous compounds, with constant ΔR_M values for the methylene group, could serve to prove the constancy of the impregnation degree within the whole range of R_F values for QSAR study.

E. Mobile Phase

The mobile phase plays a predominant role in the separation process in partition thin-layer chromatography, as in liquid partition chromatography [18]. The mobile phase determines not only the separation of the components in the mixture, but also the degree of resolution, selectivity, and mobility. The mobile phase strength (polarity) determines the migration rate (R_F values) and the mobile phase composition determines the selectivity and separation efficiency (R_F differences between the compounds in the mixture).

Once the choice of the stationary phase (nonaqueous polar or unpolar) has been made according to the solute properties, the separation process can be optimalized only by the selection of the mobile phase strength (polarity) and composition. The polarity of the mobile phase must of course be opposite to that of the stationary phase.

It is not necessary in thin-layer chromatography to find complicated mixtures of solvents as mobile phases. Reasonably chosen single- or two-component mixtures are usually convenient enough and the optimum strength (polarity) can be adjusted by their proportions in the mixture. Nevertheless, it is sometimes necessary to add a third or fourth component to achieve special solute-mobile phase or stationary-mobile phase interactions. Examples will be given later.

It should be emphasized that a mobile phase composed of two or more components cannot be considered as having a constant composition along the whole flow distance from the starting line up to the solvent front. This fact is already known from paper chromatography and it is ascribed to the interaction of the mobile phase with the stationary phase [131-134]. In thin-layer chromatography the mobile phase composition is changed analogously during its flow through the untreated or impregnated thin layer of the sorbent. Some components are retained by adsorption on the active sites of the support material so that the mobile phase near the front may contain only one of the two components. The more retained component may be found mainly near the starting line. Thus, one or even more secondary solvent fronts can be formed (solvent demixing). This phenomenon is not dangerous for analytical separations. In fact, it may sometimes improve the separation [135-137]. On the other hand, it could be the source of false information in QSAR studies.

Recently, it was pointed out that the binary mobile phase is in fact a more complicated one, since association can take place between its two components. Thus, at times a binary mixture becomes a ternary system where the third component comprises the associated solvent [138].

The properties and influence of the vapor phase should be taken into consideration when discussing the problem of the mobile phase composition. The vapor phase could, if unsaturated, influence the composition of the mobile phase by evaporation of some more volatile components. Differences in absolute values were observed when R_F values obtained from normal chamber development and by overpressured thin-layer chromatography were compared [139].

Mobile Phases for Nonaqueous Polar Stationary Phases

Partition systems with formamide as stationary phase have been intensively studied in paper chromatography by Polish authors [69,70, 140-143]. In those and similar cases (dimethyl formamide, ethylene glycol, dimethyl sulfoxide, etc. [144]) a solvent of low polarity must be taken as mobile phase, i.e., starting with n-hexane, benzene, chloroform, etc., according to the eluotropic series of solvents [145]. For some new recommendations for eluotropic series, see in Ref. 146. Mixtures of two solvents with differing polarities enable the solvent strength to be changed in a continuous manner. Curves representing the increase in R_F values of individual solutes with increasing content of the more polar mobile phase component are parallel if these solutes are homologs. A change of solute sequence may be observed if the solutes are structurally different compounds [147]. Variations of solvent composition (cyclohexane + polar solvents), in addition to giving an appropriate range of partition coefficients, can also improve the selectivity of separation, particularly when selective interactions with some of the solutes to be separated (due to different electron donor or acceptor interactions) can occur.

The only requirement is the use of mobile phases which are immiscible with the particular stationary phase used.

Sometimes an additional mobile phase component must be used. It was already mentioned (see Section II.B) that buffering the system with either ammonia or formic acid is necessary in particular cases. The addition of a basic mobile phase component (pyridine, triethylamine, etc.) is often required, if bases of middle strength are chromatographed, to suppress the acidity of the support of the stationary phase (silica gel, e.g., represents an acid of pH 4-5 [148], or if these basic solutes are applied on the origin of the chromatogram as salts. It is also possible to develop the chromatograms in ammonia atmosphere (a dish with concentrated ammonia solution at the bottom of the chamber) [149]. However, the amine modifier is strongly adsorbed, especially on silica gel supports. A zone of this adsorbed amino compound is thus formed from the lower edge of the sorbent layer up to the solvent front, depending on the amount added. Displacement thin-layer chromatography [150-152] takes advantage of this phenomenon.

It was shown in Section II.D that a certain degree of the stationary phase is required to suppress the activity of the sorbent used as support. However, in some cases it is practically impossible to overload the layer with the required amount of the impregnant, and adsorption will still have considerable effect on the separation mechanism. For example, when silica gel and cellulose layers are impregnated with the same solution of formamide, and the same mobile phase is used, the absolute retention values (R_F values) will

Table 2 R_F Values of Phenoxyalkanoic Acids on Layers Impregnated with a 20% Solution of Formamide in Ethanol

Acid	Sorbent Mobile phase	Silica gel Benzene-chloroform (4:1)	Cellulose Benzene-chloroform (4:1)	Cellulose Benzene-hexane (1:1)
Phenoxyacetic		0.08	0.19	0.00
α-Phenoxypropionic		0.14	0.57	0.11
α-Phenoxybutyric		0.26	0.87	0.24
α-(3-Methylphenoxy)butyric		0.30	0.93	0.32

Source: Ref. 153.

differ considerably on both layers. A mobile phase of lower polarity is to be used on cellulose layers. This phenomenon can be illustrated with the retention of phenoxyalkanoic acids chromatographed on silica gel and cellulose layers both impregnated with a 20% solution of formamide in ethanol [153], as is shown in Table 2.

Mobile Phases for Unpolar Stationary Phases

Since the solubility of many solutes in pure water would not allow their migration in the reversed phase system, the addition of an organic solvent is necessary. The so-called organic modifier can be methanol, ethanol, acetone, acetic acid, dimethyl formamide, acetonitrile, tetrahydrofuran, etc. The addition of the organic modifier results in the increase of the migration rate of the solutes. The higher its content, the higher are the R_F values of water-insoluble solutes. By changing its content the proper range of R_F values is achieved.

It should be mentioned, however, that some proportions of methanol-water mixtures or acetone-water mixtures, when used as mobile phases on unimpregnated silica gel or cellulose layers, give the same sequence of analytes as if a nonpolar liquid were present as stationary phase [55,154]. However, there is a difference in the shape of the spots: spots on unimpregnated layers are of adsorptive character (i.e., elongated) while the spots on impregnated layers are round [155]. Some of the analytes are so lipophilic that, using paraffin oil as the stationary phase, 100% methanol, acetonitrile, or acetic acid must be used to move the solute from the starting line.

The problem is very similar to that in liquid chromatography using covalently bonded reversed phases. Here too the volume of water affects the resolution and retention of a solute mixture. This may be due to the fact that organic modifiers, which would dissolve the organic solutes, require water to achieve different solute partition coefficients.

Mostly linear correlations have been found between the R_M values and the organic modifier concentration in the mobile phase [100,156-158]. Some compounds, however, did not follow this general rule [159-165]: their R_M values do not depend linearly on the organic phase concentration. For example, some morphine derivatives or quaternary ammonia steroids moved faster with increasing quantity of the organic modifier, but after modifier reached a certain concentration (about 70-80%) their movement slowed down again. This phenomenon was explained by a dual retention mechanism: the role of the free silanol groups prevails at higher organic solvent content. On the other hand, the structure of the compounds combining extreme lipo- and hydrophilic functional groups in the molecule must also be taken into consideration.

The possible interaction of the mobile phase components with the solute molecules should also be considered. It has been pointed out that using the reversed phase system silicone oil/0-50% aqueous acetone the proportion of solute (steroid) molecules that are solvated by water may be greater than the proportion of the acetone-solvated ones. At higher acetone concentrations this proportion is reversed [167]. Similarly, a linear relationship between the capacity factors and the methanol content in methanol-water mobile phase was found with neutral and/or nonpolar compounds in reversed phase high-performance liquid chromatography (RP-HPLC), hydrophobic expulsion being the predominant retention mechanism. For ionized polar compounds the relationship between the capacity factors and the organic modifier content in the mobile phase was nearly parabolic due to a dual retention mechanism. In water-poor eluent, polar interactions played the major role in the retention mechanism while in water-rich eluents hydrophobic expulsion was predominant [168]. Often the composition of the mobile phase has been selected mainly by trial and error. Systematic approaches to the selection of the mobile phase which would give optimum separation on reversed phase thin-layer plates and columns have been published (see, e.g., Refs. 169-174). The classification of eluents in reversed phase thin-layer chromatographic systems was carried out using the method of spectral mapping. The strength of particular organic modifiers was tabulated [175].

Reversed phase systems (nonpolar stationary phase/water + organic modifier) are suited for partition chromatography of neutral organic compounds. Problems arise with ionizable compounds. Solutes which are strong electrolytes do not pose difficulties since they completely dissociate in aqueous media. Acids and bases of middle strength can form streaks when chromatographed under neutral conditions due to dissociation, hydrolysis, and interaction with the active sites of the support (silanol groups). Therefore a secondary modifier must be added to the mobile phase; for review, see Ref. 176. Acidic or basic solutes can be chromatographed either in the form of neutral undissociated molecules or in the form of ions (ion chromatography) according to what secondary modifier has been added to the mobile phase: the presence of a suitably selected acid or base can suppress ionization or, to the contrary, bring about the complete ionization of the solute. This can also be achieved by adding a suitable counterion or liquid ion exchanger [177,178] to form ion pairs.

The addition of a counterion to acidic or basic solutes results in the formation of an ion pair (for review, see Ref. 179). This counterion can be provided with more lipophilic properties, thus influencing the chromatographic behavior in a more defined manner. Owing to the different concepts involved, various names have been used to describe this type of chromatography including ion pair chromatography [180], soap chromatography [181], surfactant chromatography [182], dynamic ion exchange chromatography [183], and chromatography on sorbed ion sites [184].

An ion pair is formed between a water-soluble cation (Q^+) and a water-soluble anion (X^-). The resulting ion pair is soluble in organic solvents but insoluble in water:

$$Q^+_{aq} + X^-_{aq} \rightleftarrows QX_{org} \tag{6}$$

The equilibrium of this reaction is controlled by the extraction constant E_{QX}:

$$E_{QX} = \frac{[QX]_{org}}{[Q^+]_{aq} \cdot [X^-]_{aq}} \tag{7}$$

where $[QX]_{org}$ is the molar concentration of the ion pair in the organic phase, and $[Q^+]_{aq}$ and $[X^-]_{aq}$ are the molar concentrations of the cation and anion in the aqueous phase. The magnitude of the equilibrium (i.e., the magnitude of E_{QX}) is controlled by the chemical nature of the ion pair, the pH value, the chemical structure of the starting materials, and the organic phase. The real situation is more complicated than shown by Eq. 7 due to additional side reactions such as dissociation, hydrolysis, dimerization, etc., in the organic phase [185].

The influence of E_{QX} on the retention for a system with a lipophilic stationary phase and a hydrophilic mobile phase (reversed phase system) is determined by:

$$K' = E_{QX} \cdot [X^-] \cdot \frac{V_S}{V_M} \qquad (8)$$

where K' is the capacity factor, V_S the volume of the stationary phase, and V_M the volume of the mobile phase. For a system with a hydrophilic stationary phase and a lipophilic mobile phase (normal or straight phase system), the following is valid:

$$K' = \frac{1}{E_{QX} \cdot [X^-]} \cdot \frac{V_M}{V_S} \qquad (9)$$

The relationship between K' and R_F can be described by:

$$K' = \frac{1 - R_F}{R_F} \qquad (10)$$

This means that in a reversed phase system increasing concentrations of the counterion result in an increase of K', i.e., a decrease of R_F. Suitable counterions for bases are sulfonic acids and sulfates. For acids the counterions are quaternary ammonium compounds. The separation of bases requires an acidic pH range (pH 1-3.5), whereas a basic medium (pH 7-9) is required for separation of acids.

This type of chromatography, known as ion pair chromatography, has become quite common, especially in RP-HPLC. It has been shown [186-188] that the retention of analytes is determined by the type, size, and concentration of the counterion, the pH, the type and concentration of organic modifier in the mobile phase, the type of stationary phase, and temperature. For some new aspects, see Refs. 189-192.

Ion-pairing reagents such as cetrimide, which possess a hydrophilic trimethylamino group attached to a long lipophilic alkyl chain, differ from ion pair reagents such as tetrabutylammonium iodide in showing pH-dependent effects on R_F values [193-198]. Procedures using such counterions or ionic surfactants of the same charge as the solute ion cannot always be considered as simple ion pair chromatography. The formation of micellar mobile phase is to be expected under defined conditions [199-203]. Aqueous micellar mobile phases produce unique separations when used in reversed phase thin-layer chromatography on chemically bonded phases. The most unusual feature of this technique is the existence of two solvent fronts (see above). It is apparent that the stationary phase strongly adsorbs the surfactant from the mobile phase. As a result, there are no micelles in the initial solution moving up the thin-layer plate. Once the stationary phase is saturated with the surfactant, the micellar mobile phase begins to move up the plate. Relatively water-insoluble molecules cannot migrate further than the surfactant front. While separations in this region are very good, one is limited to using 50% or less of the length of the chromatogram for the separation of hydrophobic compounds. Hydrophilic compounds, however, tend to migrate in the region between the upper "solvent" front and the "surfactant" front. Thus, one has a potentially useful technique for separating hydrophilic from hydrophobic compounds. Furthermore, hydrophilic compounds (like hydrophobic ones) are separated from one another in their own section of the thin-layer plate [199].

An inclusion phenomenon can be illustrated with the use of α- and β-cyclodextrins in reversed phase thin-layer chromatography. α- and β-Cyclodextrins (CD) are torus-shaped cyclic oligosaccharides made up of six to seven α-1,4-linked D-glucopyranose units, respectively. The inside of the CD cavities is relatively hydrophobic, while the hydrophilic groups are on the outside; this structure gives rise to the remarkable ability of CD to form inclusion compounds with various molecules and ions [204]. The fit of the total or at least part of the guest molecules in the CD (host) cavity determines the stability of the inclusion compounds and the selectivity of the complexation process. Some examples of applications of CD in mobile phases in reversed phase chromatography can be found in Refs. 205-215. For a review of the use of cyclodextrins for the separation of optical, geometrical, and structural isomers, see Refs. 216 and 217.

The use of buffer solutions as mobile phases in reversed phase planar chromatography can be demonstrated by the so-called pH chromatography on papers impregnated with olive oil [218,219]. Buffered mobile phases have been also used frequently in reversed phase thin-layer chromatographic systems, either for modifying analytical separations (see, e.g., Ref. 220) or for lipofilicity determinations [221]. It has been observed that the retention of analytes is affected not only by the pH of the buffer solution used but also by the quality of its ions [222] and that the achieved pH of the mobile phase is considerably dependent on the quantity of the organic modifier [223]. The acid-base properties of the support must also be taken into account. It has been already stated that the composition of multicomponent mobile phases is changed during their flow through the support. This is also the case with buffer solutions. It was found [224] that using a veronal buffer of pH 8.8 as mobile phase on silica gel layers the pH was 8.8 at the starting line, decreasing successively to pH 4 near the solvent front at a distance of 10 cm from the start. The pH value of undeveloped silica gel was also 4. The analogous difference on cellulose layers under identical conditions was only 1 pH unit. It was established [225] that the buffering capacity of veronal, phosphate, and acetate buffers is different in reversed phase thin-layer chromatography.

The movement of the basic front depends strongly on the organic phase ratio and to a lesser extent on the buffer concentration of the eluent. The retention of buffer ions increases in each case with the increasing extent of impregnation of the layer with unpolar phases.

The use of buffered mobile phases, alone or in mixture with an organic modifier, can give reversed phase separation of analytes even if used on bare, i.e., unimpregnated, silica gel [155]. These observations are in good agreement with those made in HPLC [226].

Ions present in the buffer solutions can act as counterions for the ionic solute. This was shown by the change from sodium to different kinds of ammonium ions in the buffer which caused increased retention of anionic samples, while the retention of cationic or uncharged compounds was decreased. The change in retention increased with the increasing hydrophobic character of the substituents at the nitrogen atom [226].

Salts as mobile phase components also cause interactions with solutes. It was pointed out [227] that in very dilute solutions both ion exchange and ion exclusion play a role, but that in solutions above about 1 M these effects become negligible. The so-called salting-out chromatography [228], using untreated cellulose layers, has been explained by adsorption mechanism as a kind of hydrophobic interaction between the hydrated cellulose and the aqueous solution [229]. For further examples, see Refs. 230-234.

F. The Solute

It was mentioned several times in the preceding chapters that the most important property of the solute is its solubility. This holds true when the solute is a nonelectrolyte, i.e., an un-ionized molecule. The partition coefficient of such a compound is given by the ratio of its molar concentrations in both phases. The solubility of the solute will thus influence our choice of the proper partition system.

Retention of ionizable compounds must be interpreted with special care [235]. The standard partition coefficient P refers to the concentration ratio of neutral, nonionized compound in each phase. In practice, the so-called distribution coefficients are observed. These are defined [236] as the ratio of the concentration of a compound in the organic phase to the concentration of all species, both neutral and ionized, in the aqueous phase at a given pH. The organic phase is believed to contain only un-ionized species. It should be admitted, however, that there may be some solution of the ionized form in the organic phase due to the formation of ion pairs in the presence of proper counterions (buffer ions, for example). The distribution coefficient D is used in such cases instead of the partition coefficient P but correction for relative differences in hydrogen-bonding effects when comparing different experimental procedures [237] must be done. The standard partition coefficient P of neutral species is according to:

$$P = \frac{D}{1 - \alpha} \qquad (11)$$

where α is the degree of ionization. Thus, for acids Eq. 12 can be derived, where K_a is the dissociation constant and $[H^+]$ is the hydrogen ion concentration.

$$P = D\left(\frac{K_a}{[H^+]} + 1\right) \qquad (12)$$

The retention of an un-ionized electrolyte-type compound in reversed phase systems is considerably stronger than that of its ionized form unless the solute ion has been paired with a strongly hydrophobic counterion. On the other hand, differences in pK values between the solutes can be utilized advantageously using pH chromatography: by changing continuously the pH of the mobile phase, compounds of different pK values are successively converted into ionized forms of higher mobility [218,219].

Furthermore, as has already been mentioned, the thin-layer chromatographic system cannot be considered simply as a system of the mobile and the stationary phases. All possible mutual interactions between the support, the mobile and stationary phases, and their components and with the solute forms must be taken into consideration. The interactions of the solute with the support of the stationary phase, mostly due to adsorption, seem to be the most important ones. It has been shown in the case of formamide- or dimethyl formamide-impregnated layers that a certain loading of the layer with the stationary phase is necessary to establish conditions for a liquid-liquid partition system. It is the strength of the solute intermolecular interactions which determines its competition for the active sites of the sorbent and the stationary phase.

Also, in the case of reversed phase chromatography, i.e., on layers impregnated with paraffin oil or 1-octanol, the interaction of the polar solutes with active sites of the support can be stronger than with the unpolar stationary phase. It was shown already in paper chromatography that sulfonated azo dyes chromatographed on papers impregnated with 1-dodecanol migrated on the chromatograms in the same sequence as on untreated paper [22]. The same phenomenon was observed in thin-layer chromatography of sulfonamides [238] and sulfonated azo dyes [25]: their migration on silica gel or cellulose layers both bare and impregnated with 1-octanol was identical.

Table 3 R_F Values of Alkylpyridinium Chlorides Using
1-Butanol Saturated with Ammonia as the Mobile Phase

Compound	R_F	
	C	Ac
Ethylpyridinium chloride	0.12	0.16
1-Butylpyridinium chloride	0.31	0.28
1-Octylpyridinium chloride	0.62	0.08
1-Dodecylpyridinium chloride	0.73	0.08
1-Cetylpyridinium chloride	0.89	0.06

Note: C = cellulose paper; Ac = acetylated paper.
Source: From Ref. 55, according to data from Ref. 239.

Recently, the same observation was made in chromatographing sulfonated and unsulfonated azo dyes on commercial acetylcellulose materials, which are known to be incompletely acetylated. Using the same mobile phases, the unsulfonated dyes migrated as in reversed phase chromatography, whereas the sulfonated ones as in normal phase chromatography on cellulose [55]. Similar observations were made with amino acids [230].

Very illustrative is the behavior of homologous series of compounds where the lower members are hydrophilic and the higher lipophilic, as in the case of alkylpyridinium halides separated on acetylcellulose plates. This continuously changing hydrophobicity from C_1 up to C_{18} homologs resulted in increasing R_F values of C_1-C_6 alkylated compounds and decreasing R_F values with further increase in the alkyl chain. On cellulose plates, under the same conditions, the R_F values are increased with the number of carbon atoms (Table 3). The same phenomenon can be observed when chromatographing those compounds on cellulose layers impregnated with 1-dodecanol and using ethanol-1 M HCl (1:1) as mobile phase [240].

The already-mentioned anomalous behavior of quaternary ammonio steroids and morphine derivatives, which contain functional groups of extreme lipophility and hydrophility in one molecule, can serve as a further example [159-165]. This phenomenon will not be restricted to thin-layer chromatography, but will also have a general importance in RP-HPLC. As a recent example, the partitioning of sulfonyl-containing drugs [241] can be mentioned.

III. APPLICATIONS

A. Analytical Separations

The separations based on partition or adsorption mechanisms are controlled by different chemical structure-chromatographic behavior relationships. Thus, it is the structure of the analytes to be separated which forces us to decide whether separations based on the partition or adsorption mechanism should be used in our particular case. Many examples can be found in the literature showing that the separation of complicated mixtures can be achieved by the combination of both separation principles, e.g., using two-dimensional thin-layer chromatography. Adsorption chromatography is often used to separate such mixtures into classes of compounds and the separation within these classes is achieved using partition thin-layer chromatography (see, e.g., separation of lipids [8], aliphatic aldehydes and ketones [241], etc.). In any case, separation of homologous or substituted compounds achieved by partition thin-layer chromatography is more efficient than by adsorption chromatography. This fact can be illustrated by ΔR_M values obtained from literature data for different substituents; see Tables 4-6.

Some of the practical applications of partition chromatographic separations of different types of compounds on layers impregnated with organic stationary phases described in the literature have been summarized in Table 7.

B. Lipophilicity Determination and QSAR Studies

It has been recognized that the lipophilic character of a molecule plays an important role in determining its biological affinity. 1-Octanol-water partition coefficient has become an important measure of the hydrophobicity of a compound [65,473].

Table 4   Mean ΔR_M Values for the Methylene Group of Six Lower Aliphatic Alcohols Chromatographed in the Form of Their N,N-Dimethyl-p-aminobenzeneazobenzoates on Silica Gel

Silica gel/mobile phase	ΔR_M
Impregnated with dimethyl formamide/cyclohexane-benzene (25:1)	0.128
Impregnated with paraffin oil/dimethyl formamide-methanol-water (4:1:1)	0.126
Untreated/cyclohexane-ethyl acetate (4:1)	0.066

Source: Ref. 243.

Table 5 Mean $\Delta R_{M(CH_2)}$ Values of Homologous Morpholinoethyl Esters of Alkoxy-Substituted Carbanilic Acids

Compounds	$\Delta R_{M(CH_2)}$	
	Adsorption chromatography[a]	Partition chromatography[b]
2-Alkoxy isomers	0.057	0.112
3-Alkoxy isomers	0.050	0.227
4-Alkoxy isomers	0.039	0.300

[a]Silica gel (Silufol)/light petroleum-diethylamine (4:1).
[b]Cellulose (Lucefol) impregnated with a 30% solution of formamide/n-heptane-diethylamine (20:1).
Source: Ref. 244.

Today, the partition coefficient, log P (Eqs. 1-3), in the 1-octanol-water system is used to predict not only the migration of nonelectrolytes to and from the cells in drug research, but also the bioconcentration of lipophilic pollutants, the soil adsorption, the mobility in soil, and the bioaccumulation by aquatic organisms as well as the transfer, via food chains, of aquatic organisms and the depuration of organic chemicals of a nonionic character [623]. The determination of the 1-octanol-water partition coefficient has become compulsory for every new commercial chemical in some countries [624]. Correlation between structure-activity relationships and the 1-octanol-water partition coefficient have been shown for a series of chemical compounds.

Various methods are in use for the determination of 1-octanol-water partition coefficients, such as the shake-flask, the countercurrent distribution, or the continuous partitioning method [625,626]. The direct determination by the shake-flask method is tedious, requires a relatively large amount of an analytically pure compound as well as time-consuming measurements of the concentration of compounds in both phases, and it presents difficulties when the compound under investigation is highly insoluble in either solvent phase. At present, the most promising method seems to be reversed phase partition chromatography.

To avoid the practical difficulties in determining the partition coefficients directly, the use of the chromatographic R_M values (Eq. 2) was suggested [11,434,528]. These values are related to the logarithm of the partition coefficient between the polar and nonpolar phases of a chromatographic system. On the other hand, if there

Table 6 ΔR_M Values of Different Substituents on Phenols Chromatographed in the Form of Antipyrylquinoneimine Dyes

Substituent	I[a]	II[b]
2-CH_3	-0.548	-0.148
2-C_2H_5	-0.488	-0.065
2-n-C_3H_7	-0.300	-0.082
3-CH_3	-0.298	-0.057
3-C_2H_5	-0.436	-0.044
2,6-Cl_2	-0.598	-0.333
2,6-Br_2	-0.646	-0.371
2,6-I_2	-0.815	-0.550

[a]Cellulose impregnated with formamide/heptane-benzene (1:1).
[b]Silica gel/benzene-acetone (3:1).
Source: Ref. 245.

are no group interactions, the ΔR_M value, defined as the difference between the R_M value of the substituted and unsubstituted compounds, can be a parameter for estimating lipophilicity of compounds. It is, of course, important to make R_M values determinations in systems where partition is either the sole or the predominating process taking place [11].

A general mathematical relationship between the penetration of a molecule into cells and a substitution constant π has been proposed. This parameter was defined as the difference in the logarithm of the 1-octanol-water partition coefficient P of the substituted (log P_X) and the unsubstituted (log P_H) compounds:

$$\pi = \log P_X - \log P_H \tag{13}$$

For a functional group, π is constant, provided that the introduction of the substituent into a molecule does not cause group interactions that would affect the partition coefficients of the molecule itself [64, 627]. Therefore, because of possible interaction, calculated π values cannot be completely substituted for the experimentally determined ones. They can, however, quickly focus on possible errors in measurement and can indicate where other procedures may not be delivering values actually related to 1-octanol-water distribution [628]

Table 7 Examples of Practical Applications of Chromatographic Systems Using Impregnated Layers

Stationary phase	Compounds chromatographed	Sorbent	Method of impregnation[a]	Mobile phase	Type of study[b]	Ref.
Formamide	Ergot alkaloids	Cellulose	20% in acetone (dipping)	Hexane–benzene (5:6) Hexane–benzene–chloroform (5:6:3)	A	246
	Rauwolfia Belladona Morphine Lobelia alkaloids	Cellulose	20% in acetone (dipping)	Hexane–methyl ethyl ketone (1:1) in NH_3 atm. Benzene–heptane–chloroform–diethylamine (6:5:1:0.02) Benzene–heptane–diethylamine (1:6:0.02)	A	247
	Steroidal sapogenins	Silica gel			A	248
	Rauwolfia alkaloids	Silica gel	20% in acetone (developing 2 hr)	Hexane–methyl ethyl ketone (2:1) satd. with water	A	249
	Steroids	Celite No. 545	(Spraying)	Hexane, benzene, chloroform, and their mixtures	A	250
	Cardenolide glycosides	Talcum	25% in acetone (spraying)	Dioxane–chloroform–1-butanol (20:70:5) Ethyl acetate–chloroform (1:4) Methyl ethyl ketone–p-xylene (1:1)	A	251
	Corticosteroids	Silica gel	40% in acetone (developing)	Hexane, benzene, chloroform, and their mixtures	A	252
	Lanatoside	Talcum	Slurry–talcum–formamide–ethanol 4:1:7	Chloroform Ethyl acetate	A	253

Corticosteroids	Celite	(Spraying)	Benzene Cyclohexane–benzene	A	254
Digitalis cardenolides	Kieselguhr	10% in acetone (developing)	Chloroform–tetrahydrofuran–formamide (50:50:6) Xylene–methyl ethyl ketone–formamide (50:50:4)	A	255
Oxid. products of phenols	Cellulose	25% in acetone	Petroleum ether (3x)	A	256
Phenols as azodyes	Kieselguhr	5% in acetone (dipping)	Benzene–cyclohexane–dipropylene glycol (30:70:3) Benzene–cyclohexane–diethylamine (5:5:1)	A	257
Ergot alkaloids	Cellulose	25% in acetone (dipping)	Ethyl acetate–hexane–diethylamine (50:60:0.2)	A	258
Steroids	Silica gel	20% in acetone 25% in acetone (developing)	Chloroform–diethyl ether–water (80:20:0.5) Cyclohexane–tetrachloromethane–water (50:50:0.1)	A	259
Phenols	Silica gel	33% in acetone (dipping)	Dichloromethane–cyclohexane (55:45)	A	260
Alcohols as 3,5-dinitrobenzoates	Silica gel	33% in acetone (dipping)	Benzene–methyl acetate (150:1)	A	261
Furocoumarins	Silica gel	33% in water (slurry)	Dibutyl ether	A	262
Lanatosides A, B, C, D	Talcum	30% in acetone (slurry 1 ml + 4 g talcum + 7 ml ethanol)	Chloroform–benzene–ethanol–formamide (79:10:10:1)	A	89

Table 7 (Continued)

Stationary phase	Compounds chromatographed	Sorbent	Method of impregnation[a]	Mobile phase	Type of study[b]	Ref.
Formamide	Dian	Silica gel	30% in ethanol (slurry, 3 ml + 1.04 g silica gel + 0.16 g gypsum)	Chloroform-ethanol (99:1)	A	263
	Ergot alkaloids	Cellulose	15% in acetone + 1% conc. NH$_3$	Ethyl acetate-heptane-dimethyl formamide (250:300:1)	A	264
	Phenols	Silica gel	(Slurry)	Benzene-methanol (95:5)	A	3
	Glucocorticoids	Kieselguhr	10% in acetone (developing)	Dichloromethane-toluene (3:2), (1:1) Chloroform-toluene (1:3) Chloroform	A	265
	Corticoids	Silica gel	40% in methanol	Petroleum ether-isobutyl alcohol-tetrachloromethane (5:3:4) Chloroform-petroleum ether-isobutylalcohol-tetrachloromethane (40:20:1:10)	A	266
	Aristolochic acids	Cellulose	20% in acetone (dipping)	Benzene-heptane-chloroform-acetic acid (15:15:70:3)	A	267
	Thiocarbamic acid derivs.	Silica gel	5% in acetone	Chloroform	A	268
	Ergot alkaloids	Silica gel	(Slurry, 8-10 ml/10 g) or 30% in acetone (developing)	Benzene-hexane (3:2) Benzene-cyclohexane (1:1) Petroleum ether-ethyl acetate-1 M NH$_4$OH (65:35:1)	A	269

Ergot alkaloids	Talcum	20% in acetone (slurry, 0.8 ml/2.5 g talcum)	Hexane–tetrahydrofuran–toluene (5:4:1)	A	270
Cucurbitacin	Cellulose	50% in ethanol (spraying)	Tetrachloromethane–benzene (1:1)	A	271
Rauwolfia alkaloids	Silica gel		Heptane–methyl ethyl ketone (2:1) satd. with water	A	272
Salicylamide Carbamates Urea	Kieselguhr	5% in methanol (dipping)	Tetrachloromethane Benzene–chloroform (1:4)	A	273
Polyhydric phenols	Silica gel	3% in acetone (spraying)	Benzene–methanol (95:5)	A	274
Digoxin Digitoxin	Cellulose	20% in acetone (dipping)	Chloroform–benzene (95:5) (multiple development)	A	275
Ergot alkaloids	Silica gel	(Slurry)	Hexane–butyl acetate (1:1, 6.5:3.5) satd, with formamide and ammonia	A	276
Phenols	Cellulose	(Slurry; see Section II.C)	Hexane	A,T	88, 125
Erythromycins	Silica gel + kieselguhr (1:1)	15% in acetone (developing)	Dichloromethane–hexane–ethanol (60:35:5) Dichloromethane–ethyl acetate–hexane–ethanol (40:40:15:5)	A	277
Ergot alkaloids	Cellulose	20% in acetone	Heptane–benzene–chloroform (25:30:15) Heptane–benzene (25:30) Chloroform–ethanol (96:4)	A	278
Digitoxins	Silica gel	20% in acetone (developing)	Xylene–methyl ethyl ketone (1:1)	A	279

Table 7 (Continued)

Stationary phase	Compounds chromatographed	Sorbent	Method of impregnation[a]	Mobile phase	Type of study[b]	Ref.
Formamide	Phenols as derivatives	Silica gel	33% in acetone (developing—3 hr)	Benzene Butyl acetate Benzene-butyl acetate (9:1)	A	280
	Phenols	Silica gel	Slurry, 70 ml 0.5-6 M in acetone + 15 g cellulose)	Hexane	A	88, 125, 281
	Nitrophenols	Cellulose		Hexane	A, T	282
	Indanols	Cellulose	(Slurry)	Hexane	A, T	126
	Alkaloids	Cellulose	20%	Heptane-methyl ethyl ketone (2:1)	A	283
	Phenols	Silica gel	20% in acetone	Hexane-chloroform (3:2)	A	284
	Cardenolide glycosides	Silica gel	20% in acetone	Xylene-methyl ethyl ketone (2:3)	A	285
	Ergot alkaloids	Silica gel	40% in ethanol (slurry, 11.5 ml/5 g silica gel	Benzene-cyclohexane-diethylamine (5:2:0.01)	A	85
	Ergot alkaloids	Silica gel	(Slurry, 18 ml + 4 ml concd. ammonia + 45 ml ethanol)	Diisopropyl ether-tetrahydrofuran-toluene-diethylamine (70:15:15:0.1) Diisopropyl ether-anhydr. ethanol-toluene-diethylamine (75:5:20:0.1)	A	67
	Barbiturates	Cellulose	20% in ethanol (dipping)	Chloroform	A	286

Alkaloids Local anesthetics	Cellulose	20-40% in ethanol (dipping)	Benzene Chloroform Cyclohexane and mixtures	A	287
Digitoxin Gitoxin	Kieselguhr	6% in acetone	Methyl ethyl ketone-xylene-formamide (50:50:4)	A	288
Barbiturates	Kieselguhr	20% in water (slurry)	Tetrachloromethane-chloroform (1:1)	A	289
Drugs	Talcum	(Slurry, 1 ml + 11 ml ethanol + 8 g talcum)	Benzene Toluene Cyclohexane-chloroform mixtures	A	290
Ergot alkaloids	Silica gel	(Slurry, 18 ml + 4 ml 5% aq. NH_3 + 45 ml ethanol + 20 g silica gel)	Diisopropyl ether-tetrahydrofuran-diethylamine (80:20:0.2) Dibutyl ether-dichloromethane-diethylamine (60:40:0.2)	A	87
Cardiac glycosides	Kieselguhr	6% in acetone	Chloroform-tetrahydrofuran-formamide (30:20:0.5)	A	291
1,4-Benzodiazepines	Silica gel Cellulose	20% in acetone (dipping)	Chloroform Heptane-chloroform (3:1)	A	149
Furil oximes	Cellulose	20% in ethanol	Diisopropyl ether	A	292
Ergot alkaloids	Silica gel	18% in acetone + 0.6 ml 25% NH_3	Diisopropyl ether-toluene-ethanol-diethylamine (75:20:5:0.1) Diisopropyl ether-toluene-tetrahydrofuran-heptane-diethylamine (50:15:15:20:0.1) Diisopropyl ether-tetrahydrofuran-diethylamine (90:10:0.2)	A	68
Phenols Cannabinoids	Silica gel	(In gas phase)	Toluene-dichloromethane (1:1)	A	108

Table 7 (Continued)

Stationary phase	Compounds chromatographed	Sorbent	Method of impregnation[a]	Mobile phase	Type of study[b]	Ref.
Formamide	Digoxin Dihydrodigoxin	Cellulose	25% in acetone (dipping)	Chloroform	A	123
	Ergot alkaloids	Cellulose	15% in acetone	Ethyl acetate–heptane–diethyl-amine (5:6:0.005), (4:6:0.2)	A	293
		Silica gel	25% in ethanol + 5% NH_3	Tetrahydrofuran–diisopropyl ether–toluene–diethylamine (20:70:10:0.5)	A	294
	Ergot alkaloids	Silica gel	20% in acetone	Diethyl ether	A	295
	Carbanilic acid esters	Silica gel Cellulose	10–40% in acetone (dipping)	Heptane Heptane–diethylamine (20:2; 10:1)	A	244
	Phenols as derivatives	Cellulose	10% in ethanol (developing)	Hexane–benzene (1:1)	A	245
	Ergot alkaloids	Silica gel	10% in acetone (dipping)	Chloroform–diethyl ether (3:10)	A	296
	Alkaloids	Cellulose		Hexane–ethyl acetate 12 M NH_4OH (280:140:1)	A	297
	Testosterone derivatives	Kieselguhr	10% in acetone (developing)	Hexane–acetone (2:1)	A	298
	Local anesthetics	Silica gel	40% in ethanol + 1% Tris	Heptane–diethylamine (10:1)	L	299
		Cellulose	10% in ethanol 30% in ethanol (developing)	Heptane Heptane–diethylamine (20:1)		

Carbanilic acid esters	Cellulose	10% in ethanol 30% in ethanol 40% in ethanol + 1% Tris	Heptane Heptane–diethylamine (20:1) Heptane–diethylamine (9:1)	L	300
	Silica gel	40% in ethanol + 1% Tris	Heptane–diethylamine (10:1)		
Ergot alkaloids hydrogenated	Silica gel	18% in acetone	Diisopropyl ether–tetrahydrofuran–diethylamine (80:20:0.2; 90:10:0.2) Dibutyl ether–dichloromethane–diethylamine (60:40:0.2)	A	301
Digitalis cardenolides	Silica gel	20% in acetone (developing)	Xylene–methyl ethyl ketone (7:10.5), (2:3) Chloroform–tetrahydrofuran–formamide (5:11:0.5)	A	302, 303
	Kieselguhr		Xylene–methyl ethyl ketone–formamide (50:50:4)	A	303
Cardioactive glycosides	Kieselguhr	10% in acetone (developing)	Methyl ethyl ketone–xylene (1:1) Chloroform–tetrahydrofuran–formamide (50:50:6.5)	A	304
Phenylurea herbicides	Silica gel silanized	20% in acetone	Hexane–chloroform–diethyl ether (2:1:1)	A	305
Triazine herbicides	Silica gel	20%	Hexane–chloroform–diethyl ether (3:1:1)	A	306
Digilanides A, B, C	Silica gel	In acetone	Tetrahydrofuran–chloroform–formamide (50:50:6.5)	A	307
Alkoxyphenyl-carbamic acid esters	Cellulose	40% in acetone + 2.5% HCOO-NH4 (dipping)	Heptane–diethylamine (50:0.05) Xylene–diethylamine (50:0.5) Pentane–diethylamine (50:0.5)	A	308
Furocoumarins	Silica gel	In methanol	Hexane–benzene–methanol (7:2:1)	A	309

Table 7 (Continued)

Stationary phase	Compounds chromatographed	Sorbent	Method of impregnation[a]	Mobile phase	Type of study[b]	Ref.
Formamide	Phenols Aromatic amines	Silica gel	5-30% in acetone (slurry, 50 ml/20 g silica gel)	Heptane in mixtures with methyl pentyl ketone or 1-heptanol Benzene, toluene, trichloroethylene, chloroform	T	86
	Ergot alkaloids	Silica gel	18% in ethanol + 0.6% NH$_4$OH	Diethyl ether-toluene-ethanol-diethylamine (75:20:5:0.1)	A	310
	Crotonolactones	Cellulose	20-50% in ethanol	Cyclohexane	L	124
	Dimethylamino-ethylalkoxycar-banilate chlorides	Cellulose	40% contg. 1% Tris	1-Propanol	L	311
	Glyceollin isomers	Silica gel	5% in acetone (developing 2x)	Diethyl ether-hexane (3:1) (4x)	A	312
	Phenylcarbamic acid esters	Cellulose	40% in ethanol + 1% Tris (dipping)	Hexane-ethanol (1:1) Ethanol-diethyl ether (9:1) 2-Propanol	A	313
	Nitrophenols	Silica gel Cellulose	20% in ethanol (dipping)	Benzene Benzene-acetic acid (95:5)	T	314, 366
	Alkoxyphenyl-carbamic acid esters	Cellulose	40% in ethanol + 1% Tris (dipping)	Pentane Pentane-diethylamine (9:0.05)	L	315
	1,4-Benzodiazepines	Silica gel	(Slurry, 57%)	Toluene or chloroform in mixtures with methyl amyl ketone or 1-heptanol	A	316

	Compound	Support	Reagent	Solvent	Detection	Ref.
	Alcohols as 3,5-dinitrobenzoates	Silica gel	20% in acetone (dipping)	Hexane	A	317
	Color couplers	Silica gel	20% in ethanol (dipping)	Benzene-hexane (1:1) Benzene-ethyl acetate (10:1)	A	318
Monomethyl formamide	Phenols	Cellulose	0.5-6.0 M in acetone (slurry)	Hexane	T	281
	Indanols	Cellulose	0.5-6.0 M in acetone (slurry)	Hexane	A, T	126
Dimethyl formamide	4-Arylthiazoles	Kieselguhr	10% in acetone	Cyclohexane	A	319
	Polycyclic hydrocarbons	Alumina	0.5-10% in diethyl ether (dipping)	Petroleum ether with 2% pyridine or 1% diethyl ether	A	320
	Thiophosphate pesticides	Alumina	15-25% in diethyl ether	Methylcyclohexane	A	321
	Cannabinols	Silica gel	60% in CCl₄ (developing)	Cyclohexane	A	322
	Pesticides	Alumina	25% in diethyl ether (dipping)	Isooctane	A	323
	Polycyclic hydrocarbons	Silica gel	(Spraying)	Cyclohexane	A	324
	2,4-Dinitrophenylhydrazones of aliphatic oxo compounds	Starch	30% in diethyl ether	Hexane	A	325
	Chlorinated pesticides	Silica gel		Hexane	A	326

Table 7 (Continued)

Stationary phase	Compounds chromatographed	Sorbent	Method of impregnation[a]	Mobile phase	Type of study[b]	Ref.
	Cannabis	Silica gel	50% in CCl$_4$ (developing)	Cyclohexane	A	327
	Polycyclic hydrocarbons	Cellulose	20% in diethyl ether (dipping)	Isooctane	A	328
	Dian	Kieselguhr	10% in acetone	Cyclohexane–ethyl acetate–acetic acid (125:25:4.5)	A	329
	Phenols	Silica gel	5% in acetone (spraying)		A	274
	Cannabidiol	Silica gel		Petroleum ether–diethyl ether (4:1)	A	330
	Chlorinated pesticides	Alumina	25% in diethyl ether (dipping)	Isooctane	A	331
	Phenols	Cellulose	0.5–6.0 M in acetone (slurry)	Hexane	A, T	281
	3,5-Dinitrobenzoates of alcohols	Alumina Silica gel Kieselguhr	0–90%	Cyclohexane	A, T	127, 128
	Cannabis	Silica gel	50% in CCl$_4$	Cyclohexane	A	332
	2,4-Dinitrophenylhydrazones of benzophenones	Silica gel	25–50% in acetone (developing)	Cyclohexane (6x) Cyclohexane–benzene (2:1)	A	333
	Aliphatic acids as derivatives	Silica gel	20 or 40% in methanol (developing)	Petroleum ether–toluene (5:1)	A	334

Alcohols as colored derivatives	Silica gel	40% in methanol (developing)	Cyclohexane–benzene (25:1)	A	243
Indanols	Cellulose	0.5–6.0 M in acetone (slurry)	Hexane	A, T	125, 126
Cannabinols	Silica gel	(In gas phase)	Petroleum ether–diethyl ether (4:1)	A	335
Cannabis	Silica gel		Cyclohexane	A	336, 337
2,4-Dinitrophenyl hydrazones of aromatic aldehydes	Kieselguhr		Cyclohexane–diethyl ether (7:3) I. direct. 90% methanol–acetic acid–benzene–tetraline (10:1: 1:1); II. direct. (two-dimens.)	A	338
Aromatic acids as 2,4-dinitrobenzyl esters	Silica gel	40% in ethanol	Cyclohexane–benzene (25:1)	A	339
Hexachlorobenzene	Alumina	25%	Isooctane	A	340
Hydroxyanthraquinones	Cellulose	5, 10% in acetone	Cyclohexane Cyclohexane–tetrachloromethane (9:1)	A	341
Phenols	Silica gel	(In gas phase)	Cyclohexane–dichloromethane (4:1)	A	108
Cannabinoids			Petroleum ether–diethyl ether (4:1)		
Thiobenzamides	Cellulose	In acetone 30% (dipping)	Cyclohexane–benzene (1:1)	L	342
Anthraquinone derivatives	Silica gel	30% in ethanol	Toluene	A	343
Cannabinoids	Silica gel	20% in CCl$_4$	Cyclohexane	A	344

Table 7 (Continued)

Stationary phase	Compounds chromatographed	Sorbent	Method of impregnation[a]	Mobile phase	Type of study[b]	Ref.
Formamide–dimethyl formamide mixture	Acrylic acid esters	Cellulose	40% in ethanol	Cyclohexane–benzene (24:1) Petroleum ether	A	345
	Rauwolfia alkaloids	Cellulose	Formamide–dimethyl formamide–acetone (20:15:65) (developing)	Hexane–methyl ethyl ketone (2:1) atm. NH_3	A	346
Polyamide	Polyalcohols	Kieselguhr	(Slurry)	Chloroform	A	91
	Phenols	Cellulose	(Slurry, 11 ml 0.13 g/1 ml + 13.5 g cellulose + 74 ml formic acid)	Cyclohexane–acetic acid (93:7) 10% Aqueous acetic acid	A, T	90 347
	Chlorinated phenols				A, T	348
	Nitrophenols				A, T	349, 350
	p-Acetanilides			Water–acetone–dioxane (2:1:1)	A	351
	Polyhydric phenols	Silica gel	20%	Cyclohexane–ethanol (1:1) Benzene–acetic acid–water (45:4:8)	A	352, 353
	Sugars	Glass powder		Acetone–acetic acid–water (3:1:1) Acetone–HCl–water (50:4:4) Ethyl acetate–acetic acid–methanol–water (30:7.5:7.5:10)	A	354
	Polyhydric phenols	Glass powder	(In slurry)	Benzene–methanol–acetic acid (80:13:7)	A	355

Amino acids	Silica gel	(Slurry 10 g/ 50 g silica gel in water)	Methanol–butyl acetate–acetic acid–pyridine (20:20:10:5)	A	356
	Glass powder	Slurry		T	357
Sterols	Silica gel	15% in acetone (dipping)	Heptane	A	358
2,4-Dinitrophenyl hydrazones of aliphatic oxo compounds	Kieselguhr Alumina	10% in acetone (dipping)	Petroleum ether Two-dimensional development	A	359
2,4-Dinitrophenyl hydrazones of aldehydes	Kieselguhr	In ethanol (dipping)	Heptane	A	360
3,5-Dinitrobenzoates of alcohols	Kieselguhr	20% in acetone (dipping)	Petroleum ether (b.p. 100–140°C)	A	95
Chlorinated benzoquinones	Kieselguhr	In benzene (developing)	Hexane	A	361
2,4-Dinitrophenyl hydrazones of aliphatic aldehydes	Kieselguhr	7.8% in methanol	Hexane	A	362
Dimethyl sulfoxide	Sugar acetate anomers	Silica gel	50%	Diethyl ether, diisopropyl ether and mixtures	A	363
Ethylene glycol	Steroids	Silica gel	30% in methanol (developing)	Methylcyclohexane–dichloromethane (97.5:2.5) Benzene–methylcyclohexane (1:1) Benzene–methylcyclohexane–chloroform (1:1:1)	A	364

Table 7 (Continued)

Stationary phase	Compounds chromatographed	Sorbent	Method of impregnation[a]	Mobile phase	Type of study[b]	Ref.
Ethylene glycol	Alkaloids	Silica gel	20% in methanol	Petroleum ether–chloroform (9:1)	A	365
	Aromatic amines Heterocyclic bases	Alumina		Decalin, tetrachloromethane, benzene, toluene	A	367
	Nitrophenols	Silica gel Cellulose	20% in ethanol (dipping)	Benzene Benzene–acetic acid (95:5)	A, T	314, 366
	1- and 2-Naphthols and naphthylamines	Cellulose		Cyclohexane, tetrachloromethane, toluene, benzene	T	368
	Aliphatic acids	Silica gel	10% in water	Petroleum ether–acetone (2:1)–monocarboxylic Diethyl ether–formic acid (50:1)–dicarboxylic	A	369
Propylene glycol	*Rauwolfia* alkaloids	Cellulose	Propylene glycol–methanol–acetic acid (50:100:1) (developing)	Benzene–cyclohexane (1:1)	A	346
	Steroids	Kieselguhr	10% in acetone (developing)	Toluene Cyclohexane–toluene (4:1) Petroleum ether	A	370
	Bruceolides	Kieselguhr	20% in acetone	Benzene–ethyl acetate (9:1; 85:15; 4:1)	A	371
	Steroids	Celite No. 545	(Spraying)	Ligroin	A	250

Impregnating agent	Substances	Support	Impregnation	Mobile phase	Detection	Ref.
	Corticosteroids	Celite	(Spraying)	Toluene	A	254
	Digoxin Digitoxin	Cellulose	20% in acetone (dipping)	Benzene–ethyl acetate (9:1)	A	275
Diethylene glycol	Phenylurea herbicides	Silica gel silanized	20% in acetone	Hexane–benzene–acetone (1:1:1)	A	303
	Triazine herbicides	Silica gel silanized	20% in acetone	Hexane–benzene–tetrahydrofuran (4:1:1)	A	306
Polyglycol (PEG)	2,4-Dinitrophenylhydrazones of oxo compounds	Kieselguhr		Petroleum ether	A	372
	2,4-Dinitrophenylhydrazones of aromatic aldehydes	Kieselguhr	20% in acetone	Diethyl ether	A	338
PEG 200	Fat-soluble vitamins	Silica gel	(Slurry in acetone 0.4 ml/g)	Benzene–petroleum ether (1:1)	A	373
	Vitamins A, D	Silica gel Cellulose	(Slurry in water 1 ml/4 g)	Benzene–petroleum ether (1:1)	A	93
PEG 400	2,4-Dinitrophenylhydrazones	Kieselguhr	2.5%	Petroleum ether	A	374
	2,4-Dinitrophenylhydrazones of aliphatic oxo compounds	Basic zinc carbonate		Petroleum ether	A	375
	2,4-Dinitrophenylamines	Kieselguhr + MgO	(Slurry, 14 g MgO + 7 g kieselguhr + 48 ml ethanol + 8 ml PEG)	Heptane	A	376

Table 7 (Continued)

Stationary phase	Compounds chromatographed	Sorbent	Method of impregnation[a]	Mobile phase	Type of study[b]	Ref.
	2,4-Dinitrophenylhydrazones of aldehydes	Kieselguhr	33%	Petroleum ether	A	377
	Tetracyclines	Kieselguhr	(Slurry, 20% in glycerine)	Ethyl acetate + 1.0 M EDTA adj. to pH 7.0	L	378
				Dichloromethane-ethyl formiate-ethanol (9:9:2) satd. with McIlvaine buffer pH 4.7	L	58
	2,4-Dinitrophenylhydrazones of carbonyls	MgO + microcell T38	20% in chloroform (developing)	Chloroform-SVR (4:1)	A	379
	4-Dimethylamino-3,5-dinitrobenzamides	Kieselguhr	(Slurry, 10 g + 80 ml water + 40 g kieselguhr)	Hexane-ethyl acetate (7:3)	A	94, 380
	Mono- and dicarboxylic acids	Silica gel	10% in water	Petroleum ether-acetone (2:1)—monocarb. Diethyl ether-formic acid (50:1)—dicarb.	A	368
PEG 600, 1000, 1500	2,4-Dinitrophenylhydrazones of aldehydes and ketones	Silica gel	5% in water (slurry, 15 ml + 6.5 g silica gel)	Petroleum ether-pyridine (4:1; 8:1)	A	381
PEG 600	2,4-Dinitrophenylhydrazones of aldehydes and ketones	Silica gel	20% in acetone (dipping)	Hexane-benzene (4:1)	A	382

	3,5-Dinitrobenzoates of alcohols	Kieselguhr	(Slurry 15 g + 30 g silica gel + 45 ml water)	Petroleum ether	A	95
PEG 1000	Dicarboxylic	Kieselguhr	(Slurry, 15 g + 45 ml water + 30 g kieselguhr)	Diisopropyl ether-formic acid-water (90:7:3)	A	383, 384
	Softeners	Silica gel			A	261
PEG 4000	2,4-Dinitrophenylhydrazones of aldehydes	Silica gel	(Slurry, 3 g + 10 g silica gel + 35 g Tinopal WG + 15 ml water + 5 ml 1 M NaOH)	Hexane-benzene (35:65)	A	385
β,β'-Oxydipropionitrile	Quinone 2,4-dinitrophenylhydrazones	Silica gel Alumina	5–20% in acetone (developing)	Benzene-petroleum ether (1:1)	A	92
	Fat-soluble vitamins	Silica gel Alumina	(Slurry, 0–0.2 ml/g)	Benzene-petroleum ether	A	372
1-Butanol 1-Pentanol	Basic drugs	Acetylcellulose	(Developing)	Aqueous salt solutions	A	386
1-Pentanol	Phenolic acids	RP-2 plates	10% in acetone	Phosphate buffer pH 6.0 + 0.003 M tetrabutylammonium chloride	A	387
1-Octanol	Penicillins	Cellulose	10% in acetone (developing)	0.5 M 6-aminohexanoic acid (pH adj. by HCl)	A	388, 389
	Acetanilides	Silica gel	5% in heptane (developing)	10% aqueous acetone	L	23, 390
	Cardenolides	Silica gel	5% in diethyl ether	10–50% aqueous acetone	L	391

Table 7 (Continued)

Stationary phase	Compounds chromatographed	Sorbent	Method of impregnation[a]	Mobile phase	Type of study[b]	Ref.
	Sulfonamides	Silica gel	5, 10, 20% in diethyl ether (developing)	Sodium acetate-veronal buffer pH 7.4	L	392
	Thiobenzamides	Silica gel		Aqueous methanol Phosphate buffer + methanol	L	393
	Penicillins	Cellulose	15% in acetone (developing)	0.3 M NaCl in citric acid-phosphate buffer pH 3.5 or 4.5	L	394
	1,4-Benzodiazepines	Silica gel	5% in diethyl ether	35-40% acetone in water or 1/7 M acetate-veronal buffer	L	395
	Sulfonamides Azo dyes	Silica gel Cellulose	5% in acetone (dipping)	Acetone in acetate-veronal buffer	T	25
	Diethylphenyl phosphates	Silica gel	1.5% in hexane	40% aqueous acetone	A	396
	Nitrophenols	Silica gel Cellulose	10% in ethanol (dipping)	Ethanol-buffer pH 3 (1:2) Ethanol-buffer pH 8 (1:2) Ethanol-buffer pH 10 (1:2)	A, T	314, 366
1-Dodecanol	Anion-active tensides	Kieselguhr	2% in petroleum ether	Methanol-formic acid-ammonia (50:1:50)	A	397
	Surfactants	Silica gel Alumina Cellulose	3-5% in ethanol (dipping, spraying)	Methanol-ammonia (2:3 to 3:1)	A	122
	Nitrophenols	Silica gel Cellulose	10% in ethanol (dipping)	Ethanol-buffer solutions pH 3, 8, 10 (1:2)	A, T	314, 366

Oleylalcohol	Local anesthetics	Cellulose	10% in petroleum ether (developing)	0.5 M sodium phosphate buffer pH 1-10	L	398
	1,4-Benzodiazepines	Kieselguhr	(Slurry)	Methanol-aqueous buffer solutions	L	26
	Phenothiazines	Kieselguhr Cellulose	(Slurry or developing)	Methanol-aqueous buffer solution pH 7-9	L	98, 399
	4-Hydroxycoumarin derivatives	Kieselguhr Cellulose Silica gel Kieselguhr silanized		Aqueous methanol	T	40
	Different compounds	Celite No. 545 silanized	(Slurry)	30-90% aqueous methanol	T	400
Ethyl oleate	Nitrophenols	Cellulose	(Slurry)	25 or 37.5% aqueous alcohol	A,T	97
	Alkylated phenols					401
	Chlorophenols					402
	Phenols					403
	Indanols	Cellulose	(Slurry)	37.5% aqueous ethanol	A,T	404
	Dinitrophenols	Cellulose	(Slurry, 2 ml 98 ml diethyl ether + 25 g cellulose)	40 and 60% aqueous ethanol	A	405, 406
Dimethylphthalate	Ergot alkaloids hydrogenated	Cellulose			A	407
Dinonylphthalate	Alkyl sulfoxides	Alumina	5% in acetone	Methanol-water-pyridine (5:1:1)	A	408

Table 7 (Continued)

Stationary phase	Compounds chromatographed	Sorbent	Method of impregnation[a]	Mobile phase	Type of study[b]	Ref.
Polyesters	Substituted aceto-acetamides	Kieselguhr	(Slurry)	Diisopropyl ether–petroleum ether–tetrachloromethane–water–formic acid (50:20:20:1:8)	A	96
	Hydroxamic acids 2-Hydroxybenzo-phenones	Kieselguhr	(Slurry)		A A	409 410
	Substituted benzoic acids	Kieselguhr	(Slurry)	Diisopropyl ether–petroleum ether–tetrachloromethane–undecane–formic acid–water (40:20:10:20:8:1)	A	411
Triglycerides	Chloroplast pigments	Kieselguhr	7% in petroleum ether (developing)	Methanol–acetone–water (20:4:3)	A	99
	Bacteriochlorophylls	Cellulose		Acetonitrile–water	A	412
Triolein	Bacteriochlorophylls	Kieselguhr	7% in petroleum ether	Methanol–acetone–water (20:4:3)	A	413
Castor oil	Chlorophyll derivatives	Kieselguhr	10% in methanol (dipping)	Methanol–acetone–2-propanol–water–benzene (80:2.5:2.5:15:2)	A	414
	Barbiturates	Cellulose	5% soln. of a mixture (8:2) with Ricilan B (developing)	0.1 M HCl	L	415

1,4-Benzodiazepines		Cellulose	8% in acetone (developing)	9 or 25% citric acid 5 M Acetic acid	A	416
				Phosphate-citric acid buffers pH 2-8	A	417
Peanut oil	Chlorophylls	Kieselguhr + silica gel mixture	14% in isooctane	Methanol-acetone-water (20:4:3)	A	418
Soybean or cottonseed oil	Chlorophylls	Kieselguhr	7% in petroleum ether (developing)	Methanol-acetone-water (20:4:3)	A	419
Olive oil	Chlorophylls a, b	Cellulose	4% in petroleum ether (dipping)	Acetonitrile-tetrahydrofuran-tetrachloromethane-water (70:15:10:5)	A	420
	Plastide pigments	Kieselguhr	8% in diethyl ether (developing)	Acetone-methanol-water (10:38:2)	A	421
	Aromatic polyamines	Cellulose	(Slurry, 65 ml 5% in CCl$_4$ + 25 g cellulose)	0.6-1.2 M HCl-2 M sodium perchlorate (1:10)	A	422
2-Ethylhexyl phosphate	Fatty acids methyl esters Higher fatty acids	Silica gel	5% in diethyl ether	15 and 25% aqueous acetic acid-esters Acetonitrile-acetic acid-water (70:10:25)-acids	A	423
Silicone oil	Triglycerides	Kieselguhr	7% in petroleum ether (dipping)	Methanol-acetonitrile (5:4)-synth. trigl. Methanol-acetonitrile-propionitrile (5:4:1.5)-unsaturated trigl.	A	424

Table 7 (Continued)

Stationary phase	Compounds chromatographed	Sorbent	Method of impregnation[a]	Mobile phase	Type of study[b]	Ref.
Silicone oil	N-Benzoyl-triacetyl derivs. of bases	Silica gel	5% in diethyl ether (dipping)	Acetonitrile–acetic acid–water (70:10:25)	A	425
	Pesticides	Silica gel	5% in diethyl ether (dipping)	Ethanol–acetone–water (1:1:2)	A	426
	Fatty acids and esters Fatty alcohols	Silica gel	5% (developing)	Acetonitrile–acetic acid–water (70:10:25) 70, 80, 90% aqueous acetic acid	A	427
	Porphyrin esters	Silica gel	12.5% in petroleum ether (dipping)	95% aqueous methanol	A	428
	Fatty acid methyl esters	Silica gel	5% in petroleum ether (developing)	95% aqueous methanol	A	429
	2,4-Dinitrophenylhydrazones of aldehydes	Kieselguhr	10% in petroleum ether (dipping)	85% aqueous acetone	A	430
	Pesticides and alkaloids	Cellulose	5% in acetone (dipping)	Water–ethanol–chloroform (56:42:2)	A	431
	Antibiotics	Silica gel		Sodium acetate–veronal buffer pH 7.4	L	432
	Phenolic antioxidants	Silica gel		60% aqueous ethanol	A	433

Compound	Support	Solvent system	Mode	Ref.
Penicillins	Silica gel	5% in diethyl ether (developing—2 hr)		
		30-50% acetone in sodium acetate-veronal buffer pH 7.4	L	434
Cephalosporins		0-25% acetone in sodium acetate-veronal buffer pH 7.4	L	435
Oligosaccharide antibiotics		Aqueous sodium acetate-veronal buffer pH 7.4	L	436
Penicillins Cephalosporins	Silica gel	5% in diethyl ether (developing)		
		Aqueous sodium acetate-veronal buffer pH 7.4	L	221 438
Testosterone esters		54% aqueous acetone or methanol	L	437, 439
Rifamycins		30% aqueous acetone	L	440, 441
Bisdichloro-acetamides		50% aqueous acetone	L	442
Polyenic antibiotics		30% aqueous acetone	L	443
Phenothiazines		Aqueous acetone	L	444
Corticosteroids		24-52% aqueous acetone	L	445
Androgens		16-23.5% aqueous acetone	L	445
Testosterone derivatives		54% aqueous acetone	L	446
Rifamycin derivatives		30 and 40% aqueous acetone	L	447, 448
Phenols		5-55% aqueous acetone	L	449
Steroids		15-65% aqueous acetone	L	450

Table 7 (Continued)

Stationary phase	Compounds chromatographed	Sorbent	Method of impregnation[a]	Mobile phase	Type of study[b]	Ref.
	Sulfonamides	Silica gel Polyamide	5% in diethyl ether (developing)	Sodium acetate-veronal buffer pH 7.4	L	392, 451
	Penicillins	Silica gel		Aqueous acetone	L	389, 452
	1,4-Benzodiazepines			Aqueous acetone	L	395
	3-Aryl-n-butyric acids			60% aqueous acetone	L	453
	Palmitic acid	Silica gel		Acetic acid-acetonitrile (1:1)	A	454
	Phenols	Silica gel	5% in diethyl ether	Acetone-phosphate buffer pH 7.4 mixtures	L	455
	Cardenolides	Silica gel	5% in diethyl ether (developing)	10-50% aqueous acetone	L	391
	Arylpropionic acids	Silica gel	5% in dioxane	50% aqueous acetone	L	456
	Arylaliphatic acids	Silica gel	2.5-7.5% in dioxane	Acetone-buffer pH 3.4 (1:1)	L	457
	Naphthols Acetophenones	Silica gel		25-60% acetone in sodium acetate-veronal buffer pH 7.4	L	458
	Arylaliphatic acids	Silica gel	(Slurry)	Acetone-citrate buffer pH 3.4	L	103
	Alkylphenols	Silica gel		Acetone-phosphate buffer pH 7.4 mixtures	L	459

Arylacetic acids	Silica gel	5% in dioxane	50% aqueous acetone	L	460, 461
5-Nitroimidazoles	Silica gel	5% in diethyl ether	5-30% methanol in 1 M ammonium chloride buffer pH 9 or sodium acetate-veronal buffer pH 3.6	L	462
Xanthone derivatives	Silica gel	5% in diethyl ether	40-85% methanol in 0.1 M glycine buffer pH 13	L	463
Benzoyloxyaryl-aliphatic acids	Silica gel		50% aqueous acetone	L	464, 465
Arylaliphatic acids	Silica gel	(Slurry)	Acetone-citrate buffer pH 3.4 50% aqueous acetone	L	466
Diethylphenyl phosphates	Silica gel	1.5%	40% aqueous acetone	L	396
Dermorphine oligopeptides	Silica gel	5% in diethyl ether (developing)	Acetone in sodium acetate-veronal buffer pH 7.0	L	467
Arylaliphatic acids	Silica gel silanized	5% in diethyl ether	Acetone-citrate buffer pH 3.4 (1:1)	L	468
Nitroimidoazo-thiazoles	Silica gel	5% in diethyl ether	10-70% acetone in 0.1 M gly-cocol buffer pH 13 10-90% methanol in 0.1 M gly-cocol buffer pH 13	L	469
Prostaglandins				L	470
Iodinated aromatic and aryl-aliphatic acids	Silica gel	5% in diethyl ether (developing overnight)	20-50% acetone in citrate buffer pH 2.11	L	471
Colchicine alkaloids	Silica gel	5% in diethyl ether (developing overnight)	Acetone-phosphate buffer pH 7.3 mixtures	L	472

Table 7 (Continued)

Stationary phase	Compounds chromatographed	Sorbent	Method of impregnation[a]	Mobile phase	Type of study[b]	Ref.
	Oligopeptides			Methanol-sodium acetate-veronal buffer pH 7.4 mixtures	L	473
	N,N-Diakyldi-thioxamides			20-30% acetone in phosphate buffer pH 6.64	L	474
	Thiohydrazides			45-70% aqueous methanol	L	475
	1,4-Benzodiazepines	Cellulose	5% in toluene (developing)	Ethanol-acetone-water (2:1:7)	L	416
	Benzophenones		5% (dipping)	65% aqueous methanol	L	476
Decane	Fatty acids and alcohols	Silica gel	15% in light petroleum (developing)	Acetic acid-acetonitrile (1:1)	A	477
Undecane	Fatty acids	Silica gel	15% in petroleum ether (dipping)	Acetic acid-acetonitrile (1:1) 96% aqueous acetic acid	A	478
	Fatty alcohols			Acetic acid-acetonitrile (1:3)		
	Epoxy acids			80% aqueous acetic acid		
	Diglycerides			Chloroform-methanol-water (5:15:1)		
	Triglycerides			Acetone-acetonitrile (7:3)		
	Sterols	Silica gel	15% in petroleum ether (dipping)	Methanol	A	358

Fatty acids	Kieselguhr	10% in petroleum ether (dipping)	Acetic acid–acetonitrile (1:1)	A	479
Fatty acids	Kieselguhr	10% in petroleum ether	Acetic acid–acetonitrile (3:2) Acetic acid–acetonitrile (3:7) + 0.5% Br	A	480
Fatty aldehydes and ketones	Silica gel	15% in petroleum ether (dipping)	70–100% aqueous methanol Acetonitrile 75% aqueous acetonitrile	A	481
Triglycerides Free fatty acids	Calcium sulfate	10% in petroleum ether (dipping)	Acetone–acetonitrile (4:1) Acetic acid–acetonitrile (1:1) Acetic acid–acetonitrile + 0.5% Br	A	482
Cholesterol Desmosterol	Silica gel	15% in petroleum ether	Acetic acid–acetonitrile (1:1)	A	483
Sterol acetates	Kieselguhr	10% in petroleum ether (dipping)	92% aqueous acetic acid	L	484, 485
Sterols			Acetic acid–acetonitrile (1:3) 90% aqueous acetic acid	L L	486 484
Sulfonamides	Polyamide Silica gel	5, 10, 20% in light petroleum (developing 20 hr or slurry)	10% acetone in veronal-acetate buffer pH 7.4	L	451
5-Nitroimidazoles	Silica gel	5% in petroleum ether	5–30% methanol in 1 M ammonium chloride buffer pH 9.0	L	462
2,4-Dinitrophenylhydrazones	Silica gel	10% in petroleum ether (dipping)	25% aqueous methanol	A	360
Steroids	Silica gel	(Developing)	Acetic acid–acetonitrile (4:1)	A	487
Phenylazobenzoates of alcohols (Dodecane)	Silica gel	5% in hexane (dipping)	Acetonitrile–2-butanone (4:1)	A	488

Table 7 (Continued)

Stationary phase	Compounds chromatographed	Sorbent	Method of impregnation[a]	Mobile phase	Type of study[b]	Ref.
Dodecane	Fatty acid methyl esters	Silica gel	10% in hexane	Two-dimensional development	A	489
Tetradecane	Triglycerides	Kieselguhr	5% in petroleum ether (dipping)	Acetone–acetonitrile (4:1)	A	424
	Fatty hydroxy acids	Kieselguhr	5% in petroleum ether	90% aqueous acetic acid	A	490
	Triglycerides	Kieselguhr		Acetone–acetonitrile (4:1; 7:4)	A	491
	Triglycerides Waxes	Kieselguhr	5% in petroleum ether (dipping)	2-Propanol–ethanol–acetic acid–water (8:3:4:2; 8:3:4:1.5)	A	492
	Triglycerides	Kieselguhr	5% in petroleum ether	Acetone–acetonitrile (4:1)	A	480, 493
	Glycerides	Kieselguhr	5% in petroleum ether (dipping)	Acetone–acetonitrile (4:1)	A	494
Hexadecane	Organic thiols and thioethers	Alumina	5% in petroleum ether (dipping)	Methanol–chloroform–water (5:15:1) Acetic acid–acetonitrile (1:3) Chloroform–methanol (1:3)	A	495
	Thiols	Alumina	5% in hexane	Acetonitrile–1-propanol (7:2)	A	496
Paraffin oil	Carotene aldehydes	Silica gel	5% in petroleum ether (dipping)	Methanol	A	497

Compound class	Adsorbent	Pretreatment	Solvent	Method	Ref.
Cholesterol esters	Silica gel		Methyl ethyl ketone-acetonitrile (7:3)	A	498
Ubiquinones K vitamins	Silica gel	5% in petroleum ether	95% aqueous acetone	A	499
Fat-soluble vitamins	Silica gel + kieselguhr (1:1)	5% in petroleum ether	Methanol-2-propanol (9:1)	A	500
Steroids	Celite No. 545	(Spraying)	95% aqueous methanol	A	250
Fat-soluble vitamins	Silica gel		80 and 90% aqueous acetone	A	501
Vitamin E	Kieselguhr	5% in petroleum ether	80% aqueous acetone	A	502
Cholesterol esters	Silica gel	1% in petroleum ether	Acetic acid	A	477
Plastide pigments	Kieselguhr	8% in diethyl ether (development)	Acetone-methanol-water (50:47:3)	A	421
Phenolic antioxidants	Silica gel	10% in petroleum ether (developing)	75% aqueous methanol	A	503
Triglycerides Cholesterol esters	Silica gel	0.5% in ether (dipping)	Acetic acid	A	504
Carotenoids	Silica gel		Methanol-acetone (5:2)	A	505
Plant phloroglucinol butanones	Silica gel		Methanol-formic acid-water (15:2:3)	A	506
Ubiquinones	Silica gel	5% in diethyl ether (developing)	90% aqueous acetone	A	507

Table 7 (Continued)

Stationary phase	Compounds chromatographed	Sorbent	Method of impregnation[a]	Mobile phase	Type of study[b]	Ref.
	Fatty acids	Kieselguhr	10% in ligroin (developing)	50-90% aqueous acetic acid	A	121
	Glycerides			90-100% aqueous acetic acid	A	121
	Resins	Silica gel	5% in petroleum ether	70% aqueous methanol	A	508
	Fatty acids Triglycerides	Kieselguhr	10% in petroleum ether	85% aqueous methanol Acetone-acetonitrile (4:1)	A	509
	Wax alcohols	Kieselguhr	10% in petroleum ether	85% aqueous acetic acid 90% aqueous acetone	A	510
	Fatty acid esters	Kieselguhr	10% in petroleum ether	90% aqueous acetone	A	511
	Ubiquinones Vitamin K	Silica gel + kieselguhr (1:1)	5% in petroleum ether	Methanol-2-propanol (9:1)	A	512
	Plastoquinone Ubiquinone	Silica gel		95% aqueous acetone	A	513
	Phosphorus pesticides	Silica gel	5% in hexane (dipping)	Ethanol-acetone-water (1:1:2)	A	426
	Lutein esters	Cellulose		Acetone-methanol (66:34)	A	514
	2,4-Dinitrophenylhydrazones of aldehydes and ketones	Silica gel	10% (dipping)	65% aqueous acetone	A	515

Fatty acid methyl esters	Kieselguhr	10% in petroleum ether (dipping)	Nitromethane–acetonitrile–acetic acid (75:10:10)	A	516
Fat-soluble food dyes	Starch	10% in petroleum ether (slurry)	Methanol–water–acetic acid (80:15:5; 8:1:1)	A	102
Fat-soluble food dyes	Starch + calcium carbonate	10% in petroleum ether (dipping)	Methanol–water–ammonia (20:5:1)	A	517
Terpene alcohols	Kieselguhr	5% in petroleum ether (dipping)	Acetone–water–paraffin oil (65:35:0.5)	A	518
2,4-Dinitrophenylosazones of α-dicarbonyls	Silica gel	10% in petroleum ether (dipping)	60% aqueous acetone	A	519
Thiocarbamic acid derivatives	Silica gel	5% in diethyl ether	50% aqueous dimethyl formamide	A	268
Vitamins E	Silica gel + kieselguhr (1:1)		Methanol–2-propanol (9:1)	A	520
Ubiquinones	Silica gel	5% in petroleum ether	95% aqueous acetone	A	521
Adonispigments Ketokarotenoids	Cellulose		93% aqueous ethanol	A	522
Isoprenoid quinones	Cellulose + silica gel (66:34)		Acetone–methanol–water (65:33:2)	A	523
Sterol esters	Silica gel	1% in petroleum ether	Acetic acid	A	524

Table 7 (Continued)

Stationary phase	Compounds chromatographed	Sorbent	Method of impregnation[a]	Mobile phase	Type of study[b]	Ref.
	Fatty acid esters	Kieselguhr	In heptane (dipping)	Acetic acid–acetone (2:3) Acetic acid–acetonitrile (2:3)	A	525
	Natural quinones	Silica gel	5% in hexane (developing)	90% aqueous acetone	A	526, 527
	Alkyltrityl amines	Silica gel	5% in hexane (developing)	70% aqueous acetone	L	528
	Fatty acids	Kieselguhr	8% in benzene	Acetic acid–paraffin oil–water (8:1:1)	A	529
	Terpene alcohols	Silica gel		70% aqueous methanol	A	530
	Triglycerides	Kieselguhr	5% in petroleum ether	Acetone–methanol (4:1, 7:4)	A	491, 493, 531
	Triglycerides	Kieselguhr		Acetone–acetonitrile (4:1)	A	532
	Dinitrophenols	Cellulose	10% in diethyl ether 5% in diethyl ether + 1% acetic acid (dipping)	Methanol–acetonitrile–water (30:25:45) as methyl ethers Methanol–acetic acid–water (73:2:25)	A	533
	Isoprenoid alcohols	Kieselguhr	5% in petroleum ether (dipping)	90% aqueous acetone	A	534, 535
	S-Alkylphenyl-thiouronium picrates	Silica gel	10% in petroleum ether	50% aqueous acetone	A	536

3,5-Dinitroben-zoates of alcohols	Silica gel	10% in petro-leum ether (dipping)	80% aqueous acetic acid 65% aqueous acetic acid	A	537
Steroids	Kieselguhr	10% in ligroin (developing)	Aqueous acetic acid	A	370
Polyprenols	Kieselguhr	5% in petro-leum ether	92% aqueous acetone	A	538, 539
Alkaloids	Silica gel	5% in petro-leum ether	60% aqueous methanol	A	365
Glycerides	Kieselguhr		Acetone-methanol-acetic acid (60:40:0.5)	A	540
Dimethylamino-benzeneazobenzo-ates of alcohols	Silica gel	10% in petro-leum ether (developing)	20 or 25% aqueous dimethyl formamide Dimethyl formamide-methanol-water (4:1:1; 8:2:1)	A	243
Alkylmercuridi-thizonates	Corn starch or Avicel SF	(Slurry)	70% aqueous ethanol 75% aqueous methylcellosolve	A	541
Pesticides	Cellulose	5% in hexane (spraying)	50% aqueous acetone	A	542
Fat-soluble food dyes	Cellulose	10% in petro-leum ether	2-Methoxyethanol-methanol-water (55:15:30)	A	543
Polychlorinated biphenyls	Kieselguhr	8%	Acetonitrile-acetone-methanol-water (20:9:20:1) 3x development	A	544
Acetanilides	Silica gel	5% in heptane (developing)	10% aqueous acetone	L	390
Menaquinones	Silica gel Kieselguhr	5% in petro-leum ether (dipping)	97% aqueous dimethyl formamide 93% aqueous acetone	A	545

Table 7 (Continued)

Stationary phase	Compounds chromatographed	Sorbent	Method of impregnation[a]	Mobile phase	Type of study[b]	Ref.
	Androgen esters	Silica gel	5% in hexane	50-80% aqueous acetone	L	546
	Isoprenoid quinones	Kieselguhr	5% in petroleum ether (dipping)	97.5% aqueous acetone	A	547
	Substituted phenylhydrazones	Silica gel	10% in chloroform	Acetone-water-dioxane (1:2:1)	L	548, 549
	Triazinone herbicides	Silica gel		Water-dioxane-acetone (13:10:7)	A	550
	Organomercuri-dithizonates	Cellulose	10% in chloroform (dipping)	65-75% aqueous methylcellosolve	A	551
	Chlorinated pesticides	Silica gel		Aqueous methanol	A	552
	Terpene alcohols	Avicel	5% in petroleum ether (dipping)	65% aqueous acetone	A	553
	Dimethylamino-benzeneazobenzo-ates of higher alcohols and diols	Silica gel	10% (developing)	75% aqueous dimethyl formamide Dimethyl formamide-methanol-water (8:2:1)	A	554
	Chlorinated pesticides	Kieselguhr	8% in petroleum ether (dipping)	Water-acetone-acetonitrile-methanol (1:3:8:8)	A	555
	Triglycerides	Kieselguhr	7.5% in petroleum ether (dipping)	Acetone-acetonitrile (4:1; 7:3)	A	556

Compound	Layer	Reagent	Solvent	Detection	Ref.
Polychlorinated biphenyls	Alumina	8% in petroleum ether (developing)	Acetonitrile-acetone-methanol-water (20:9:20:1)	A	557
Disulfides Xanthates Thioureas	Silica gel	5% (dipping)	Methanol (disulfides) 5% ammonia or 10% sodium acetate (xanthates) Water (thioureas)	A	558
Chlorophyll derivatives	Kieselguhr	8%	Methanol-acetone-2-propanol-water-benzene (35:50:10:10:2)	A	414
Dimethylamino-benzeneazobenzoates of isoprenoid alcohols	Cellulose	5% in hexane	Dimethyl formamide	A	559
Dimethylamino-benzeneazobenzoates of lower alcohols	Silica gel	5%	80% aqueous dimethyl formamide	A	560
Dimethylamino-benzeneazobenzoates of higher alcohols	Silica gel Cellulose	5% in diethyl ether (dipping)	Dimethyl formamide-methanol-water (7:12:0.1; 10:10:1)	A	561
Diphenhydramine	Silica gel	10% in petroleum ether (dipping)	Ethanol-water-conc. ammonia (55:43:2)	A	562
Capsaicinoids	Silica gel	5% in petroleum ether	60% aqueous methanol	A	563
Benzenesulfon-amidopyrimidines	Silica gel	5%	Acetone-phosphate buffer pH 7.4	L	564
Guanidinoalkane-sulfonic acids	Silica gel	5% in hexane	50% aqueous acetone	L	565

Table 7 (Continued)

Stationary phase	Compounds chromatographed	Sorbent	Method of impregnation[a]	Mobile phase	Type of study[b]	Ref.
	Polyprenols	Kieselguhr	5% in petroleum ether	92% aqueous acetone	A	566
	Chlorinated biphenyls	Kieselguhr	8% in hexane (dipping)	Acetonitrile-methanol-acetone-water (20:20:9:1) Acetonitrile-methanol-water (8:9:3)	A	567, 568
	Polychlorinated naphthalenes	Kieselguhr		Acetonitrile-methanol-acetone-water (20:20:9:1) Acetonitrile-methanol-water (8:9:3)	A	569
	Isonicotinic acid hydrazide-type compounds	Silica gel	In hexane	Aqueous acetone	L	570
	Tranquilizers	Silica gel	5% in diethyl ether	90% aqueous acetone	L	571
	Fat-soluble vitamins	Talcum	(Slurry, 2 ml + 45 ml diethyl ether + 30 g talcum)	Water-dioxane-acetone-formaldehyde (85:20:15:25) Acetone-acetic acid (3:2)	A	101
		Cellulose	(Slurry, 3 ml + 100 ml diethyl ether + 15 g cellulose)			
	Polybrominated biphenyls	Kieselguhr		Acetonitrile-methanol-acetone-water (20:20:9:1) Acetonitrile-methanol-water (8:9:3)	A	572

Polyunsaturated esters	Silica gel + kieselguhr (3:7)	10% in petroleum ether (developing)	Formic acid–acetonitrile–acetone (8:8:5)	A	573
ω-Amino acid derivatives	Silica gel	5% in hexane (developing)	20–70% aqueous acetone	L	574
Polyprenols	Kieselguhr		90% aqueous acetone	A	575
Polyprenols	Kieselguhr		92% aqueous acetone	A	576
Pyrazine carbothioamide derivs.	Silica gel		Water (extrapolated to 100% water)	L	577
2-Cyanomethylbenzimidazoles	Silica gel		Water (extrapolated to 100% water)	L	578
Higher aliphatic amines	Kieselguhr	5% in acetone (dipping)	Acetone–17% NH$_4$OH (55:45; 7:3)	A	579
O-Alkyl-O-arylphenylphosphonothioates	Silica gel	5% in hexane	60% aqueous acetone	L	580
Polychlorinated biphenyls	Silica gel	8% in petroleum ether (dipping)	Acetonitrile–acetone–methanol–water (20:9:20:1)	A	581
Vitamin K derivatives	Kieselguhr	5% in petroleum ether (dipping)	50% aqueous acetone	A	502
3-Acyloxy-1,4-benzodiazepines	Silica gel	5% in petroleum ether	60–75% aqueous methanol	T	583
Phenolic acids	Silica gel	(Slurry, 3 g + 70 ml acetone–benzene (1:1) + 20 g silica gel)	5–30% aqueous methanol with 2% acetic acid 5–25% aqueous acetonitrile with 2% acetic acid	A	584

Table 7 (Continued)

Stationary phase	Compounds chromatographed	Sorbent	Method of impregnation[a]	Mobile phase	Type of study[b]	Ref.
	Hypoglycemic sulfonamides	Silica gel	5% in hexane	60-80% aqueous acetone	L	585
	Oxyethylated nonylphenol	Silica gel	5% in hexane	Aqueous acetone	L	586
	Antimalarial sulfonamides	Silica gel	5% in hexane	Acetone-phosphate buffer pH 5	L	587
	Pesticides	Silica gel	5% in hexane	75% aqueous acetone	L	588
	5-Nitroimidazoles	Silica gel	5% in petroleum ether	5-30% methanol in 1 M ammonium chloride buffer pH 9	L	462
	Barbiturates	Silica gel	(Slurry, 3 g + 60 ml (1:1) benzene-acetone + 20 g silica gel)	Aqueous acetone, methanol, or acetonitrile	L	100
	Sterols	Silica gel	5% in hexane	80% aqueous acetone	L	234
	Dothistromin	Silica gel	5% in hexane (developing)	Methanol-water (2:1) contg. 4% of formic acid	A	589
	Naphthalene derivatives	Kieselguhr	15% in benzene	Aqueous methanol	T	590
	Oxyethylated nonylphenol	Silica gel Alumina	5% in hexane	75% aqueous acetone Acetone-1 M NaCl (3:1)	L	591, 592
	Thionophosphates	Silica gel	In hexane	50-80% aqueous methanol	A	593
	Hydroxycinnamic acids	Silica gel	(Slurry)	Buffer solutions pH 3-10	A	594

Ecdysteroids	Silica gel	7.5% in dichloromethane (developing)	50% aqueous methanol	A	595
Triazine herbicides	Silica gel	5% in hexane	Aqueous alcohols, ketones, cellosolves	L	596
n-Alkylphenyl ketones	Silica gel	5% in petroleum ether (dipping)	Aqueous mixtures of acetone, methanol, or dimethyl formamide	L	597
Triazines Quaternary ammonium compounds Phenol derivatives 2,4-Dinitrophenylhydrazones	Silica gel Alumina Cellulose	1, 5, 10% in hexane (developing overnight)	Water 50% aqueous methanol 33 and 50% aqueous acetonitrile	L	598
Carboxamide derivatives	Silica gel	5% in hexane (overnight developing)	50% aqueous methanol	L	599
s-Triazine derivatives	Silica gel	5% in hexane (dipping overnight)	Different aqueous organic solvents	T, L	175
Ecdysteroids Aminophenols Antiinflammatory drugs	Silica gel	7.5% in hexane or dichloromethane (developing)	50% aqueous methanol 50% aqueous ethanol 50% aqueous ethanol	A	600, 601
Acrylic and methacrylic acid esters	Silica gel	5%	40-90% aqueous dimethyl formamide Dimethyl formamide-methanol-water (9:5:1)	A	602
Benzophenones	Silica gel		60-65% aqueous acetone	L	603

Table 7 (Continued)

Stationary phase	Compounds chromatographed	Sorbent	Method of impregnation[a]	Mobile phase	Type of study[b]	Ref.
	Dinitrooctylphenyl crotonate	Silica gel	5% in hexane	50-75% aqueous acetone or methanol	L	604
				20-70% aqueous 1-propanol		
				30-70% aqueous 2-propanol		
				50-70% Tetrahydrofuran		
	Phenol derivatives	Silica gel	5% in hexane	1 M calcium chloride	L	605
	Naphthols	Kieselguhr	15% in benzene	90% aqueous methanol	T	606
	Crown ethers	Silica gel	5% in hexane (dipping overnight)	Aqueous acetone, acetonitrile, methanol	L	607
	Dioleylphosphatidylcholine	Cellulose	5% in hexane (dipping overnight)	Water-ethanol (12:13) mixtures contg. nonionic tensides	T	608
	Antimycin A	Silica gel	5% in hexane	30-50% aqueous methanol	L	609
				50% aqueous acetonitrile		
	Higher alcohols and acids	Kieselguhr	15% in benzene	Aqueous acetic acid	L	610
	Different compounds	Celite No. 545 silanized	(Slurry)	70-90% aqueous methanol	T	400
	Amino acids	Silica gel Alumina Kieselguhr	10% in hexane	Aqueous methanol	L	611

Organic acids		Silica gel	7.5% in 0.07 M solution of the ion pair reagent	50% aqueous methanol	T	612
Acrylic acid esters		Silica gel	5% in petroleum ether	Dimethyl formamide–water–methanol (2:1:1; 2:2:1)	A	345
Aniline derivatives		Silica gel	5% in hexane (dipping)	Aqueous methanol or salt solutions	L	613
Fatty oils		Kieselguhr Cellulose	In petroleum ether (developing)	Acetic acid	A	614
Phospholipid adducts		Silica gel Kieselguhr (10:9)		30–70% aqueous 2-propanol	L	615
Fully protected peptides		Silica gel	5% in hexane (dipping overnight)	25–65% aqueous methanol 40% methanol in aqueous salt solutions	L	616
5-Nitroimidazoles	Squalan	Silica gel	5% in petroleum ether	1 M ammonium chloride buffer pH 9 in 30% methanol	L	462
Vitamin A compounds		Kieselguhr		85% aqueous ethanol	A	617
Alcohols as o-nitrophenyl urethans	Vaseline	Silica gel	5% in heptane (dipping)	95% aqueous methanol	A	618
Carbazole halogen derivatives	Apiezon	Silica gel	5% in hexane (dipping)	Dioxane–tetrahydrofuran–water (3:2:5) Pyridine–ethanol–water (7:4:10)	A	619
Metallocarboranes	Nujol	Cellulose	10% in benzene (developing)	94% aqueous methanol	A	620

Table 7 (Continued)

Stationary phase	Compounds chromatographed	Sorbent	Method of impregnation[a]	Mobile phase	Type of study[b]	Ref.
Nujol	Chlorophyll derivatives	Cellulose	8% in dichloromethane (dipping)	Methanol-acetone-2-propanol-water (35:50:10:2) Methanol-acetone-water-benzene (60:20:15:10) Methanol-acetone-2-propanol-water-benzene (30:10:5:5:1; 60:5:5:10:10)	A	621
	Ecdysteroids	Silica gel	7.5% in dichloromethane	50% aqueous methanol	A	622
1-Bromonaphthalene	Nitrophenols	Silica gel Cellulose	10% in chloroform	Ethanol-buffers pH 3, 8, 10 (1:2) 80% aqueous acetic acid	A, T	314, 366
Tetralin	2,4-Dinitrophenylhydrazones of aldehydes	Silica gel		Methanol-water-acetic acid-benzene-tetralin (9:1:1:1:1)	A	338

[a] $X\% = X\%$ solution.
[b] A = analytical separation; L = lipophilicity or QSAR studies; T = theoretical study.

(see also Section II.B). Comparison of stationary phases for reversed phase thin-layer chromatography for correlation between structure and biological response of probiotics was also compiled [629].

Recently, a number of f values (hydrophobic fragmental constants), which also allow the calculation of log P values of organic chemicals, were presented [13,630].

The chromatographic methods have some advantages. They are rapid and relatively simple, and use very small quantities of the compounds, which need not be very pure. However, in reversed phase thin-layer chromatography the actual values of lipophilicity (R_M values) depend on the chromatographic conditions. The support particles partially retain their original adsorptive characteristics even after coating with the stationary phase [40,462,598], and the R_M values can change with the quantity [598] and the quality [306,462] of the stationary phase used. When the compound also contains one or more dissociable substituents, the pH value of the eluent [156,631] and the salt concentration [164,222] modify its lipophilicity. The organic modifier in the mobile phase also influences the retention [596,632,633].

Papers in which lipophilicity measurements of QSAR studies for different groups of compounds have been described are referred to in Table 7.

Analogous to thin-layer chromatography, lipophilicity measurements and QSAR studies are carried out using partition HPLC [51, 235,634-643] and gas chromatographic [644-648] methods.

IV. CONCLUSIONS

The principles, techniques, and applications of partition thin-layer chromatography on layers impregnated with organic stationary liquids, collected from the literature and from the author's own experience, have been summarized in this chapter. This type of thin-layer chromatography played an important role in achieving very significant analytical separations. Since the introduction of chemically bonded layers, thin-layer chromatography on impregnated plates has been largely ignored. However, the impregnated plates can give just as good chromatographic results as their bonded equivalents. They can be prepared very simply in the laboratory and their great advantage is the low cost of materials [649,650]. At present, it is used very often in lipophilicity and QSAR studies.

Factors playing a significant role in creating conditions for a partition process in thin-layer chromatography, using organic liquid stationary phases, have been summarized and discussed. The sorbent used as support of the stationary liquid and its activity, the quality and quantity of the stationary phase, the method of covering the layer with the stationary phase, the quality and composition of the mobile phase, the role of the vapor phase, and the properties of the solute all represent individual parts of this complex system. To establish conditions for a proper migration rate of the solutes in the form of well-defined spots with sufficient R_F differences, the original purely partition process can be modified by adsorption, ion pairing, ion exchange, salting out, inclusion, and by complex formation. These variations can bring about an improvement of the analytical separations. It can, however, represent a complication in using the chromatographic data for lipophilicity measurements. The partition thin-layer chromatographic system must be considered to be very complex, e.g., as illustrated in Fig. 6 showing the mutual interactions of its individual parts. In fact, Fig. 6 is a simplification, because it was shown in Section II.E that the mobile phase can be composed of two or more solvents which can also interact, forming species which are different from the originally present components.

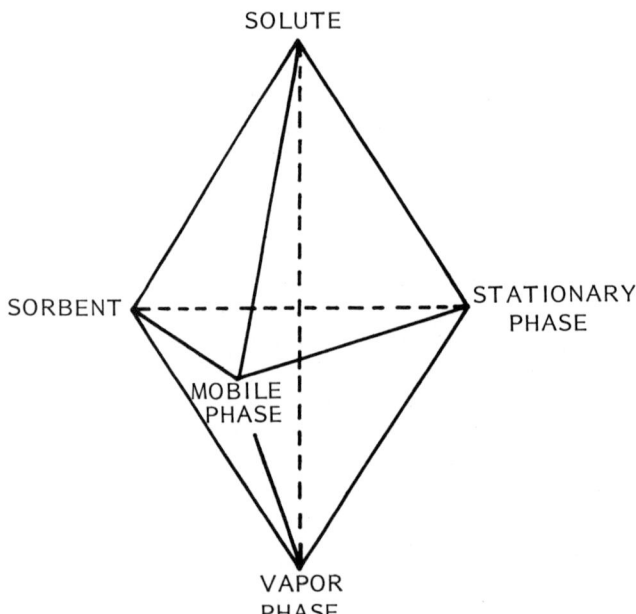

Fig. 6 Schematic representation of the chromatographic system using thin layers impregnated with organic stationary phases.

REFERENCES

1. Ž. Procházka, in *Stationary Phase in Paper and Thin-Layer Chromatography* (K. Macek and I. M. Hais, eds.), Publ. House Czech. Acad. Sci., Prague, 1965, pp. 180, 218.
2. J. W. Copius-Peereboom, in *Stationary Phase in Paper and Thin-Layer Chromatography* (K. Macek and I. M. Hais, eds.), Publ. House Czech. Acad. Sci., Prague, 1965, p. 224.
3. A. Waksmundzki and R. Manko, in *Stationary Phase in Paper and Thin-Layer Chromatography* (K. Macek and I. M. Hais, eds.), Publ. House Czech. Acad. Sci., Prague, 1965, p. 221.
4. R. Neher, *Steroid Chromatography*, Elsevier, Amsterdam, 1964.
5. E. Heftmann, *Chromatography of Steroids*, J. Chromatogr. Library, Vol. 8, Elsevier, Amsterdam, 1976.
6. B. P. Lisboa, Meth. Enzymol. *15*, 3 (1969).
7. A. Baerheim Svendsen and R. Verpoorte, *Chromatography of Alkaloids. Thin-Layer Chromatography*, J. Chromatogr. Library, Vol. 23, Elsevier, Amsterdam, 1983.
8. H. K. Mangold, J. Am. Oil Chem. Soc. *38*, 708 (1961); in E. Stahl, ed., *Thin-Layer Chromatography: A Laboratory Handbook*, 2nd ed., Springer-Verlag, Berlin, 1969, pp. 374, 409.
9. U. A. T. Brinkman and G. de Vries, J. High Resolut. Chromatogr. Chromatogr. Commun. *2*, 79 (1979).
10. G. L. Biagi, A. M. Barbaro, and M. C. Guerra, Adv. Chem. Ser. *114*, 61 (1972).
11. E. Tomlinson, J. Chromatogr. *113*, 1 (1975).
12. R. Kaliszan, J. Chromatogr. *220*, 71 (1981).
13. R. F. Rekker, J. Chromatogr. *300*, 109 (1984).
14. G. A. Howard and A. J. P. Martin, Biochem. J. *46*, 532 (1950).
15. R. M. Scott, J. Chromatogr. Sci. *11*, 129 (1973).
16. V. Rodwell, J. Chromatogr. Sci. *15*, 332 (1977).
17. K. E. Bij, Cs. Horváth, W. R. Melander, and A. Nahum, J. Chromatogr. *203*, 65 (1981).
18. H. J. Issaq, in *New Approaches in Liquid Chromatography*, Proc. Symp. Adv. Liquid Chromatogr. (H. Kalász, ed.), Szeged, 1982. Akadémiai Kiadó, Budapest, 1984, p. 109.
19. R. Consden, A. H. Gordon, and A. J. P. Martin, Biochem. J. *38*, 224 (1944).
20. E. C. Bate-Smith and R. G. Westall, Biochim. Biophys. Acta *4*, 427 (1950).
21. E. Soczewinski, A. Waksmundzki, and R. Manko, in *Stationary Phase in Paper and Thin-Layer Chromatography* (K. Macek and I. M. Hais, eds.), Publ. House Czech. Acad. Sci., Prague, 1965, p. 278.
22. J. Gasparič, J. Chromatogr. *47*, 51 (1970).
23. E. Tomlinson and J. C. Dearden, J. Chromatogr. *106*, 481 (1975).

24. Z. Juvancz, M. Vajda, T. Cserháti, and B. Bordás, Proc. Int. Symp. TLC with Special Emphasis on Overpressured Layer Chromatography, Szeged, Sept. 10-12, 1984 (E. Tyihák, ed.), p. 152.
25. J. Gasparič, J. Chromatogr. 196, 391 (1980).
26. A. Hulshoff and J. H. Perrin, J. Chromatogr. 129, 263 (1976).
27. K. K. Unger, Porous Silica: Its Properties and Use as Support in Column Liquid Chromatography, J. Chromatogr. Library, Vol. 16, Elsevier, Amsterdam, 1979.
28. J. J. Fripiat, Soluble Silicates (J. S. Falcone, Jr., ed.), ACS Symposium Series, No. 194, Washington, D.C., 1982, p. 165.
29. L. R. Snyder, Principles of Adsorption Chromatography. The Separation of Nonionic Organic Compounds, Marcel Dekker, New York, 1968.
30. L. R. Snyder, Anal. Chem. 46, 1384 (1974).
31. R. P. W. Scott and P. Kučera, J. Chromatogr. 171, 37 (1979)
32. S. Hara, M. Hirasawa, S. Miyamoto, and A. Ohsawa, J. Chromatogr. 169, 117 (1979).
33. L. R. Snyder and H. Poppe, J. Chromatogr. 184, 363 (1980).
34. R. P. W. Scott, J. Chromatogr. 122, 35 (1976); Faraday Symp. Chem. Soc. 15, 49 (1981).
35. L. R. Snyder and J. L. Glajch, J. Chromatogr. 248, 165 (1982).
36. J. H. Knox, J. N. Done, A. F. Fell, M. T. Gilbert, A. Pryde, and R. A. Wall, High-Performance Liquid Chromatography, Edinburgh University Press, Edinburgh, 1979.
37. J. Oscik and J. K. Rozylo, Chromatographia 4, 516 (1971).
38. J. K. Rozylo, G. Chojnacka, and H. Kolodziejczyk, J. High Resolut. Chromatogr. Chromatogr. Commun. 3, 139 (1982).
39. J. Pitra, in Stationary Phase in Paper and Thin-Layer Chromatography (K. Macek and I. M. Hais, eds.), Publ. House Czech. Acad. Sci., Prague, 1965, p. 211.
40. W. F. Van der Giessen and L. H. M. Janssen, J. Chromatogr. 237, 199 (1982).
41. R. N. Nikolov, J. Chromatogr. 286, 147 (1984).
42. A. Nahum and Cs. Horváth, J. Chromatogr. 203, 53 (1981).
43. L. Nondek, B. Buszewski, and D. Berek, J. Chromatogr. 360, 241 (1986).
44. H. Colin, A. Krstulovič, G. Guiochon, and Z. Yun, J. Chromatogr. 255, 295 (1983).
45. M. Popl, Le Duy Ky, and J. Strnadová, J. Chromatogr. Sci. 23, 95 (1985).
46. E. Bayer and A. Paulus, J. Chromatogr. 400, 1 (1987).
47. T. Welsch, H. Frank, H. Zwanziger, S. Liebisch, and W. Engewald, Chromatographia 19, 457 (1984).
48. J. C. Kraak and P. Bijster, J. Chromatogr. 143, 499 (1977).
49. W. R. Melander, J. Stovenken, and Cs. Horváth, J. Chromatogr. 185, 111 (1979).

50. J. P. Crombeen, J. C. Kraak, and H. Poppe, J. Chromatogr. 167, 219 (1978).
51. J. M. McCall, J. Med. Chem. 18, 549 (1975).
52. N. H. C. Cooke and K. Olsen, J. Chromatogr. Sci. 18, 512 (1980).
53. K. Karch, I. Sebestian, and I. Halasz, J. Chromatogr. 122, 3 (1976).
54. C. Gonnet and M. Marichy, Analusis 7, 204 (1979); Proc. Third Int. Symp. Instrum. High Perform. Thin-Layer Chromatogr., 1985, p. 49.
55. J. Gasparič, J. Chromatogr. 454, 352 (1988).
56. L. Hörhammer, H. Wagner, and K. Macek, Chromatogr. Rev. 9, 103 (1967).
57. H. Wagner, L. Hörhammer, and K. Macek, J. Chromatogr. 31, 455 (1967).
58. N. D. Gyanchandani, I. J. McGilveray, and D. W. Hughes, J. Pharm. Sci. 59, 224 (1970).
59. S. M. Partridge, Nature 158, 270 (1946).
60. F. Brown and L. P. Hall, Nature 166, 66 (1950).
61. J. Boldingh, Experientia 4, 270 (1948); Rec. Trav. Chim. 69, 247 (1950).
62. J. V. Koštíř and K. Slavík, Chem. Listy 44, 17 (1950); Collect. Czech. Chem. Commun. 15, 17 (1950).
63. A. Zaffaroni, R. B. Burton, and E. H. Keutmann, Science 111, 6 (1950).
64. A. Leo, C. Hansch, and D. Elkins, Chem. Rev. 71, 526 (1971).
65. C. Hansch, J. P. Björkroth, and A. Leo, J. Pharm. Sci. 76, 663 (1987).
66. J. Reichelt, Pharmazie 13, 24 (1958).
67. J. Reichelt and S. Kudrnáč, J. Chromatogr. 87, 433 (1973).
68. J. Reichelt, Českosl. Farm. 25, 213 (1976).
69. E. Soczewinski and W. Maciejewicz, Chem. Anal. (Warsaw) 15, 1199 (1970).
70. M. Przyborowska, Chem. Anal. (Warsaw) 16, 1217 (1971); 20, 591, 1191 (1970); 21, 449 (1976); Acta Polon. Pharm. 32, 173 (1975).
71. M. Przyborowska and E. Soczewinski, J. Chromatogr. 42, 516 (1969).
72. W. Holstein and H. Hemetsberger, Chromatographia 15, 186, 251 (1982).
73. V. M. Bhatnagar, Acta Chim. Hungar. 46, 179 (1965).
74. V. M. Bhatnagar and A. Liberti, J. Chromatogr. 18, 177 (1965).
75. J. Kuczynski, Chem. Anal. (Warsaw) 26, 277 (1981).
76. L. Lepri, P. G. Desideri, and D. Heimler, J. Chromatogr. 153, 77 (1978); 155, 119 (1978); 169, 271 (1979); 173, 119 (1979); 195, 65, 339 (1980).

77. R. Däppen, H. Arm, and V. R. Meyer, J. Chromatogr. *373*, 1 (1986).
78. C. Testa, J. Chromatogr. 5, 236 (1961).
79. E. Cerrai, Chromatogr. Rev. 6, 129 (1964).
80. E. Cerrai and G. Ghersini, Adv. Chromatogr. 9, 3 (1970).
81. U. A. Th. Brinkman and G. de Vries, J. Chem. Educ. 244 (1972).
82. U. A. Th. Brinkman, Progr. Sep. Purif. 4, 241 (1971).
83. H. R. Leene, G. de Vries, and U. A. Th. Brinkman, J. Chromatogr. *80*, 221 (1973); 57, 173 (1971).
84. S. P. Srivastava and V. K. Gupta, Chromatographia *12*, 496 (1979).
85. V. Miscov, L. Rosca, and E. Nichiforescu, Farmacia (Bucharest) *21*, 499 (1973).
86. M. Przyborowska, Chem. Anal. (Warsaw) 27, 125 (1982).
87. J. Reichelt and S. Kudrnáč, Českosl. Farm. 23, 13 (1974).
88. R. J. T. Graham, L. S. Bark, and J. Daly, J. Chromatogr. *33*, 107 (1968).
89. J. Żurkowska and A. Ozarowski, Acta Polon. Pharm. *22*, 83 (1965).
90. L. S. Bark and R. J. T. Graham, J. Chromatogr. *27*, 131 (1967).
91. E. Knappe, D. Peteri, and I. Rohdewald, Z. Anal. Chem. *199*, 270 (1964).
92. B. Rittich and M. Šimek, Chem. Zvesti *30*, 208 (1976).
93. J. Richter and D. Ropte, Pharmazie *21*, 495 (1966); 24, 601 (1969).
94. I. P. G. Wirotama and K. H. Ney, J. Chromatogr. *61*, 166 (1971).
95. W. Diemair, K. Pfeilsticker, and I. Holscher, Z. Anal. Chem. *234*, 418 (1968).
96. E. Knappe and I. Rohdewald, Z. Anal. Chem. *208*, 195 (1965).
97. L. S. Bark and R. J. T. Graham, Proc. SAC Conf., Nottingham, 1965; Heffer, Cambridge, 1966, p. 112.
98. A. Hulshoff and J. H. Perrin, J. Chromatogr. *129*, 249 (1976).
99. K. Egger, Planta *58*, 664 (1962).
100. L. Ekiert, Z. Grodzinska-Zachwieja, and J. Bojarski, Chromatographia *13*, 472 (1980).
101. N. Perišić-Janjić, S. Petrović, and P. Hadzić, Chromatographia *9*, 130 (1976).
102. J. Davídek and G. Janíček, J. Chromatogr. *15*, 542 (1964).
103. M. Kuchař, V. Rejholec, B. Brůnová, and M. Jelínková, J. Chromatogr. *195*, 329 (1980).
104. E. Stahl, in *Thin-Layer Chromatography: A Laboratory Handbook*, 2nd ed. (E. Stahl, ed.), Springer-Verlag, Berlin, 1969, p. 50.

105. J. Gasparič and J. Borecký, J. Chromatogr. 5, 466 (1961).
106. J. Gasparič, Acta Chimica (Budapest) 27, 221 (1961).
107. J. Churáček, J. Chromatogr. 33, 45 (1968).
108. E. Stahl, G. Becker, and V. Brüderle, J. Chromatogr. 129, 41 (1976).
109. J. H. Dhont and G. J. C. Mulders-Dijkman, Analyst (London) 94, 1090 (1969).
110. J. Dhont, J. Chromatogr. 238, 465 (1982).
111. H. M. Chawla, S. S. Chibber, and U. Khera, J. Chromatogr. 111, 246 (1975).
112. H. M. Chawla, S. S. Chibber, and R. Saigal, J. Chromatogr. 138, 243 (1977).
113. H. M. Chawla and S. S. Chibber, Sci. Cult. 48, 105 (1982); Chem. Abstr. 97, 188383g (1982).
114. H. M. Chawla, I. Gambhir, and L. Kathuria, J. Chromatogr. 188, 289 (1980).
115. Y. Suzuki, Y. Yamazaki, and T. Takeuchi, Bunseki Kagaku 20, 824, 1158 (1971); 19, 926 (1970).
116. T. Takeuchi, Y. Suzuki, and H. Okazaki, Bunseki Kagaku 21, 1149 (1972).
117. E. von Arx, J. Chromatogr. 33, 217 (1968).
118. C. Guinchard, M. Baud, J. J. Panouse, and M. Porthault, J. Liquid Chromatogr. 5, 1103 (1982).
119. D. Volkmann, J. High Resolut. Chromatogr. Chromatogr. Commun. 5, 134 (1982).
120. E. Soczewinski and R. Manko, J. Chromatogr. 33, 40 (1968).
121. L. Anker and D. Sonanini, Pharm. Acta Helv. 37, 360 (1962).
122. A. C. Breyer, M. Fischl, and E. J. Seltzer, J. Chromatogr. 82, 37 (1973).
123. J. J. Sabatka, D. A. Brent, J. Murphy, J. Charles, J. Vance, and M. H. Gault, J. Chromatogr. 125, 523 (1976).
124. K. Dadáková, J. Hartl, and K. Waisser, J. Chromatogr. 254, 277 (1983).
125. L. S. Bark, J. Daly, and R. J. T. Graham, Int. Symp. IV Chromatogr., Electrophorese, Presses Académiques Europeénes, Brussels, 1968, p. 128.
126. R. J. T. Graham and J. Daly, J. Chromatogr. 46, 187 (1970).
127. J. Vámos and A. Brantner, Int. Symp. V Chromatogr., Electrophorese, Bruxelles, 1970 (Publ. 1971) 228.
128. J. Vámos, A. Brantner, Gy. Szász, and A. Végh, Gyógyszerészet 15, 371 (1971); 16, 61 (1972).
129. J. Vámos, A. Brantner, Gy. Szász, and A. Végh, Gyógyszerészet 18, 59 (1974).
130. J. Gasparič, Z. Kalousková, and P. Nouzovská, unpublished results.

131. K. Macek, in *Stationary Phase in Paper and Thin-Layer Chromatography* (K. Macek and I. M. Hais, eds.), Publ. House Czech. Acad. Sci., Prague, 1965, p. 166.
132. L. Horner, W. Emrich, and A. Kirschner, Z. Elektrochem. 56, 987 (1952).
133. G. Ackermann and J. Michal, in *Stationary Phase in Paper and Thin-Layer Chromatography* (K. Macek and I. M. Hais, eds.), Publ. House Czech. Acad. Sci., Prague, 1965, p. 169.
134. J. Michal and G. Ackermann, Talanta 11, 441, 451 (1964); 12, 171 (1965).
135. M. Brenner, in *Stationary Phase in Paper and Thin-Layer Chromatography* (K. Macek and I. M. Hais, eds.), Publ. House Czech. Acad. Sci., Prague, 1965, p. 263.
136. A. Niederwieser and M. Brenner, Experientia 21, 50 (1965).
137. F. Geiss and S. Sandroni, J. Chromatogr. 33, 201 (1968).
138. E. D. Katz, K. Ogan, and R. P. W. Scott, J. Chromatogr. 352, 67 (1986).
139. É. János, *Proc. Int. Symp. Special Emphasis on Overpressured Layer Chromatography (OPLC)*, Szeged, Sept. 10-12, 1984 (E. Tyihák, ed.), p. 282.
140. E. Soczewinski and C. A. Wachtmeister, J. Chromatogr. 7, 311 (1962).
141. H. Szumilo and E. Soczewinski, J. Chromatogr. 94, 219 (1974).
142. E. Soczewinski and M. Ciszewska, J. Chromatogr. 96, 163 (1974).
143. E. Soczewinski, J. Iskierko, and J. Klimek, Chromatographia 9, 328 (1976).
144. J. Gasparič, in *Stationary Phase in Paper and Thin-Layer Chromatography* (K. Macek and I. M. Hais, eds.), Publ. House Czech. Acad. Sci., Prague, 1965, p. 176.
145. J. Gasparič and J. Churáček, *Laboratory Handbook of Paper and Thin-Layer Chromatography*, Ellis Horwood Ltd., Chichester, 1978, p. 28.
146. M. L. Buchmann and U. W. Kesselring, Pharm. Acta Helv. 56, 166 (1981).
147. J. Gasparič and M. Večeřa, Collect. Czech. Chem. Commun. 22, 1426 (1957).
148. K. Macek, in *Pharmaceutical Applications of Paper and Thin-Layer Chromatography* (K. Macek, ed.), Elsevier, Amsterdam, 1972, p. 34.
149. J. Gasparič, J. Zimák, I. Nádvorníková-Zdráhalová, and J. Volke, *Proc. 2nd Symp. Chromatogr.*, Oct. 1975, Dům Techn. ČVTS, Praha, 1976, p. 61.
150. H. Kalász and Cs. Horváth, J. Chromatogr. 239, 423 (1982).
151. H. Kalász, J. Nagy, and J. Knoll, in *Chromatography and Mass Spectrometry in Biological Sciences* (A. Frigerio, ed.), Elsevier, Amsterdam, 1983, p. 203.

152. H. Kalász, L. Kerecsen, and J. Nagy, J. Chromatogr. 316, 95 (1984).
153. P. Davídková and J. Gasparič, J. Chromatogr. 410, 33 (1987).
154. T. Cserháti, Chromatographia 18, 18 (1984).
155. J. Gasparič (unpublished results).
156. B. Rittich, M. Polster, and O. Králík, J. Chromatogr. 197, 43 (1980).
157. W. Butte, C. Fooken, R. Klussmann, and D. Schuller, J. Chromatogr. 214, 59 (1981); Kontakte 3, 25 (1982).
158. T. Cserháti, Chromatographia 18, 318 (1984).
159. É. János, T. Cserháti, E. Tyihák, G. Simon, and Z. Tuba, Proc. Symp. Analysis of Steroids, Eger, Hungary, 1981, p. 447.
160. É. János, B. Bordás, T. Cserháti, and G. Simon, in Chromatograpy: The State of the Art (H. Kalász and L. S. Ettre, eds.), Akadémiai Kiadó, Budapest, p. 751.
161. T. Cserháti, B. Bordás, M. Dévai, and A. Kuszmann, in Chromatography, The State of the Art (H. Kalász and L. S. Ettre, eds.), Akadémiai Kiadó, Budapest, p. 601.
162. Y. M. Darwish, T. Cserháti, and Gy. Matolcsy, Acta Phytopat. Acad. Sci. Hung. 17, 203 (1982); Proc. 3rd Danube Symp. on Chromatography, Siófok, 1981, p. 239.
163. É. János, T. Cserháti, and E. Tyihák, J. High Resolut. Chromatogr. Chromatogr. Commun. 5, 634 (1982); in Proc. 2nd Int. Symp. Instrum. HPTLC (R. E. Kaiser, ed.), Interlaken, May 2-6, 1982, p. 186.
164. T. Cserháti, Y. M. Darwish, and Gy. Matolcsy, J. Chromatogr. 241, 223 (1982).
165. É. János, Advances in Liquid Chromatography, Int. Symp., Szeged, Sept. 10-14, 1984, Abstracts, p. 43.
166. M. Gazdag, G. Szepesi, K. Varsányi-Riedl, Z. Végh, and Zs. Pap-Sziklay, J. Chromatogr. 328, 279 (1985).
167. J. Draffehn, K. Ponsold, and B. Schönecker, J. Chromatogr. 216, 69 (1981).
168. N. El Tayar, H. van de Waterbeemd, and B. Testa, J. Chromatogr. 320, 293, 305 (1985).
169. J. L. Glajch, J. J. Kirkland, K. M. Squire, and J. M. Minor, J. Chromatogr. 199, 57 (1980).
170. P. J. Schoenmakers, A. C. J. H. Drouen, H. A. H. Billiet, and L. de Galan, Chromatographia 15, 688 (1982).
171. A. C. J. H. Drouen, H. A. H. Billiet, P. J. Schoenmakers, and L. de Galan, Chromatographia 16, 48 (1982).
172. P. Jandera, H. Colin, and G. Guiochon, Chromatographia 16, 132 (1982).
173. P. R. Haddad and S. Sekulic, J. Chromatogr. 392, 65 (1987).
174. M. Gazdag, G. Szepesi, and E. Szeleczki, J. Chromatogr. 454, 83 (1988).

175. T. Cserháti and B. Bordás, J. Chromatogr. 286, 131 (1984).
176. R. K. Gilpin, S. S. Yang, and G. Werner, J. Chromatogr. Sci. 26, 388 (1988).
177. M. Puttemans, L. Dryon, and D. L. Massart, Anal. Chim. Acta 165, 245 (1984).
178. S. Przeszlakowski, Chromatogr. Rev. 15, 29 (1971).
179. D.-G. Volkmann, Topics Curr. Chem. 126, 51 (1984).
180. B. A. J. Bidlingmeyer, J. Chromatogr. Sci. 18, 525 (1980).
181. J. H. Knox and G. R. Laird, J. Chromatogr. 122, 17 (1976).
182. E. Tomlinson, T. M. Jefferies, and C. M. Riley, J. Chromatogr. 159, 315 (1978).
183. J. C. Kraak, K. M. Jonker, and J. F. K. Huber, J. Chromatogr. 142, 671 (1977).
184. B. F. Nilsson and O. Samuelson, J. Chromatogr. 212, 1 (1981).
185. G. Schill, K. O. Borg, R. Modin, and B.-A. Persson, in Analytical Chemistry. Essays in Memory of Anders Ringbom (E. Wänninen, ed.), Pergamon Press, Oxford, 1977, p. 379.
186. R. Gloor and E. L. Johnson, J. Chromatogr. Sci. 15, 413 (1977).
187. K.-G. Wahlund and I. Beijersten, J. Chromatogr. 149, 313 (1978).
188. W. R. Melander, K. Kalghatgi, and Cs. Horváth, J. Chromatogr. 201, 201 (1980).
189. A. Bartha, Gy. Vigh, H. Billiet, and L. de Galan, J. Chromatogr. 303, 29 (1984); Chromatographia 20, 587 (1985).
190. J. Staahlberg, Chromatographia 24, 820 (1987).
191. J. Staahlberg and A. Furaengen, Chromatographia 24, 783 (1987).
192. H. Lamparczyk and R. J. Ochocka, Chromatographia 23, 337 (1987).
193. I. M. Johansson, K.-G. Wahlund, and G. Schill, J. Chromatogr. 149, 281 (1978).
194. E. Tomlinson, C. M. Riley, and T. M. Jefferies, J. Chromatogr. 173, 89 (1979).
195. C. M. Riley, E. Tomlinson, and T. M. Jefferies, J. Chromatogr. 185, 197 (1979).
196. M. Gazdag, G. Szepesi, and M. Hernyes, J. Chromatogr. 316, 267 (1984).
197. J. A. Troke and I. D. Wilson, J. Chromatogr. 360, 236 (1986).
198. R. J. Ruane, I. D. Wilson, and J. A. Troke, J. Chromatogr. 369, 168 (1986).
199. D. W. Armstrong and K. H. Bui, J. Liquid Chromatogr. 5, 1043 (1982).
200. A. Berthod and J. G. Dorsey, Analysis 16, 75 (1988).
201. A. Berthod and A. Roussel, J. Chromatogr. 449, 349 (1988).
202. Z. Mao and Q. Zhang, Fenxi Huaxne 12, 455 (1984); Chem. Abstr. 101, 78939m (1984).

203. S. Lewis and I. D. Wilson, J. Chromatogr. *312*, 133 (1984).
204. J. Szejtli, *Cyclodextrins and Their Inclusion Complexes*, Akadémiai Kiadó, Budapest, 1982.
205. W. G. Burkert, C. N. Owensby, and W. L. Hinze, J. Liquid Chromatogr. *4*, 1065 (1981).
206. T. Cserháti, É. Fenyvesi, and J. Szejtli, Acta Biochim. Biophys. Acad. Sci. Hung. *18*, 60 (1983).
207. T. Cserháti, B. Bordás, É. Fenyvesi, and J. Szejtli, J. Chromatogr. *259*, 107 (1983); J. Inclusion Phenom. *1*, 53 (1983).
208. T. Cserháti, L. Szente, and J. Szejtli, J. High Resolut. Chromatogr. Chromatogr. Commun. *7*, 635 (1984).
209. J. Debowski, G. Grassini-Strazza, and D. Sybilska, J. Chromatogr. *349*, 131 (1985).
210. T. Cserháti, J. Bojarski, É. Fenyvesi, and J. Szejtli, J. Chromatogr. *351*, 356 (1986).
211. T. Cserháti, J. Szejtli, and É. Fenyvesi, J. Chromatogr. *439*, 383 (1988).
212. D. W. Armstrong, F.-Y. He, and S. M. Han, J. Chromatogr. *448*, 345 (1988).
213. D. W. Armstrong, J. R. Faulkner, Jr., and S. M. Han *452*, 323 (1988).
214. A. Alak and D. W. Armstrong, Anal. Chem. *58*, 582 (1986).
215. M. Gazdag, G. Szepesi, and L. Huszár, J. Chromatogr. *351*, 128 (1986); *371*, 227 (1986); *436*, 31 (1988).
216. W. L. Hinze, Sep. Purif. Meth. *10*, 159 (1981).
217. S. Krýsl and E. Smolková-Keulemansová, J. Chromatogr. *349*, 167 (1985).
218. Z. Vacek and J. Staněk, Collect. Czech. Chem. Commun. *28*, 264 (1963); *29*, 3167 (1964).
219. Z. Vacek, Z. Šťota, and J. Staněk, J. Chromatogr. *19*, 572 (1965).
220. E. Sanchez-Moyano, J. M. Plá-Delfina, and M. Herráez, Ciencia i Técn. *5*, 121 (1986).
221. G. L. Biagi, A. M. Barbaro, and M. C. Guerra, J. Chromatogr. *51*, 548 (1970).
222. E. Papp and Gy. Vigh, J. Chromatogr. *259*, 49 (1983).
223. A. Leitold and Gy. Vigh, J. Chromatogr. *257*, 384 (1983).
224. J. Gasparič, *Fifth Danube Symp. on Chromatography*, Yalta, Nov. 11-16, 1985. Abstracts, Nauka, Yalta, p. 187.
225. T. Cserháti and J. Gasparič, J. Chromatogr. *394*, 368 (1987).
226. J. Crommen, J. Chromatogr. *186*, 705 (1979).
227. I. Jakubec, Collect. Czech. Chem. Commun. *24*, 617 (1959); *25*, 1736 (1960); *26*, 1072 (1961).
228. L. Hagdahl and A. Tiselius, Nature *170*, 799 (1952).
229. A. O. Kuhn and M. Lederer, J. Chromatogr. *440*, 165 (1988).
230. T. Cserháti, B. Bordás, and E. Tyihák, J. Chromatogr. *365*, 289 (1986).

231. H. H. W. Thijssen, J. Chromatogr. *133*, 355 (1977).
232. B. A. Bidlingmeyer, J. K. Del Rios, and J. Korpi, Anal. Chem. *54*, 442 (1982).
233. D.-G. Volkmann, in *Topics in Analytical Chemistry*, Vol. 126 (F. L. Boschke, ed.), Springer-Verlag, Berlin, 1984, p. 51.
234. T. Cserháti, É. János, and B. Barna, *Proc. Symp. Anal. Steroids*, Eger, Hungary, 1981, p. 441.
235. S. H. Unger, J. R. Cook, and J. S. Hollenberg, J. Pharm. Sci. *67*, 1364 (1978).
236. R. A. Scherrer and S. M. Howard, J. Med. Chem. *20*, 53 (1977).
237. T. Fujita, T. Nishioka, and M. Nakajima, J. Med. Chem. *20*, 1071 (1977).
238. J. Gasparič, *Rozvoj farmacie v rámci vědeckotechnické revoluce*, Universita Karlova, 1979, p. 105.
239. A. Kabil and V. Prey, Monatsh. Chem. *87*, 625 (1956).
240. J. Gasparič, unpublished results.
241. N. El Tayar, A. Tsantili-Kakoulidou, T. Roethlisberger, B. Testa, and J. Gal, J. Chromatogr. *439*, 237 (1988).
242. E. Stahl and H. Jork, in *Thin-Layer Chromatography: A Laboratory Handbook*, 2nd ed. (E. Stahl, ed.), Springer-Verlag, Berlin, 1969, p. 219.
243. J. Churáček, M. Hušková, H. Pechová, and J. Říha, J. Chromatogr. *49*, 511 (1970).
244. M. Bachratá, M. Blešová, Ž. Bezáková, and A. Lukáš, Českosl. Farm. *26*, 198 (1977).
245. J. Gasparič and D. Svobodová, J. Chromatogr. *153*, 153 (1978).
246. K. Teichert, E. Mutschler, and H. Rochelmeyer, Deut. Apoth.-Ztg. *100*, 283 (1960).
247. K. Teichert, E. Mutschler, and H. Rochelmeyer, Deut. Apoth.-Ztg. *100*, 477 (1960).
248. L. Carreras Matas, Anales Real Acad. Farm. *26*, 371 (1960).
249. E. Ullmann and H. Kassalitzky, Arch. Pharm. *295*, 37 (1962).
250. J. Vaedtke and A. Gajewska, J. Chromatogr. *9*, 345 (1962).
251. J. Žurkowska, M. Lukaszewski, and A. Ozarowski, Acta Polon. Pharm. *20*, 115 (1963).
252. L. Göldel, W. Zimmermann, and D. Lommer, Z. Physiol. Chem. *333*, 35 (1963).
253. J. Žurkowska and A. Ozarowski, Planta Med. *12*, 222 (1964).
254. M. Ywata and E. M. Gold, Steroids *3*, 435 (1964).
255. D. Sonanini, Pharm. Acta Helv. *39*, 673 (1964).
256. F. Takacs, Monatsh. Chem. *95*, 961 (1964).
257. G. A. L. Smith and P. J. Sullivan, Analyst (London) *89*, 312 (1964).
258. T. Hohmann and H. Rochelmeyer, Arch. Pharm. *297*, 186 (1964).

259. C. J. Clifford, J. V. Wilkinson, and J. S. Wragg, J. Pharm. Pharmacol. *16*, Suppl. 11T (1964).
260. D. Braun and G. Vorendohre, Z. Anal. Chem. *207*, 26 (1965).
261. D. Braun, Chimia *19*, 77 (1965).
262. Th. Beyrich, Planta Med. *13*, 439 (1965).
263. H. Zowall and T. Lewandowska, Chem. Anal. (Warsaw) *10*, 947 (1965).
264. S. Agurell, Acta Pharm. Suec. *2*, 357 (1965).
265. D. Sonanini, R. Hofstetter, L. Anker, and H. Mühlemann, Pharm. Acta Helv. *40*, 302 (1965).
266. R. Stainier, J. Pharm. Belg. *20*, 89 (1965).
267. W. Schunack, E. Mutschler, and H. Rochelmeyer, Pharmazie *20*, 685 (1965).
268. L. Fishbein and J. Fawkes, J. Chromatogr. *19*, 364 (1965).
269. V. Procházka, F. Kavka, M. Průcha, and J. Pitra, Českosl. Farm. *15*, 363 (1966).
270. I. Zarebska and A. Ozarowski, Farm. Polska *22*, 518 (1966).
271. J. C. Schabort and D. J. J. Potgieter, J. Chromatogr. *31*, 235 (1967).
272. E. Ullmann and H. Kassalitzki, Deut. Apoth.-Ztg. *107*, 152 (1967).
273. T. W. MacConnell Davis, J. Chromatogr. *29*, 283 (1967).
274. A. N. Crabtree and A. E. J. McGill, Mikrochim. Acta (Wien) 85 (1967).
275. G. Rabitzsch, J. Chromatogr. *35*, 122 (1968).
276. P. Horák, Českosl. Farm. *17*, 37 (1968).
277. A. Banaszek, E. Krowicki, and A. Zamojski, J. Chromatogr. *32*, 581 (1968).
278. K. C. Güven and L. Eroglu, Eczacilik Bul. *10*, 53 (1968).
279. W. Hauser, Th. Kartnig, and G. Verdino, Sci. Pharm. *36*, 237 (1968).
280. H. Thielemann, Z. Chem. *9*, 350 (1969); Pharmazie *24*, 483 (1969); *25*, 418 (1970); Mikrochim. Acta (Wien), 994 (1970).
281. R. J. T. Graham and J. Daly, J. Chromatogr. *48*, 67, 78 (1970).
282. R. J. T. Graham, A. E. Nya, and D. A. Tinsley, *Int. Symp. VI Chromatogr. Electrophorese,* Brussels, 1970 (Publ. 1971), p. 105.
283. V. P. Georgievskii, N. Ya. Tsarenko, M. S. Schraiber, and G. Ya. Khait, Herba Pol. *17*, 258 (1971).
284. D. Braun and J. Arndt, Kunststoffe *62*, 41 (1972).
285. F. Hammerstein and F. Kaiser, Planta Med. *21*, 5 (1972).
286. J. Večerková and M. Vacková, Českosl. Farm. *22*, 162 (1973).
287. E. Nováková and J. Večerková, Českosl. Farm. *22*, 347 (1973).
288. Ch. B. Lugt, Planta Med. *23*, 176 (1973).
289. M. C. Bonjean, J. Alary, and M. C. Luu Duc, Chim. Thér. 93 (1973).

290. D. Radulovic, Z. Blagojevic, and D. Zivanov-Stakić, Acta Pharm. Jugosl. 24, 173 (1974).
291. Ch. B. Lugt and L. Noordhoek-Ananias, Planta Med. 25, 267 (1974).
292. J. Touj and M. Mouková, Proc. 2nd Symp. Chromatogr., Oct. 1975, Dům Techn. ČVTS, Praha, 1976, p. 258.
293. M. Prošek, E. Kučan, M. Katič, and M. Bano, Chromatographia 9, 273, 325 (1976); 10, 147 (1977).
294. M. Prošek, E. Kučan, M. Katič, M. Bano, and A. Medja, Chromatographia 11, 578 (1978).
295. E. Anghelescu, O. Bojor, S. M. Gruia, E. Ciobanu, and M. Pelea, Rom. Pat. 70 914 (Oct. 15, 1980); Chem. Abstr. 96, 228263f (1982).
296. S. M. Hassan, Pharmazie 33, 237 (1978).
297. P. Hatinguais, D. Beziat, P. Negol, and R. Tarroux, Trav. Soc. Pharm. Montpellier 38, 329 (1978).
298. A. Kutner, J. Smolińska, and W. Cieślik, Chem. Anal. (Warsaw) 23, 727 (1978).
299. M. Bachratá, M. Blešová, A. Schultzová, L. Grolichová, Ž. Bezáková, and A. Lukáš, J. Chromatogr. 171, 29 (1979).
300. M. Blešová, M. Bachratá, Z. Bezáková, J. Čižmárik, and D. Kováčová, Acta Fac. Pharm. Univ. Comenianae 33, 85 (1979).
301. B. Kakáč and M. Šaršúnová, Zborník prednášok X. konf. Syntéza a Analýza Léčiv, Tatr. Lomnica, 1980, p. 103.
302. Th. Kartnig and P. Kobosil, J. Chromatogr. 138, 238 (1977).
303. K. Winsauer and W. Buchberger, Chromatographia 14, 623 (1981).
304. A. Janssen and D. Sopczak, Chromatographia 13, 479 (1980).
305. L. Ogierman and G. Brysz, Z. Anal. Chem. 308, 463 (1981).
306. L. Ogierman and A. Silowiecki, J. High Resolut. Chromatogr. Chromatogr. Commun. 4, 357 (1981).
307. T. Tomova and A. Kolusheva, Tr. Nauchnoizsled. Khim.-Farm. Inst. 11, 200 (1981). Chem. Abstr. 96, 168805x (1982).
308. J. Čižmárik, M. Blešová, M. Bachratá, Ž. Bezáková, and A. Borovanský, Pharmazie 37, 554 (1982).
309. S. H. Hilal, A. S. Radwan, M. Y. Haggag, F. R. Melek, and S. M. Abdel Khalek, Egypt. J. Pharm. Sci. 23, 365 (1982, Publ. 1984); Chem. Abstr. 102, 84476v (1985).
310. D. A. Voloshina and S. S. Shain, Khim. Farm. Zh. 17, 438 (1983).
311. M. Bachratá, J. Čižmárik, Ž. Bezáková, M. Blešová, and A. Borovanský, Chem. Zvesti 37, 217 (1983).
312. T. Kömives, J. Chromatogr. 261, 423 (1983).
313. M. Bachratá, M. Blešová, M. Stankovičová, Ž. Bezáková, Z. Bugáň, J. Csöllei, L. Beneš, and A. Borovanský, Farm. Obzor 53, 543 (1984).

314. J. Gasparič, J. Skutil and N.-H. Mai, *Proc. Int. Symp. TLC with Special Emphasis on OPLC*, Szeged, Sept. 10-12, 1984, p. 72.
315. M. Blešová, J. Čižmárik, M. Bacharatá, Ž. Bezáková, and A. Borovanský, Collect. Czech. Chem. Commun. *50*, 1133 (1985).
316. H. Szumilo and M. Przyborowska, Chem. Anal. (Warsaw) *30*, 267 (1985).
317. E. Sanchez-Moyano, J. M. Plá-Delfina, and M. Herraez, Cienc. Ind. Farm. *6*, 252 (1987).
318 P. Davídková, J. Inf. Rec. Mater. *16*, 121 (1988).
319. G. Vernin and J. Metzger, Chim. Anal. *46*, 487 (1964).
320. A. Berg and J. Lam, J. Chromatogr. *16*, 157 (1964).
321. M. F. Kovács, J. Assoc. Offic. Agr. Chemists *47*, 1097 (1964).
322. F. Korte and H. Sieper, J. Chromatogr. *13*, 90 (1964); *14*, 178 (1964).
323. M. F. Kovács, Jr., J. Assoc. Offic. Agr. Chemists *48*, 1018 (1965).
324. H. Woggon and D. Jehle, Plaste Kaut. *13*, 460 (1966).
325. J. Davídek, Nahrung *10*, 713, 717 (1966).
326. M. F. Kovács, Jr., J. Assoc. Offic. Anal. Chemists *49*, 365 (1966).
327. T. J. Betts and P. J. Holloway, J. Pharm. Pharmacol. *19*, Suppl. 97S (1967).
328. R. H. White and J. W. Howard, J. Chromatogr. *29*, 108 (1967).
329. P. K. Ghosh and A. N. Saha, Indian J. Technol. *5*, 301 (1967); Chem. Abstr. *68*, 114199f (1968).
330. S. Agurell, J. M. Nilsson, and F. Sandberg, Svensk Farm. Tidskr. *72*, 662 (1968).
331. E. J. Thomas, J. A. Burke, and J. H. Lawrence, J. Chromatogr. *35*, 119 (1968).
332. P. Chambon and R. Chambon-Mougenot, Ann. Pharm. Franc. *27*, 739 (1969).
333. H.-J. Petrowitz, J. Chromatogr. *40*, 462 (1969).
334. J. Churáček and H. Pechová, J. Chromatogr. *48*, 250 (1970).
335. G. Machbert and A. von Lukowicz, Pharm. Ztg. *116*, 517 (1971).
336. F. W. H. M. Merkus, Pharm. Weekbl. *106*, 49 (1971).
337. F. W. H. M. Merkus and J. F. C. Roovers-Bollen, *Int. Symp. VI Chromatogr. Electrophorese*, Brussels, 1970 (Publ. 1971) p. 355.
338. Y. Yamazaki, Y. Suzuki, and T. Takeuchi, Japan Analyst *21*, 1223 (1972).
339. H. Pechová, J. Churáček, V. Červinka, and M. Vondruška, Sb. Věd. Pr. Vys. Sk. Chemtechn. Pardubice *28*, 51 (1972).
340. J. Sherma, in *Analytical Methods for Pesticides and Plant Growth Regulation*, Vol. 7, *Thin-Layer and Liquid*

Chromatography and Pesticides of International Importance (J. Sherma and G. Zweig, eds.), Academic Press, New York, 1973, Chap. 1.
341. B. Rittich and M. Šimek, Chem. Zvesti 30, 200 (1976).
342. K. Waisser, H. Synková, and M. Čeladník, Českosl. Farm. 32, 3 (1983).
343. J. Franc, Czech. Pat. 204 746; Chem. Abstr. 98, 209400y (1983).
344. P. Oroszlán, G. Verzár-Petri, E. Mincsovics, and T. Székely, Proc. Int. Symp. Special Emphasis Overpressured Layer Chromatography (OPLC), Szeged, Sept. 10-12, 1984, p. 343.
345. A. Horna, H. Pechová, A. Pikulová, L. Hornová, and J. Churáček, J. Chromatogr. 367, 155 (1986).
346. H. Wullen and E. Stainier, J. Pharm. Belg. 22, 291 (1967).
347. L. S. Bark and R. J. T. Graham, J. Chromatogr. 27, 116 (1967).
348. L. S. Bark and R. J. T. Graham, J. Chromatogr. 27, 109 (1967).
349. L. S. Bark and R. J. T. Graham, Int. Symp. IV Chromatogr. Electrophorese, Brussels, 1966. Presses Academiques Europeenes, Brussels, 1968, p. 105.
350. R. J. T. Graham, J. Chromatogr. 33, 118 (1968).
351. J. C. Dearden, A. M. Patel, and J. H. Tubby, J. Pharm. Pharmacol. 26, 74P (1974).
352. S. Gocan and C. Konnert, Stud. Univ. Babes-Bolyai (Ser.) Chem. 22, 39 (1977).
353. S. Gocan, C. Konnert, and H. Voik, Rev. Rouman. Chim. 23, 133 (1978).
354. S. Gocan and C. Konnert, Stud. Univ. Babes-Bolyai (Ser.) Chem. 24, 39 (1979); Chem. Abstr. 91, 222015b (1979).
355. S. Gocan and C. Konnert, Rev. Chim. (Bucharest) 30, 82 (1979).
356. S. P. Srivastava, R. Bhushan, and R. S. Chauhan, J. Liquid Chromatogr. 7, 1359 (1984).
357. S. Gocan and C. Konnert, Stud. Univ. Babes-Bolyai (Ser.) Chem. 31, 13 (1986); Chem. Abstr. 107, 51104x (1987).
358. D. I. Cargill, Analyst (London) 87, 865 (1962).
359. G. Urbach, J. Chromatogr. 12, 196 (1963).
360. H. M. Edwards, Jr., J. Chromatogr. 22, 29 (1966).
361. P. Švec, R. Seifert, and M. Zbirovský, J. Chromatogr. 53, 587 (1970).
362. H. Czarniecka and J. Czarniecki, Proc. Int. Symp. TLC with Special Emphasis on Overpressured Layer Chromatography (OPLC), Szeged, Sept. 10-12, 1984, p. 126.
363. G. R. Inglis, J. Chromatogr. 20, 417 (1965).
364. E. Chang, Steroids 4, 237 (1964).

365. J. Halpaap, J. Chromatogr. *33*, 144 (1968).
366. J. Gasparič and J. Skutil, J. Chromatogr. (in press).
367. G. Chojnacka, Ann. Univ. Mariae Curie-Sklodowska, Sect. AA: Chem. *33*, 51 (1978, publ. 1979).
368. G. Chojnacka, Biul. Lubel. Tow. Nauk, Mat.-Fiz. Chem. *21*, 11 (1979).
369. C. J. Lupton, J. Chromatogr. *104*, 223 (1975).
370. D. Sonanini and L. Anker, Pharm. Acta Helv. *42*, 54 (1967).
371. J. D. Phillipson and F. A. Darwish, Planta Med. *41*, 209 (1981).
372. D. Sloot, J. Chromatogr. *24*, 451 (1966).
373. B. Rittich, M. Krška, M. Šimek, and J. Čoupek, Rocz. Nauk Zootech. *5*, 33 (1978).
374. H. T. Badings and J. G. Wassink, Neth. Milk Dairy J. *17*, 132 (1963).
375. H. T. Badings, J. Chromatogr. *14*, 265 (1964).
376. D. P. Schwartz, R. Brewington, and O. W. Parks, Microchem. J. *8*, 402 (1964).
377. P. W. Meijboom, J. Chromatogr. *24*, 427 (1966).
378. P. P. Ascione, J. B. Zagar, and G. P. Chrekian, J. Pharm. Sci. *56*, 1393 (1967).
379. J. D. Craske and R. A. Edwards, J. Chromatogr. *51*, 237 (1970).
380. A. Zeman and I. P. G. Wirotama, Z. Anal. Chem. *259*, 351 (1972).
381. S. Grodzka-Zapytowska and W. Janitz, Chem. Anal. (Warsaw) *22*, 549 (1977).
382. J. H. Tumlinson, J. P. Minyard, P. A. Hedin, and A. C. Thompson, J. Chromatogr. *29*, 80 (1967).
383. E. Knappe and D. Peteri, Z. Anal. Chem. *188*, 184, 352 (1962); *190*, 380 (1962).
384. E. Knappe and I. Rohdewald, Z. Anal. Chem. *210*, 183 (1965).
385. E. Bloem, J. Chromatogr. *35*, 108 (1968).
386. K. Gröningsson, Acta Pharm. Suec. *7*, 635 (1970).
387. J. Kuczynski and M. Bieganowska, Chem. Anal. (Warsaw) *28*, 709 (1983).
388. A. E. Bird and A. C. Marshall, J. Chromatogr. *63*, 313 (1971).
389. T. Yamana, A. Tsuji, E. Miyamoto, and O. Kubo, J. Pharm. Sci. *66*, 747 (1977).
390. J. C. Dearden and E. Tomlinson, J. Pharm. Pharmacol. *24*, Suppl. 115P (1972).
391. E. Cohnen, H. Flasch, N. Heinz, and F. W. Hempelmann, Arzneim.-Forsch. *28*, 2179 (1978).
392. G. L. Biagi, A. M. Barbaro, M. C. Guerra, G. Cantelli-Forti, and M. E. Fracasso, J. Med. Chem. *17*, 28 (1974).

393. J. Hlavatá, K. Waisser, K. Palát, and R. Karlíček, Rozvoj farmacie v rámci vědeckotechnické revoluce, Univerzita Karlova, 1979, p. 108.
394. H. H. W. Thijssen, Arzenim.-Forsch. 28, 1065 (1978); Eur. J. Med. Chem. 16, 449 (1981).
395. G. L. Biagi, A. M. Barbaro, M. C. Guerra, M. Babbini, M. Gaiardi, M. Bartoletti, and P. A. Borea, J. Med. Chem. 23, 193 (1980).
396. B. R. Gandhe, R. K. Danikhel, and D. Purnanand, J. Chromatogr. 288, 233 (1984).
397. K. Bey, Fette, Seifen, Anstrichm. 67, 217 (1965).
398. J. Büchi and J. A. Fresen, Pharm. Acta Helv. 41, 551 (1966).
399. A. Hulshoff and J. H. Perrin, J. Chromatogr. 120, 65 (1976); J. Med. Chem. 20, 430 (1977).
400. M. Waksmundzka-Hajnos and M. L. Bieganowska, Chem. Anal. (Warsaw) 31, 475 (1986).
401. L. S. Bark and R. J. T. Graham, J. Chromatogr. 23, 417 (1966).
402. L. S. Bark and R. J. T. Graham, J. Chromatogr. 25, 357 (1966).
403. L. S. Bark and R. J. T. Graham, Talanta 13, 1281 (1966).
404. R. J. T. Graham, J. Chromatogr. 33, 125 (1968).
405. D. R. Clifford, D. M. Fieldgate, and D. A. M. Watkins, J. Chromatogr. 43, 110 (1969).
406. D. R. Clifford, A. C. Deacon, and M. E. Holgate, Ann. Appl. Biol. 64, 131 (1969).
407. B. Trtík, Czech. Pat. 195 151; Chem. Abstr. 97, 115409k (1982).
408. H. W. Prinzler, H. Tauchmann, and C. Tzscharnke, J. Chromatogr. 29, 151 (1967).
409. E. Knappe and K. G. Yekundi, Z. Anal. Chem. 203, 87 (1964).
410. E. Knappe, D. Peteri, and I. Rohdewald, Z. Anal. Chem. 197, 364 (1963).
411. E. Knappe and J.-I. Stuck, Z. Anal. Chem. 227, 353 (1967).
412. B. Scholz and K. Ballschmiter, J. Chromatogr. 252, 269 (1982).
413. W. S. Kim, Biochim. Biophys. Acta 112, 392 (1966).
414. R. J. Daley, C. B. J. Gray, and S. R. Brown, J. Chromatogr. 76, 175 (1973).
415. J. M. Plá-Delfina, J. Moreno, J. Durán, and A. Del Pozo, J. Pharmacokinet. Biopharmaceut. 3, 115 (1975).
416. E. Sánchez-Moyano, M. Herráez, and J. M. Plá-Delfina, Cienc. Ind. Farm. 6, 252 (1987).
417. E. Sánchez-Moyano, J. M. Plá-Delfina, and M. Herráez, Ciencia i Téchnica 5, 123 (1986).
418. I. D. Jones, L. S. Butler, E. Gibbs, and R. C. White, J. Chromatogr. 70, 87 (1972).

419. J. Sherma, J. Chromatogr. 52, 177 (1970).
420. B. Scholz, K. D. Willaschek, H. Müller, and K. Ballschmiter, J. Chromatogr. 208, 156 (1981).
421. K. Egger, Chromatogr. Symp. 2nd, Brussels, 1961, p. 75.
422. G. Duncan, L. Kitching, and R. J. T. Graham, J. Chromatogr. 47, 232 (1970).
423. D. C. Malins and H. K. Mangold, J. Am. Oil Chemists Soc. 37, 576 (1960).
424. H. P. Kaufmann, Z. Makus, and B. Das, Fette, Seifen, Anstrichm. 63, 807 (1961).
425. H. E. Carter and H. S. Hendrickson, Biochemistry 2, 389 (1963).
426. R. A. Conkin, Residue Rev. 6, 136 (1964).
427. M. W. Roomi, M. R. Subbaram, and K. T. Achaya, J. Chromatogr. 16, 106 (1964).
428. T. C. Chu and E. J.-H. Chu, J. Chromatogr. 21, 46 (1966).
429. M. M. Paulose, J. Chromatogr. 21, 141 (1966).
430. V. Mahadevan, C. V. Viswanathan, and W. O. Lundberg, J. Chromatogr. 24, 357 (1966).
431. J. J. Menn and J. B. McBain, Nature 209, 1351 (1966).
432. G. L. Biagi and M. F. Gamba, Boll. Soc. Ital. Biol. Sper. 44, 189 (1968).
433. R. S. Dobies, J. Chromatogr. 40, 110 (1969).
434. G. L. Biagi, A. M. Barbaro, M. F. Gamba, and M. C. Guerra, J. Chromatogr. 41, 371 (1969).
435. G. L. Biagi, A. M. Barbaro, M. C. Guerra, and M. F. Gamba, J. Chromatogr. 44, 195 (1969).
436. G. L. Biagi, A. M. Barbaro, and M. C. Guerra, Pharmacol. Res. Commun. 2, 121 (1970).
437. G. L. Biagi, A. M. Barbaro, and M. C. Guerra, Experientia 27, 918 (1971).
438. G. L. Biagi, M. C. Guerra, A. M. Barbaro, and M. F. Gamba, J. Med. Chem. 13, 511 (1970).
439. G. L. Biagi, M. C. Guerra, and A. M. Barbaro, J. Med. Chem. 13, 944 (1970).
440. G. L. Biagi, M. C. Guerra, and A. M. Barbaro, Farmaco, Ed. Sci. 25, 755 (1970).
441. A. M. Barbaro, M. C. Guerra, and G. L. Biagi, Boll. Soc. Ital. Biol. Sper. 47, 556 (1971).
442. J. D. Turnbull, G. L. Biagi, A. J. Merola, and D. G. Cornwell, Biochem. Pharmacol. 20, 1383 (1971).
443. M. C. Guerra, A. M. Barbaro, and G. L. Biagi, Boll. Soc. Ital. Biol. Sper. 47, 553 (1971).
444. M. C. Guerra, A. M. Barbaro, and G. L. Biagi, Boll. Soc. Ital. Biol. Sper. 48, 1225 (1972).

445. O. Gandolfi, A. M. Barbaro, and G. L. Biagi, Experientia 29, 689 (1973).
446. O. Gandolfi, A. M. Barbaro, M. C. Guerra, G. Cantelli-Forti, and G. L. Biagi, Boll. Soc. Ital. Biol. Sper. 49, 1022 (1973).
447. G. Pelizza, G. C. Lancini, G. C. Allievi, and G. G. Gallo, Farmaco, Ed. Sci. 28, 298 (1973).
448. A. N. Tischler, F. M. Thompson, L. J. Libertini, and M. Calvin, J. Med. Chem. 17, 948 (1974).
449. G. L. Biagi, O. Gandolfi, M. C. Guerra, A. M. Barbaro, and G. Cantelli-Forti, J. Med. Chem. 18, 868 (1975).
450. G. L. Biagi, A. M. Barbaro, O. Gandolfi, M. C. Guerra, and G. Cantelli-Forti, J. Med. Chem. 18, 873 (1975).
451. G. L. Biagi, A. M. Barbaro, M. C. Guerra, G. Cantelli-Forti, and O. Gandolfi, J. Chromatogr. 106, 349 (1975).
452. G. L. Biagi, M. C. Guerra, A. M. Barbaro, G. Cantelli-Forti, and T. Rossi, Boll. Soc. Ital. Biol. Sper. 51, 403 (1975).
453. M. Kuchař, B. Brůnová, and V. Rejholec, Českosl. Farm. 26, 239 (1977).
454. D. N. Dhar, S. C. Suri, and P. Dwivedi, Planta Med. 31, 33 (1977).
455. K. Waisser, E. Spálenská, V. Šimánek, and F. Šantavý, Českosl. Farm. 28, 321 (1979).
456. M. Kuchař, V. Rejholec, Z. Roubal, and O. Němeček, Collect. Czech. Chem. Commun. 44, 183 (1979).
457. M. Kuchař, V. Rejholec, M. Jelínková, V. Rábek, and O. Němeček, J. Chromatogr. 162, 197 (1979).
458. G. L. Biagi, A. M. Barbaro, M. C. Guerra, G. Hakim, G. C. Solaini, and P. A. Borea, J. Chromatogr. 177, 35 (1979).
459. K. Waisser and E. Spálenská, Proc. 3rd Congr. Hungarian Pharmacol. Soc. Budapest, Akadémiai Kiadó, Budapest, 1980, p. 241.
460. M. Kuchař, B. Brůnová, J. Grimová, V. Rejholec, V. Čepelák, and O. Němeček, Českosl. Farm. 29, 276 (1980).
461. M. Kuchař, B. Brůnová, Z. Roubal, J. Schlanger, and O. Němeček, Collect. Czech. Chem. Commun. 45, 1401 (1980).
462. M. C. Guerra, A. M. Barbaro, G. Cantelli-Forti, M. T. Foffani, G. L. Biagi, P. A. Borea, and A. Fini, J. Chromatogr. 216, 93 (1981).
463. A. M. Barbaro, M. C. Guerra, G. Cantelli-Forti, G. Aicardi, G. L. Biagi, P. da Re, P. Valenti, and P. A. Borea, J. Chromatogr. 242, 1 (1982).
464. M. Kuchař, V. Rejholec, B. Brůnová, J. Grimová, G. Matoušová, O. Němeček, and H. Čepeláková, Collect. Czech. Chem. Commun. 47, 2514 (1982).
465. M. Kuchař, V. Rejholec, Z. Roubal, and O. Matoušková, Collect. Czech. Chem. Commun. 48, 1077 (1983).

466. M. Kuchař, V. Rejholec, E. Kraus, V. Miller, and V. Rábek, J. Chromatogr. *280*, 279 (1983).
467. A. M. Barbaro, M. C. Pietrogrande, M. C. Guerra, G. Cantelli-Forti, P. A. Borea, and G. L. Biagi, J. Chromatogr. *287*, 259 (1984).
468. M. Kuchař, B. Brůnová, V. Rejholec, M. Jelínková, J. Holubek, and O. Němeček, Collect. Czech. Chem. Commun. *49*, 122 (1984).
469. M. C. Guerra, A. M. Barbaro, G. L. Biagi, M. C. Pietrogrande, P. A. Borea, A. Andreani, and G. Cantelli-Forti, J. Chromatogr. *320*, 281 (1985).
470. A. M. Barbaro, M. C. Guerra, G. L. Biagi, M. C. Pietrogrande, and P. A. Borea, J. Chromatogr. 347, 209 (1985).
471. M. Lázníček, P. Beňo, K. Waisser, and J. Květina, Českosl. Farm. *34*, 353 (1985).
472. S. Dvořáčková, D. Guénard, F. Picot, V. Šimánek, and K. Waisser, Acta Univ. Palackianae Olomucensis, Fac. Med. *111*, 13 (1985).
473. G. L. Biagi, A. M. Barbaro, M. C. Guerra, G. Cantelli-Forti, P. A. Borea, M. C. Pietrogrande, S. Salvadori, and R. Tomatis, in *Chromatography: The State of the Art* (H. Kalasz and L. S. Ettre, eds.), Akadémiai Kiadó, Budapest, 1985, p. 719.
474. K. Waisser, Ž. Odlerová, W. Thiel, and R. Mayer, Pharmazie *43*, 794 (1988).
475. K. Waisser, Ž. Odlerová, N. Houngbedji, W. Thiel, and R. Mayer, Zentralbl. Mikrobiol. *144*, 355 (1989).
476. G. J. Bijloo and R. F. Rekker, J. Chromatogr. *351*, 122 (1986).
477. S. J. Purdy and E. V. Truter, Analyst (London) *87*, 802 (1962).
478. H. P. Kaufmann and Z. Makus, Fette, Seifen, Anstrichm. *62*, 1014 (1960).
479. H. P. Kaufmann, Z. Makus, and T. H. Khoe, Fette, Seifen, Anstrichm. *63*, 689 (1961).
480. H. P. Kaufmann, Z. Makus, and T. H. Khoe, Fette, Seifen, Anstrichm. *64*, 1 (1962).
481. R. Marcuse, U. Mobech-Hanssen, and P.-O. Göthe, Fette, Seifen, Anstrichm. *66*, 192 (1964).
482. H. P. Kaufmann and T. H. Khoe, Fette, Seifen, Anstrichm. *64*, 81 (1962).
483. L. Wolfman and B. A. Sachs, J. Lipid Res. *5*, 127 (1964).
484. J. W. Copius-Peereboom, J. Chromatogr. *9*, 316 (1962).
485. J. W. Copius-Peereboom and H. W. Beekes, J. Chromatogr. *17*, 99 (1965).
486. J. W. Copius-Peereboom, Z. Anal. Chem. *205*, 325 (1964).
487. A. S. Truswell and W. D. Mitchell, J. Lipid Res. *6*, 438 (1965).

488. I. Katz and M. Keeney, Anal. Chem. 36, 231 (1964).
489. L. D. Bergelson, E. V. Dyatlovitskaya, and V. V. Voronkova, J. Chromatogr. 15, 191 (1964).
490. H. P. Kaufmann and Y. S. Ko, Fette, Seifen, Anstrichm. 63, 828 (1961).
491. H. P. Kaufmann and B. Das, Fette, Seifen, Anstrichm. 64, 214 (1962).
492. H. P. Kaufmann and B. Das, Fette, Seifen, Anstrichm. 65, 398 (1963).
493. H. P. Kaufmann and H. Wessels, Fette, Seifen, Anstrichm. 66, 81 (1964).
494. D. Lefort, R. Perron, A. Pourchez, C. Madelmont, and J. Petit, J. Chromatogr. 22, 266 (1966).
495. H. W. Prinzler, D. Pape, and M. Teppke, J. Chromatogr. 19, 375 (1965).
496. W. Goworek, Chem. Anal. (Warsaw) 30, 669 (1985).
497. A. Winterstein, A. Studer, and R. Rüegg, Chem. Ber. 93, 2951 (1960).
498. H. P. Kaufmann, Z. Makus, and F. Deicke, Fette, Seifen, Anstrichm. 63, 235 (1961).
499. R. Rüegg and O. Isler, Planta Med. 9, 386 (1961).
500. H. R. Bolliger, in Dünnschicht-Chromatographie (E. Stahl, ed.), Springer-Verlag, Berlin, 1962, p. 217.
501. H. R. Bolliger and A. König, in Thin-Layer Chromatography: A Laboratory Handbook (E. Stahl, ed.), Springer-Verlag, Berlin, 1969, p. 264.
502. J. F. Pennock and P. J. Dunphy, in Thin-Layer Chromatography: A Laboratory Handbook, 2nd ed. (E. Stahl, ed.), Springer-Verlag, Berlin, 1969, p. 285.
503. J. Jonas, J. Pharm. Belg. 17, 103 (1962).
504. Č. Michalec, M. Šulc, and J. Měšťan, Nature 193, 63 (1962).
505. K. Randerath, Dünnschicht-Chromatographie, Verlag Chemie, Weinheim, 1962, p. 155.
506. E. Stahl and P. J. Schorn, Naturwissenschaften 49, 14 (1962).
507. H. Wagner, L. Hörhammer, and B. Dengler, J. Chromatogr. 7, 211 (1962).
508. H. Jork, Deut. Apoth.-Ztg. 102, 1263 (1961); J. Pharm. Belg. 18, 429 (1963).
509. H. P. Kaufmann and C. V. Viswanathan, Fette, Seifen, Anstrichm. 65, 538 (1963).
510. H. P. Kaufmann and C. V. Viswanathan, Fette, Seifen, Anstrichm. 65, 607 (1963).
511. H. P. Kaufmann and C. V. Viswanathan, Fette, Seifen, Anstrichm. 65, 925 (1963).
512. E. Stahl, H. R. Bolliger, and L. Lehnert, Wiss. Veroeffentl. Deut. Ges. Ernährung 9, 129 (1963).

513. D. R. Threlfall and T. W. Goodwin, Biochim. Biophys. Acta 78, 532 (1963).
514. K. Egger, Ber. Deut. Botan. Ges. 77, 145 (1964).
515. L. M. Libbey and E. A. Day, J. Chromatogr. 14, 273 (1964).
516. T. W. Hammonds and G. Shone, J. Chromatogr. 15, 200 (1964).
517. M. K. Ramamurthy and V. R. Bhalerao, Analyst (London) 89, 740 (1964).
518. G. P. McSweeney, J. Chromatogr. 17, 183 (1965).
519. W. Y. Cobb, L. M. Libbey, and E. A. Day, J. Chromatogr. 17, 606 (1965).
520. J. G. Bieri and E. L. Prival, Proc. Soc. Expl. Biol. Med. 120, 554 (1965).
521. H. M. Yusef, D. R. Threlfall, and T. W. Goodwin, Phytochemistry 4, 551 (1965).
522. K. Egger, Phytochemistry 4, 609 (1965).
523. K. Egger and H. Kleinig, Z. Anal. Chem. 211, 187 (1965).
524. D. R. Fraser and E. Kodicek, Biochem. J. 96, 59p (1965).
525. H. P. Kaufmann and K. D. Mukherjee, Fette, Seifen, Anstrichm. 67, 753 (1965).
526. H. Wagner and B. Dengler, Biochem. Ztg. 336, 380 (1962).
527. G. Rebbel and P. Mandel, Biochim. Biophys. Acta 98, 380 (1965).
528. C. B. C. Boyce and B. V. Milborrow, Nature 208, 537 (1965).
529. J. Sliwiok and Z. Kwapniewski, Mikrochim. Acta (Wien) 657 (1965).
530. E. Stahl and H. Vollmann, Talanta 12, 525 (1965).
531. M. M. Chakrabarty, D. Bhattacharyya, and A. Gupta, J. Chromatogr. 22, 84 (1966).
532. H. P. Kaufmann and H. Wessels, Fette, Seifen, Anstrichm. 68, 249 (1966).
533. G. Yip and S. F. Howard, J. Assoc. Offic. Anal. Chemists 49, 1166 (1966).
534. P. J. Dunphy, J. D. Kerr, J. F. Pennock, and K. J. Whittle, Chem. Ind. 1549 (1966).
535. P. J. Dunphy, J. D. Kerr, J. F. Pennock, K. J. Whittle, and J. Feeney, Biochim. Biophys. Acta 136, 136 (1967).
536. M. Rink, R. Lenhard, and H. Jäger, J. Chromatogr. 29, 396 (1967).
537. M. Severin, J. Chromatogr. 26, 101 (1967).
538. A. R. Wellburn, J. Stevenson, F. W. Hemming, and R. A. Morton, Biochem. J. 102, 313 (1967).
539. K. J. Stone, A. R. Wellburn, F. W. Hemming, and J. F. Pennock, Biochem. J. 102, 325 (1967).
540. M. M. Chakrabarty, D. Bhattacharyya, and A. K. Gayen, J. Chromatogr. 44, 116 (1969).

541. R. Takeshita, H. Akagi, M. Fujita, and Y. Sakagami, J. Chromatogr. *51*, 283 (1970).
542. V. Mallet and R. W. Frei, J. Chromatogr. *54*, 251 (1971).
543. R. A. Hoodless, J. Thomson, and J. E. Arnold, J. Chromatogr. *56*, 332 (1971).
544. R. H. De Vos and E. W. Peet, Bull. Environ. Contam. Toxicol. *6*, 164 (1971).
545. P. J. Dunphy, P. G. Phillips, and A. F. Brodie, J. Lipid Res. *12*, 442 (1971).
546. K. C. James, G. T. Richards, and T. D. Turner, J. Chromatogr. *69*, 141 (1972).
547. J. A. S. Rokos, J. Chromatogr. *74*, 357 (1972).
548. K. H. Buechel and W. Draber, Advan. Chem. Ser. *114*, 141 (1972).
549. W. Draber, K. H. Buechel, and G. Schäfer, Z. Naturforsch. *27b*, 159 (1972).
550. W. Draber, K. H. Buechel, and K. Dickore, Pestic. Chem., *Proc. 2nd Int. Congr. Pestic. Chem.*, Vol. 1, 5, 1972, p. 153; Chem. Abstr. *80*, 67292s (1974).
551. S. Ishikura, M. Yonaha, I. Sunaga, Y. Watanabe, N. Kikuchi, and R. Takeshita, Eisei Kagaku *18*, 281 (1972); Chem. Abstr. *78*, 1644k (1972).
552. R. Takeshita and T. Yamashita, Eisei Kagaku *18*, 388 (1972); Chem. Abstr. *78*, 144191w (1973)
553. K. Ogura, T. Shinka, and S. Seto, J. Biochem. *72*, 1101 (1972).
554. J. Churáček, H. Pechová, and M. Hušková, Sborník Věd. Prac. VŠChT, Pardubice *29*, 37 (1973).
555. D. L. Stalling and J. N. Huckins, J. Assoc. Offic. Anal. Chemists *56*, 367 (1973).
556. H. Wessels, Fette, Seifen, Anstrichm. *75*, 478 (1973).
557. R. J. Ismail and F. L. Bonner, J. Assoc. Offic. Agr. Chemists *57*, 1026 (1974).
558. H. Nakamura and Z. Tamura, J. Chromatogr. *96*, 211 (1974).
559. Č. Michalec, T. Chojnacki, and J. Churáček, *Proc. 2nd Symp. Chromatogr.*, Oct. 1975, Dům Techn. ČVTS, Praha, 1976, p. 93.
560. J. Churáček, H. Pechová, and D. Tocksteinová, *Proc. 2nd Symp. Chromatogr.*, Oct. 1975, Dům Techn. ČVTS, Praha, 1976, p. 225.
561. J. Reinišová, Č. Michalec, and J. Churáček, *Proc. 2nd Symp. Chromatogr.*, Oct. 1975, Dům Techn. ČVTS, Praha, 1976, p. 160.
562. S. A. Fusari, M. Terhalle, I. J. Holcomb, and E. C. Alix, J. Chromatogr. Sci. *13*, 563 (1975).

563. P. Todd, Jr., M. Bensinger, and T. Biftu, J. Chromatogr. Sci. *13*, 577 (1975).
564. J. K. Seydel, H. Ahrens, and W. Losert, J. Med. Chem. *18* 234 (1975).
565. A. Fujii and E. S. Cook, J. Med. Chem. *18*, 502 (1975).
566. C. M. Allen, Jr., M. V. Keenan, and J. Sack, Arch. Biochem. Biophys. *175*, 236 (1976).
567. U. A. Th. Brinkman, A. de Kok, G. de Vries, and H. G. M. Reymer, J. Chromatogr. *128*, 101 (1976).
568. H. Thielemann, Z. Anal. Chem. *282*, 144 (1976).
569. U. A. Th. Brinkman, A. de Kok, H. G. M. Reymer, and G. de Vries, J. Chromatogr. *129*, 193 (1976).
570. J. K. Seydel, K.-J. Schaper, E. Wempe, and H. P. Cordes, J. Med. Chem. *19*, 483 (1976).
571. D. Sharples, J. Pharm. Pharmacol. *28*, 100 (1976).
572. J. J. de Kok, A. de Kok, U. A. Th. Brinkman, and R. M. Kok, J. Chromatogr. *142*, 367 (1977).
573. B. A. El-Zeany and A.-K. S. Ahmed, Egypt. J. Food Sci. 5(1-2), 1 (1977).
574. A. Fujii, J. H. Bush, K. E. Shores, R. G. Johnson, R. J. Garascia, and E. S. Cook, J. Pharm. Sci. *66*, 844 (1977).
575. W. Sasak and T. Chojnacki, Arch. Biochem. Biophys. *181*, 402 (1977).
576. T. Baba and C. M. Allen, Jr., Biochemistry *17*, 5598 (1978).
577. R. Kaliszan, H. Foks, and M. Janowiec, Pol. J. Pharmacol. Pharm. *30*, 579 (1978).
578. R. Kaliszan, B. Milczarska, B. Lega, P. Szefer, and M. Janowiec, Pol. J. Pharmacol. Pharm. *30*, 585 (1978).
579. C. Prandi, J. Chromatogr. *155*, 149 (1978).
580. W. Steurbaut, W. Dejonckheere, and R. H. Kips, J. Chromatogr. *160*, 37 (1978).
581. H. Thielemann, Pharmazie *33*, 466 (1978); Z. Gesamte Hyg. Ihre Grenzgeb. *26*, 463 (1980).
582. P. L. Donnahey, V. T. Burt, H. H. Rees, and J. F. Pennock, J. Chromatogr. *170*, 272 (1979).
583. G. Maksay, Zs. Tegyey, and L. Ötvös, J. Chromatogr. *174*, 447 (1979).
584. Z. Grodzińska-Zachwieja, M. Bieganowska, and T. Dzido, Chromatographia *12*, 555 (1979).
585. R. Kaliszan, I. Kozakiewicz, F. Gajewski, and A. Madrala, Pharmazie *34*, 246 (1979).
586. T. Cserháti and É. János, in *Chemical Structure-Biological Activity Relationships: Quantitative Approaches* (I. Knoll, ed.), Akadémiai Kiadó, Budapest, 1980, p. 221.
587. A. K. Saxena and J. K. Seydel, Eur. J. Med. Chem. *15*, 241 (1980).

588. G. G. Briggs, J. Agr. Food Chem. 29, 1050 (1981).
589. R. A. Franich, J. Chromatogr. 209, 117 (1981).
590. J. Sliwock, A. Maciosczyk, and T. Kowalska, Chromatographia 14, 138 (1981).
591. T. Cserháti, M. Szögyi, and F. Tölgyesi, Proc. 2nd Egyptian-Hungarian Conf. Plant Prot., Alexandria, 1982, p. 335.
592. T. Cserháti, M. Szögyi, and B. Bordás, Gen. Physiol. Biophys. 1, 225 (1982).
593. T. Cserháti and F. Örsi, Periodica Polytechn. Chem. Engn. 26, 111 (1982).
594. Z. Grodzińska-Zachwieja, J. Chromatogr. 241, 217 (1982).
595. I. D. Wilson, C. R. Bielby, and E. D. Morgan, J. Chromatogr. 242, 202 (1982).
596. É. János and T. Cserháti, Acta Phytopathol. Acad. Sci. Hung. 17, 343 (1983).
597. H. J. M. Grünbauer, G. J. Bijloo, and T. Bultsma, J. Chromatogr. 270, 87 (1983).
598. T. Cserháti, Y. M. Darwish, and Gy. Matolcsy, J. Chromatogr. 270, 97 (1983).
599. É. János, B. Bordás, and T. Cserháti, J. Chromatogr. 286, 63 (1984).
600. I. D. Wilson, J. Chromatogr. 287, 183 (1984).
601. I. D. Wilson, J. Chromatogr. 291, 241 (1984).
602. A. Horna, H. Pechová, A. Tůmová, and J. Churáček, J. Chromatogr. 288, 230 (1984).
603. A. Kakoulidou and R. F. Rekker, J. Chromatogr. 295, 341 (1984).
604. T. Cserháti, A. Kuszmann-Borbély, and A. Dévai, Acta Phytopathol. Acad. Sci. Hung. 19, 177 (1984).
605. M. Dévai, A. Borbély-Kuszmann, T. Cserháti, and B. Bordás, Proc. Int. Symp. TLC with Special Emphasis on Overpressured Layer Chromatography (OPLC), Szeged, Sept. 10-12, 1984, p. 101.
606. B. Kocjan and J. Sliwiok, Proc. Budapest Chromatogr. Symp., Budapest, June 11-14, 1985.
607. T. Cserháti, M. Szögyi, and L. Györfi, Chromatographia 20, 253 (1985).
608. T. Cserháti, M. Szögyi, and L. Györfi, J. Chromatogr. 349, 295 (1985).
609. J. K. Pauncz, J. Chromatogr. 322, 386 (1985).
610. H. Czarniecka and J. Sliwiok, Budapest Chromatogr. Symp., Budapest, June 11-14, 1985.
611. G. Gullner, T. Cserháti, and M. Szögyi, Budapest Chromatogr. Symp., Budapest, June 11-14, 1985.
612. I. D. Wilson, J. Chromatogr. 354, 99 (1986).

613. T. Cserháti, B. Bordás, and M. Szögyi, Chromatographia 21, 312 (1986).
614. P. Pachaly, Oesterr. Apoth.-Ztg. 40, 809 (1986).
615. T. Cserháti and M. Szögyi, J. Biochem. Biophys. Meth. 14, 101 (1987).
616. T. Cserháti, B. Bordás, and G. Ösapay, Chromatographia, 23, 184 (1987).
617. O. R. Braekkan, Int. Z. Vitamin-Forsch. 33, 293 (1963).
618. J. P. Minyard, J. H. Tumlinson, A. C. Thompson, and P. A. Hedin, J. Chromatogr. 29, 88 (1967).
619. J. B. Kyziol and J. Pielichowski, Chem. Anal. (Warsaw) 19, 1253 (1974).
620. J. S. Plotkin and L. G. Sneddon, J. Chromatogr. 153, 289 (1978).
621. P. M. Schaber, J. E. Hunt, R. Fries, and J. J. Katz, J. Chromatogr. 316, 25 (1984).
622. I. D. Wilson, J. Chromatogr. 318, 373 (1985).
623. H. Ellgehausen, C. D'Hondt, and T. Fuerer, Pest. Sci. 12, No. 2, 219 (1981).
624. OECD Guideline for Testing of Chemicals 107, Partition Coefficient (1-Octanol/Water)—Flask Shaking Method, OECD, Paris, 1981.
625. H. Reinhardt and J. Rydberg, Chem. Ind. (London) 488 (1970).
626. S. S. Davis, G. Elson, E. Tomlinson, G. Harrison, and J. C. Dearden, Chem. Ind. (London) 677 (1976).
627. C. Hansch and A. J. Leo, *Substituent Constants for Correlation Analysis in Chemistry and Biology*, J. Wiley and Sons, New York, 1979.
628. A. J. Leo, J. Pharm. Sci. 76, 166 (1987).
629. A. Fujii, K. E. Shores, J. H. Bush, R. J. Garascia, and E. S. Cook, J. Pharm. Sci. 67, 713 (1978).
630. R. F. Rekker, *The Hydrophobic Fragmental Constant, Its Derivation and Application. A Means of Characterizing Membrane Systems*, Elsevier, Amsterdam, 1977.
631. Gy. Vigh, Z. Varga-Puchony, J. Hlavay, and E. Papp-Hites, J. Chromatogr. 236, 51 (1982).
632. P. Jandera, J. Churáček, and H. Colin, J. Chromatogr. 214, 35 (1981).
633. I. D. Wilson, C. R. Bielby, and E. D. Morgan, J. Chromatogr. 238, 97 (1982).
634. M. S. Mirrlees, S. J. Moulton, Ch. T. Murphy, and P. J. Taylor, J. Med. Chem. 19, 615 (1976).
635. R. M. Carlson, R. E. Carlson, and H. Kopperman, J. Chromatogr. 107, 219 (1975).

636. K. Miyako and H. Terada, J. Chromatogr. *157*, 387 (1978).
637. D. A. Brent, J. J. Sabatka, D. J. Minick, and D. W. Henry, J. Med. Chem. *26*, 1014 (1983).
638. K. Valkó, J. Liquid Chromatogr. *7*, 1405 (1984).
639. R. Kaliszan, J. Chromatogr. Sci. *22*, 362 (1984).
640. J. E. Haky and A. M. Young, J. Liquid Chromatogr. *7*, 675 (1984).
641. Th. Braumann, J. Chromatogr. *373*, 191 (1986).
642. J. C. Kraak, H. H. van Rooij, and J. L. G. Thus, J. Chromatogr. *352*, 455 (1986).
643. P. M. Sherblom and R. P. Eganhouse, J. Chromatogr. *454*, 37 (1988).
644. C. Papp, K. Valkó, Gy. Szász, I. Hermecz, J. Vámos, K. Hankó, and Zs. Ignáth-Halász, J. Chromatogr. *252*, 67 (1982).
645. K. Valkó, O. Papp, and F. Darvas, J. Chromatogr. *301*, 355 (1984).
646. K. Valkó and A. Lopata, J. Chromatogr. *252*, 77 (1982).
647. É. János, B. Bordás, and T. Cserháti, J. Chromatogr. *286*, 63 (1984).
648. É. János, J. Chromatogr. *365*, 117 (1986).
649. I. D. Wilson, Methodol. Surv. Biochem. Anal. *14*, 91 (1984).
650. I. D. Wilson and S. Lewis, J. Pharm. Biomed. Anal. *3*, 491 (1985).

4

Countercurrent Chromatography for the Purification of Peptides

Martha Knight *Peptide Technologies Corporation, Washington, D.C.*

I.	INTRODUCTION	254
II.	EARLY INSTRUMENTATION FOR THE COUNTERCURRENT CHROMATOGRAPHY OF PEPTIDES	255
	A. Countercurrent Distribution	257
	B. Countercurrent Chromatogaphy	259
III.	THEORY OF COUNTERCURRENT CHROMATOGRAPHY	262
	A. Partition Coefficient Relationships	262
	B. Resolution	264
IV.	METHODS	265
	A. Operation of the Coil Planet Centrifuge	265
	B. Analysis of the Countercurrent Chromatographic Separation	267
V.	USE OF THE HORIZONTAL FLOW-THROUGH COIL PLANET CENTRIFUGE	268
VI.	MULTILAYER COIL PLANET CENTRIFUGE	274
VII.	MULTICOIL COUNTERCURRENT CHROMATOGRAPH (MC-CCC)	276
	A. Operation	278
	B. Capability	283

VIII.	POTENTIAL INSTRUMENTS FOR CCC OF PEPTIDES	285
	A. Cross-Axis Coil Planet Centrifuge (CPC)	285
	B. "Foam" Chromatography	287
IX.	SUMMARY	289
	REFERENCES	290

I. INTRODUCTION

Today peptides are playing an important role in all areas of biological research. In biomedical, biotechnological, and agricultural research, peptide fragments are required in the investigation of proteins, especially for immunological techniques. In the promising area of diagnostics, peptides are a component. Peptide-derived compounds are emerging as useful drugs such as captopril. Future central nervous system drugs will evolve from the structure-activity studies of neuropeptides. Biotechnology products coming from recombinant sources are long peptides such as growth hormone and tissue plasminogen activator. Thus, there is much activity centering on peptide synthesis and isolation in research laboratories. Among the most exciting discoveries in recent years have been the discoveries of the endogenous opiate receptor ligand, enkephalin, and, more recently, with important implications in cardiovascular biology, the endothelin family of peptides.

For many purposes, peptides are required to be chromatographically pure in pharmacological and physiological in vivo and in vitro experiments, for conformational studies and, of course, for the ultimate application as therapeutics. Depending on the sequence, length, and structural modifications, the solubility of peptides can vary, ranging from peptides which have the characteristics of small organic molecules to large polymers with strong ionic charge or strong hydrophobicity. Thus, a variety of chromatographic techniques are needed. Reversed phase (RP) chromatography on hydrocarbon-bonded silica has proven suitable for purifying most peptides. However, not all types of peptides can be chromatographed because of their solubility characteristics. Silica is not suited for highly cationic compounds or very highly hydrophobic water-insoluble compounds. Even though RP supports provide high resolution and are used in analytical up to pilot scale chromatography, alternate modes of chromatography are needed for peptides.

Instrumentation for silica support-filled column chromatography has improved steadily over the last 20 years. A lucrative industry has developed in supplying filled columns and much research and development continues in finding higher resolution and more selective materials for the separation of biochemicals. The cost of these supports is quite high and contributes to the planned high cost of future therapeutics. Of the various types of chromatography that were known in the 1940s, liquid-liquid partition on solid support chromatography [1] has been exploited by the instrumentation industry. Left behind to be developed in research institutions has been the technological alternative, countercurrent chromatography, which is simply liquid-liquid partition chromatography. The success of reversed phase high-performance liquid chromatography (RP-HPLC) came about due to the suitability of aqueous acetonitrile solvent systems for the separation of peptides as well as other small molecular weight biochemicals. However, there are few useful solvent combinations for peptides. Thus, there remain types of peptides that cannot be chromatographed easily by RP-HPLC. Despite its great success as a technology, limitations remain. Thus, a technology that offers greater possibilities of solvent combinations bears consideration.

II. EARLY INSTRUMENTATION FOR THE COUNTERCURRENT CHROMATOGRAPHY OF PEPTIDES

Throughout recent years, the technology of countercurrent chromatography has largely been investigated in the Laboratory of Technical Development, National Heart, Lung, and Blood Institute of the National Institutes of Health, by Yoichiro Ito and his collaborators. Prior to coming to the United States, Ito had collaborated with K. Nunogaki in devising the first instrument in which centrifugal force was applied to countercurrent chromatography for stationary phase retention [2]. The centrifugal droplet countercurrent chromatograph evolved from this instrument and is presently commercialized and imported from Japan by Sanki Engineering Co. Ltd. Also, shortly after this time, a high-resolution device for countercurrent chromatography, the droplet countercurrent chromatograph, was developed [3]. These were the first new devices introduced after the popular countercurrent distribution (CCD) instrument, used since its introduction in the 1940s by Lyman Craig at the Rockefeller Institute.

At the National Institutes of Health, Ito described the hydrodynamic equilibrium behavior of a two-phase solvent system in a rotating coil and discovered that one phase could be retained and the other phase pumped through continuously [4]. Various orientations of the coiled tubing were studied. It was found that inflow tubing could be passed through a center axis of rotation and wound as a displaced and parallel coil mounted on a holder and rotated. The outflow returning through the axis of rotation

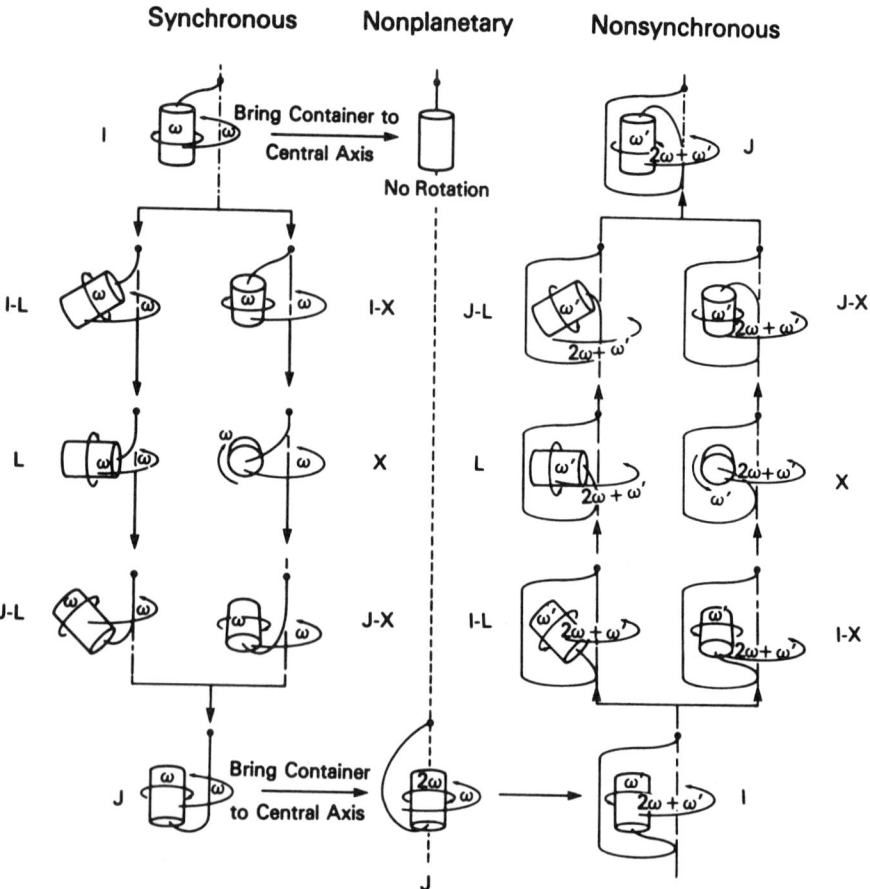

Fig. 1 The organization of seal-free flow-through centrifuge schemes of synchronous and nonsynchronous planetary motion possible with tubing coils. The various instruments studied are composed of a bundle of coiled tubing or a multilayered coil with inflow and outflow tubing. The coils revolve with their axis of revolution at the various orientations to the center axis of centrifugation. (From Ref. 46.)

could twist and untwist on itself in the motion without entangling. This was the seminal idea of the seal-free rotational coil of continuous tubing that was able to retain a stationary phase; while the mobile phase could be pumped through, and liquid-liquid partitioning could take place along the length of the tubing. The planetary motion induced by these configurations causes the mixing of the 2-phase solvent. The coil screw direction (having a head and tail) holds the stationary phase as its direction of movement is counteracted by the direction of the mobile phase being pumped in. Theoretically, a point in the coil is subjected to forces that change constantly in direction and magnitude. The forces alternate the processes of agitation and separation of the phase [46]. Thus substances are separated according to their partitioning in the solvent system.

Over the years, many different devices were designed with changed orientation of the coiled tubing relative to the center axis of revolution or rotation (Fig. 1). Due to stronger hydrodynamic forces produced by more rapid centrifugation, the efficiency of the separations increased (Fig. 2). Other instruments are being designed and studied for their capabilities. In many cases the studies are incomplete and are a fruitful area of research. Most of the newly designed instruments were tested with peptides. This chapter will describe briefly the early history of the use of countercurrent chromatography with peptides, followed by a report of the capability of present instrumentation, and concluding with a brief discussion of future possibilities of this technology to meet the needs of peptide separations.

A. Countercurrent Distribution

Countercurrent chromatography is liquid-liquid partition chromatography in which solutes are separated according to their relative solubility in a stationary liquid phase and a mobile liquid phase [5]. The early apparatus used for this process was the countercurrent distribution machine, which was a train of glass tubes. Some of these instruments contained 1000 tubes whereby two phases were mixed by rocking of the train, stopping in the vertical position and allowing the phases to separate, then moving to a position in which the upper phase is decanted to the next tube. Thus, a constant proportion of the solute is eluted with the mobile phase and is distributed throughout the fractions as a Gaussian peak. The separation units varied from 1 to 100 ml or more, and the equilibration times could be set for slow-separating solvent systems.

At present, a few laboratories use these instruments for large-scale separations of a half gram or more. The mobile phase is limited to the upper phase. A later instrument designed by Craig was the counter double-current distribution machine, which allowed mobility of both phases [6]. The mobility of the sample could be predicted from the partition coefficient. The shape of the peak, calculated by the Gaussian distribution law, indicated the purity of the substance. These instruments are still found in research and industrial

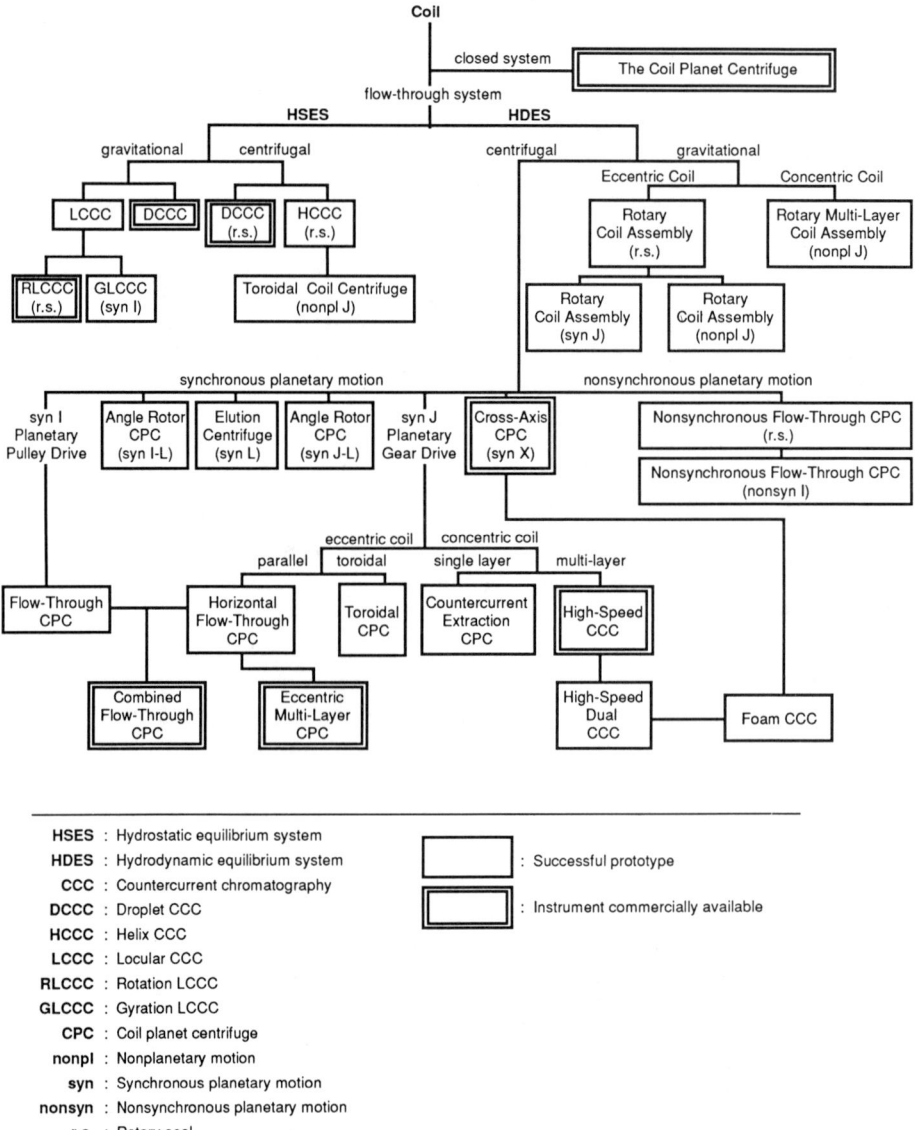

Fig. 2 Flow chart of the development of the instrumentation of CCC originating with the concept of a continuous coil.

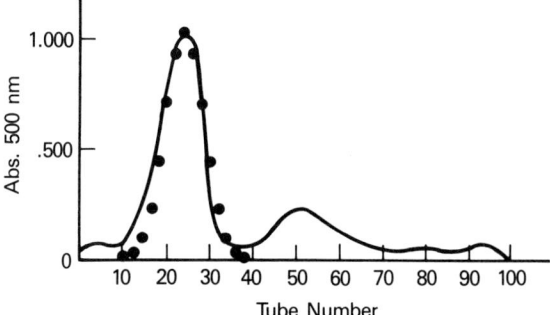

Fig. 3 Purification of the decapeptide, angiotensin I, in a countercurrent distribution machine of 10-ml-capacity tubes. A sample load of 151 mg was run in BAW for 200 transfers. Pure peptide, 63 mg, was recovered from fractions 15-30. The fractions were assayed by the Pauly reaction (solid line) that detects His- and Tyr-containing peptides. Theoretical curve (dots) was calculated. (Unpublished results of M. Knight and J. M. Stewart.)

laboratories which have been synthesizing peptides since the early 1960s. An example of the purification of angiotensin performed in a 100-tube instrument of 10-ml-capacity tubes is shown in Fig. 3. Figure 4 shows the results of a thyrotropin-releasing hormone (TRH) analog chromatographed in a 2000-tube instrument with 3-ml-capacity tubes.

Hypothalamic releasing factors [7] were isolated by CCD using various solvent systems. More recently, enkephalin-Arg was isolated from extracts of 470,000 porcine hypothalami using CCD in the course of an isolation procedure of a corticotropin-releasing factor [8]. Until the availability of laboratory scale preparative HPLC instruments, which have been available only since 1984, CCD was the main technology available for the preparative purification of peptides in the kilogram range. Some of the useful CCD solvent systems for peptides are given in Table 1.

B. Countercurrent Chromatography

During the early phase of Ito's research, a few different types of instruments were developed for CCC. These were the droplet countercurrent chromatograph [3] and other nonhelical instruments, such as rotational and gyrational locular CCC [9]. These instruments are now used primarily in the isolation of natural products [10]. The development of the coil planet centrifuge facilitated countercurrent movement of the two phases along the narrow bore of

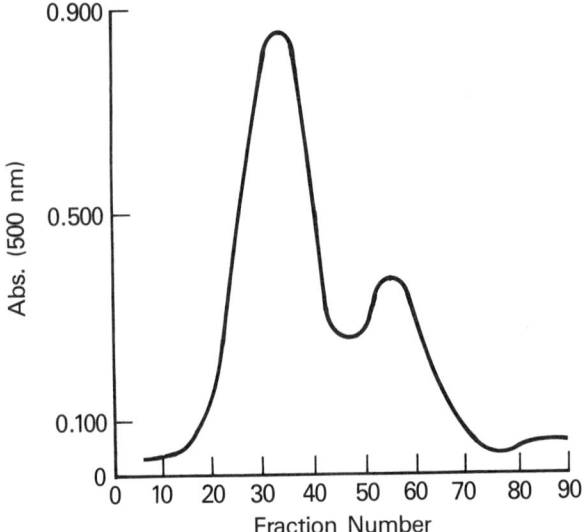

Fig. 4 Countercurrent distribution of D-Pyroglu-His-Pro-amide, an analog of thyrotropin-releasing hormone, in 0.1% acetic acid/n-butanol/pyridine (11:5:3) for 180 transfers in a 2000-tube instrument of 3-ml-capacity tubes. The pure peptide, 32 mg, was contained in fractions 16-39 and the peptide methyl ester side product was recovered from fractions 40-77. The K value calculated from the run is 0.14.

Table 1 Solvent Systems for CCD and CCC of Peptides[a]

Solvent composition	Volume ratio	Ref.
N-Butanol/acetic acid/water	4:1:5	40
1% Tfa/n-butanol	1:1	48
0.2-0.5 M NH_4acetate/n-butanol	1:1	22
0.1% Acetic acid/n-butanol/pyridine	11:5:3	7
1% DCA/sec-butanol	1:1	42
Chloroform/acetic acid/water	2:2:1	15
Methanol/chloroform/0.1 N HCl	2:2:1	43

[a]These are commonly used solvent systems. There are comprehensive listings of solvent systems used in the early literature of CCD [43,44]. CCC of peptides was reviewed previously [45].

Fig. 5 The horizontal flow-through coil planet centrifuge, a prototype built by Kontes. The upper row of coils on the gear side is made of 1.6-mm-i.d. tubing and is used for preparative chromatography. The inflow tubing enters through the center shaft from the right and extends from the left up to the holder shaft on the upper side; the tubing is wound around the rods and the outflow tubing returns from the holder shaft to the center out the same way. The tubing is clamped on the outside. When the instrument is rotated, the tubing twists and untwists and does not tangle in the rotation. On the lower side are the column coils of 0.55-mm-i.d. tubing suitable for small-scale chromatography mounted inside the covering rods. The inflow tubing comes from the pump and the outflow tubing is connected to a flow cell and fraction collector (not shown). (Reprinted with permission from Ref. 24.)

coiled tubing by rotation in a strong centrifugal force field. Study of the orientation of the coiled tubing with respect to the central axis of rotation and experimentation with the shape of the tubing itself produced systems of designs shown in Fig. 1. Many, but not all, of the schemes shown here have been tested in instruments built by Ito. An early successful instrument was based on mounting one coil on a side of the rotating unit and a counterbalancing coil on the other side with tubing entering the center shaft and coiled around column coils as in scheme J (Fig. 1). The other side was a bundle of column coils oriented as in I. The angular rotation is as shown. These are examples of synchronous planetary motion. The instrument built from this design shown in Fig. 5 is the horizontal flow-through coil planet centrifuge [11], and the coil of 1.6 or 2.6 mm PTFE tubing, mounted on one side or the pulley side, was used for many years in the semipreparative purification of synthetic peptides. The opposite column coils were of 0.5-mm-i.d. tubing for small-scale samples. Early experiments by Ito separated dipeptides and other peptides [12]. In 1979, an enkephalin analog was purified in this instrument using a solvent system of 1% aqueous trifluoroacetic acid (Tfa) and n-butanol (1:1 by volume) [13]. Since then, much work has been done with cholecystokinin and bombesin analogs, as well as various other peptides [14,15].

CCC in the coil planet centrifuge proved to be an advance over CCD. In CCC the purified material was eluted in a smaller volume than from CCD. Furthermore, either phase could be used as the mobile phase. The mobile phase of higher retention could be chosen to maximize the separation. Additionally, the instrument is simpler mechanically and much smaller than the large glass manifold.

III. THEORY OF COUNTERCURRENT CHROMATOGRAPHY

The relationships of chromatography apply to CCC as a form of liquid-liquid partition chromatography with the stationary phase held in the coiled tubing and not on a solid support [16]. The elution or retention of the solute is dependent on the ratio of its concentration in each phase. The concentrations in each phase are related to the capacity factor k':

$$k' = \frac{Q_s}{Q_m} = \frac{\text{quantity of solute in stationary phase}}{\text{quantity of solute in mobile phase}} = \frac{C_s V_s}{C_m V_m} = K \frac{V_s}{V_m} \quad (1)$$

K, the partition coefficient, can be defined as either the ratio of the concentration of the solute in the stationary phase to that in the mobile phase or the inverse of that ratio. V_s and V_m are the volumes of stationary and mobile phases retained in the coil at equilibrium during the chromatographic run.

A. Partition Coefficient Relationships

Conway [16] described partition coefficient relationships in terms of column liquid chromatography and defined K as C_s/C_m; however, former users of CCD define K as C_m/C_s. The derived functions

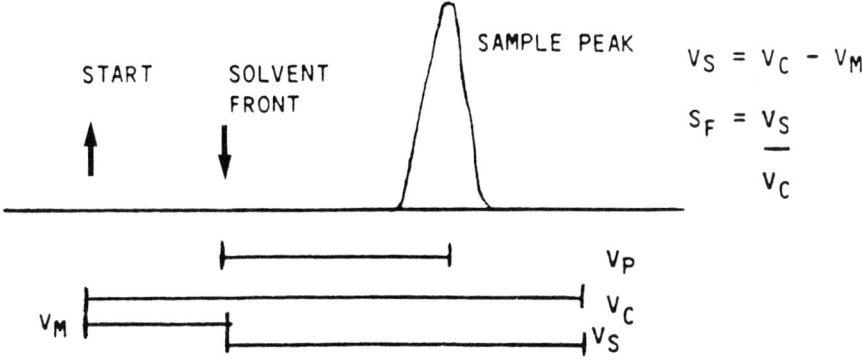

Fig. 6.

are similar to the retention-time relationship of column liquid chromatography. An important difference is the absence of the dead volume occupied by the solid support matrix.

The stationary phase fraction (S_f) is the volume of the coil occupied by the stationary phase as the mobile phase is pumped through the coil (Fig. 6). S_f is the stationary phase remaining in the coil after displacement by the mobile phase at equilibrium, or the ratio of the stationary phase volume to the total capacity of the coil. Total volume of the coil minus the mobile phase volume is the stationary phase volume.

$$V_s = V_c - V_m \tag{2}$$

$$S_f = \frac{V_s}{V_c} \tag{3}$$

$$\frac{V_s}{V_m} = \frac{S_f}{1 - S_f} \tag{4}$$

The capacity factor is related to the partition coefficient by

$$k' = K\left(\frac{S_f}{1 - S_f}\right) \tag{5}$$

The elution volume of a sample is analogous to the retention time (t_R) in a column. Thus, k' is the ratio of the time (t_s) or volume (V_s) spent by the solute in the stationary phase to the mobile phase time or volume.

$$k' = \frac{t_R - t_m}{t_m}; \quad k' = \frac{V_R - V_m}{V_m} \quad (6)$$

All chromatographic relationships, retention, separation, and resolution are described in terms of K, and from the elution volume, efficiency (N) can be determined.

B. Resolution

The separation factor $\alpha = K_2/K_1 = k_2'/k_1'$, where $K_2 > K_1$ and α is greater than 1.

The factors determining resolution according to the Knox equation:

$$R_s = \frac{1}{4}(\alpha - 1)\sqrt{N}\left(\frac{k_1'}{k_1' + 1}\right) \quad (7)$$

can be determined by the equivalent equation in countercurrent chromatography:

$$R_s = \frac{1}{4}(\alpha - 1)\sqrt{N}\,\frac{K_1}{K_1 + [(1 - S_f)/S_f]} \quad (8)$$

The separation factor α is determined by the solvent system and the resolution is increased with higher S_f.

The Ito coil planet centrifuges have S_f values of at least 0.4 and the most efficient, the multilayer coil planet centrifuge, has an S_f of 0.8. The stationary phase volume of RP-HPLC columns has been estimated to be 14%. Thus countercurrent chromatography has an obvious advantage for preparative capability.

The efficiency of N can be calculated from the elution volume of the sample and values of over 1000 are suitable for preparative separations. Interested readers are referred to more extensive descriptions of countercurrent chromatography theory [16] and resolution in CCC [17]. The equations are derived from the K defined as C_s/C_m and the comparisons of functional parameters of instruments are based on this definition; especially for the widely used multilayer coil planet centrifuge capable of high-speed countercurrent chromatography (HSCCC). However, for the work described here, which is largely preparative with the goal of maximum separation of eluents, the K as C_m/C_s is used. Solvent systems devised for higher retention (elution volume) are those with K values less than 1. Thus, later eluting peaks have a K value equal to 1 or less (0.5-0.01).

IV. METHODS

Generally, the coil planet centrifuges are operated as liquid chromatography systems. Solvent delivery is provided by an LC pump able to handle organic solvents and withstand moderate back pressures of not more than 1000 psi. Although monitoring systems and recorders for open-column chromatography can be used, in-line detection needs to be improved to remove the interference of phase droplets in the recorder tracing.

A. Operation of the Coil Planet Centrifuge

The solvent systems must be selected by determining the partition coefficient of the sample. The coefficient is determined by dissolving the dry sample in equal volumes of each phase of a solvent system and determining the concentration in the upper phase and the concentration in the lower phase. The most accurate method to use is analytical HPLC, which separates the impurities, and the peak ratio of the compound represents the K [18]. The determination can be made simply by reading the UV absorbance of the sample in each phase; however, these values include impurities. There are reviews on other techniques [19]. The ideal K is around 1 and a mobile phase should be chosen that gives the higher retention which is the lower K value. For example, if the K for U/L is 2, then it is better to run with a lower mobile phase, which would give a ratio of 0.5 with the peak fractions well behind the solvent front. If run in the other mode, then the material may be eluted too close to the solvent front. Solvent systems can be modified as described in the method development of bombesin [20] in which the percentage of Tfa or dichloroacetic acid (DCA) in a 1:1 mixture with n-butanol was changed so that K was a useful value. There is presently a need for other volatile solvent systems that can be adjusted to modify more readily the K value. The best range of K for most of the instruments is between 1 and 0.1.

For the chromatography, the solvent system is prepared by combining, in a large separatory funnel, the volumes required, equilibrating them, and then separating the phases. After the coil is filled with the stationary phase, the sample is loaded. Up to 1 g can be dissolved in equal volumes of the solvent phases in 10 ml or up to no more than 1/10 the total volume of the coil. The sample solution can simply be pushed into the coil via a syringe either directly or by way of a slide valve. Sample loading devices are also available. The rotation is started and when it is at the rpm of the run, the mobile phase is pumped at the flow rate suitable for the instrument and solvent system (see Table 2). The effluent is passed through a UV monitor to a fraction collector. Usually

Table 2 Peptide Chromatography in the Ito Coil Planet Centrifuges

Multicoil Countercurrent Chromatograph

Conditions:
1. Either phase mobile
2. Room temperature operation
3. 60 ml/hr flow, 500-800 rpm

Solvent systems	Volume ratio	Ref.
All solvent systems, examples:		
BAW	4:1:5	32
1% Tfa/n-butanol	1:1	23
Chloroform/acetic acid/water	2:2:1	23

Horizontal Flow-Through Coil Planet Centrifuge

Conditions:
1. Either phase mobile
2. Room temperature operation
3. 24 ml/hr flow, 400 rpm

Solvent systems	Volume ratio	Ref.
All solvent systems, examples:		
0.5-2% DCA/sec-butanol	1:1	45
0.2-0.4 M NH_4OAc, pH 6-9/n-butanol	1:1	14
BAW	4:1:5	12
Ethanol/n-butanol/hexane/acetic acid/water	1:3:2:1:5	14
0.1% Tfa/n-butanol	1:1	13
Dichloromethane/acetic acid/water	2:1:2	45
Chloroform/benzene/methanol/water	15:15:23:7	33

Multilayer Coil Planet Centrifuge

Conditions: (for polar solvents)
1. Lower phase mobile
2. 40-50°C
3. 60-150 ml/hr flow, 800 rpm

Solvent systems	Volume ratio	Ref.
0.5-1% Tfa/n-butanol	1:1	29
0.3-3% DCA/n-butanol	1:1	20
BAW	9:1:10	29
0.2 M NH_4OAc/n-butanol (upper ph. possible)	1:1	29

Conditions:
1. Either phase mobile
2. Room temperature
3. 100-300 ml/hr, 800-1000 rpm

Chloroform/acetic acid/water	2:2:1	15

6-15 ml fractions is collected. The run is continued for two column volumes. There are then the choice of stopping the rotation and pumping out the contents, while still collecting fractions, or changing the stationary phase to become the mobile phase. If the sample has a K near 1, then the sample will be eluted before the two column volumes.

In-line monitoring for UV-absorbing samples is possible with the LKB 6 channel chopper recorder. Recently, by modifiying the tubing leading to a UV monitor, the phase droplets were cleared, thus reducing the interference in the absorbance tracing [21]. Usually aliquots of the fractions are diluted in an alcohol-water mixture and the absorbance read manually. If the peptide is not UV-absorbing, a colorimetric determination, such as the Folin Lowry assay, can be done. These procedures are available in Ref. 22. After the peptide peak is located, it is advisable to analyze a few fractions across the peak by analytical HPLC (as in Fig. 21) to determine the purest fractions which can be combined. The other fractions can also be pooled depending on the requirements. The fractions are dried by rotary evaporation, dissolved in either water or glacial acetic acid, and lyophilized to a white powder. The smaller volume of solvent requires a shorter lyophilization time (compared to solvent removal from preparative HPLC fractions). The peptide then can be further analyzed and/or used in experiments.

To clear the coil, the contents of the coil are pushed out by air or N_2 pressure and rinsed by pumping in a wash solvent such as 50 ml acetone. The coil is finally dried with a stream of N_2.

B. Analysis of the Countercurrent Chromatographic Separation

From the elution volume of the sample, the partition coefficient (K) can be calculated. In this work the K is defined as C_m/C_s. The K is equal to the ratio of the stationary phase volume retained in the column to the retention volume of the sample after the solvent front. The stationary phase volume is the total capacity of the coil V_c minus the excluded volume V_m (determined by the solvent front tube or fraction number times the volume of the fraction tube) (Fig. 6).

$$V_s = V_c - V_m \tag{9}$$

The volume of elution or retention (V_p or V_r) is the peak fraction number times the fraction volume.

$$K = \frac{V_c - V_m}{V_p - V_m} \tag{10}$$

As an example, in an instrument of 385-ml total capacity and when fractions of 15 ml are collected during a run, for a sample eluting at fraction #35 and the solvent front having eluted at fraction #15, the K is calculated as follows:

$V_m = 15 \times 15$ ml $= 225$ ml

$V_s = 385$ ml $- 225$ ml $= 160$ ml

$V_p - V_m = 525$ ml $- 225$ ml $= 300$ ml

$$K = \frac{V_s}{V_p - V_m} = \frac{V_c - V_m}{V_p - V_m} = \frac{160 \text{ ml}}{300 \text{ ml}} = 0.53 \tag{11}$$

Additionally, the efficiency of the separation can be determined by the tangential method, which is used in gas chromatography.

$$N = (4 \text{ R/W})^2 \tag{12}$$

Theoretical plates are calculated by the shape of the peak. For this calculation R is used as the retention volume of the peak maximum (V_p) and W as the peak width expressed in the same units as R (total volume of the peak fractions). In Fig. 24 calculation of N gives a value of 1600 theoretical plates for the peptide separated in the multicoil countercurrent chromatograph [23] and over 500 in the purification of bombesin in the multilayer coil planet centrifuge [20]. N values of 500 and above are suitable for preparative chromatography, especially if there is complete removal of the impurities.

V. USE OF THE HORIZONTAL FLOW-THROUGH COIL PLANET CENTRIFUGE

The series of CCK-8 fragment peptides that were investigated for CCK inhibitory activity were synthesized by solid phase techniques and purified by CCC on the horizontal flow-through coil planet centrifuge (HFCPC). Some of the small, aromatic, negatively charged peptides were not water-soluble, but were easily loaded into the coil dissolved in the two-phase solvent system. In Fig. 7 is shown a purification of Ac-Asp-Tyr-Met-Gly-Trp-Met-Asp-NH$_2$ in n-butanol/acetic acid/water (BAW; 4:1:5 by volume) with the lower phase mobile run in the conditions of low K (0.03), which separated the impurities better [24]. If run in the high-K mode, the peptide would have eluted with the solvent front. After purification, the peptide was then sulfated by reaction with the pyridine-sulfur trioxide complex and, after drying, treated with NaHCO$_3$ and lyophilized. The reaction mixture was separated by CCC in ammonium acetate, pH 7-9,

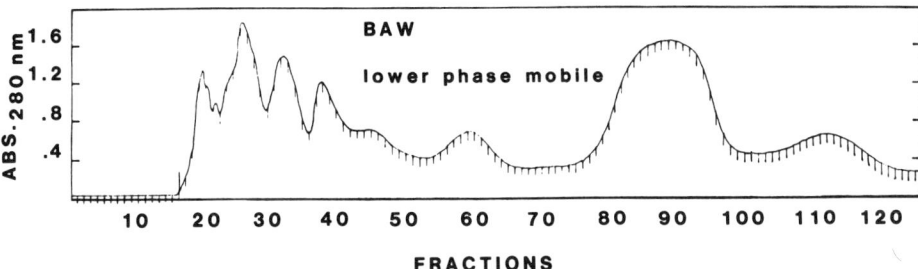

Fig. 7 Chromatography of the CCK (26-32) fragment peptide in the BAW system with the lower phase mobile in the HFCPC at 24 ml/hr and 400 rpm. The solvent front came out at fraction 17 and 60 mg of pure peptide were in fractions 76-98. All peptides were characterized for purity by HPLC and amino acid composition analysis.

and n-butanol in order to maintain the sulfate ester of the Tyr. An example of a separation of a sulfated peptide from the reaction mixture is shown in Fig. 8. The late eluting components are the side products and salt in large amounts. Details of procedures using the HFCPC for the CCK peptides were previously described [14,25]. As a result of many separations, it was found that the solvent systems that were previously used in CCD are effective with peptides which are polar compounds. Peptides that are negatively charged and have aromatic residues chromatograph well in n-butanol solvent systems using the mobile phase with the lower partition coefficient.

In Figs. 9-11 are shown successful separations of crude synthetic peptides in the BAW system with upper phase mobile with sample loads up to 1 g. These peptides are representative of peptides synthesized for making immunogens. They are highly hydrophilic sequences. And are usually 10- to 20-mers which have multiple cationic and anionic side groups with some aromatic groups. They are chromatographed in the BAW or 1-5% Tfa/n-butanol solvent systems with the mobile phase giving some retention and maximum separation of impurities. In Fig. 12 is shown a separation using the lower phase as the mobile phase.

Small, basic peptides without aromatic groups seem to be the most difficult to retain using the present solvent systems [15]. They are also not retained by RP-HPLC either. New solvent systems are needed for this purpose. In Table 1 are listed solvent systems used in the countercurrent chromatography of peptides.

In our continuing studies of neuropeptide analogs which are designed for high potency and bioavailability or for use as inhibitors,

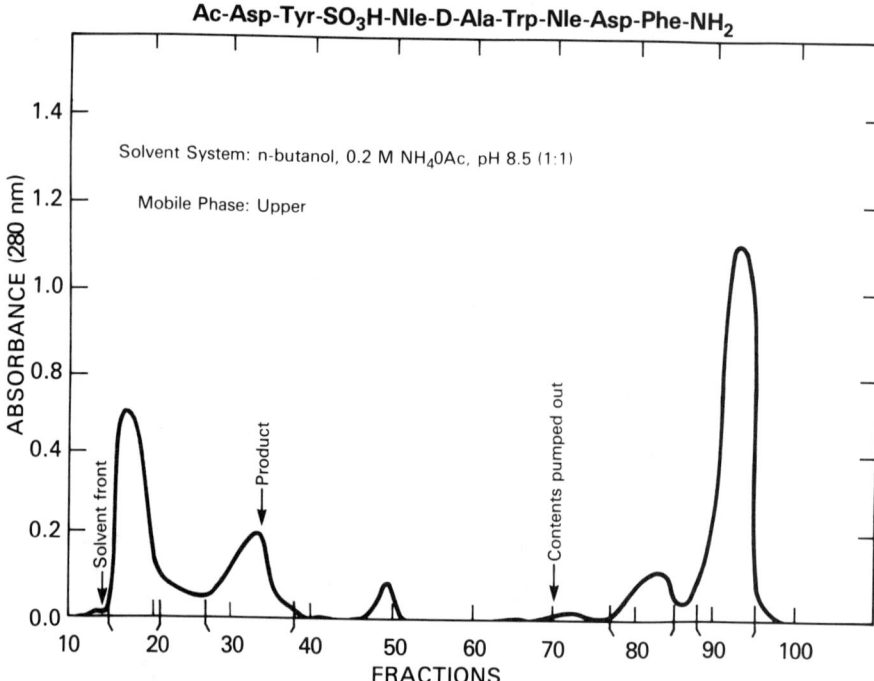

Fig. 8 Separation of a CCK analog from a sulfation reaction mixture in the HFCPC. Unsulfated peptide, 120 mg, was reacted with 176 mg of the pyridine-sulfur trioxide complex. After neutralization with NaHCO$_3$ and lyophilization, the mixture was chromatographed in 0.2 M NH$_4$ acetate, pH 8.5/n-butanol with the upper phase used as the mobile phase. The solvent front emerged at fraction 15. The rotation was stopped at fraction 70 and the contents were pumped out while collection of 6-ml fractions was continued. The peak at the solvent front contained 38 mg unreacted peptide, fractions 27-38 contained 23 mg of the sulfated octapeptide, and the final material (88-95) contained 550 mg mixed salts and side products.

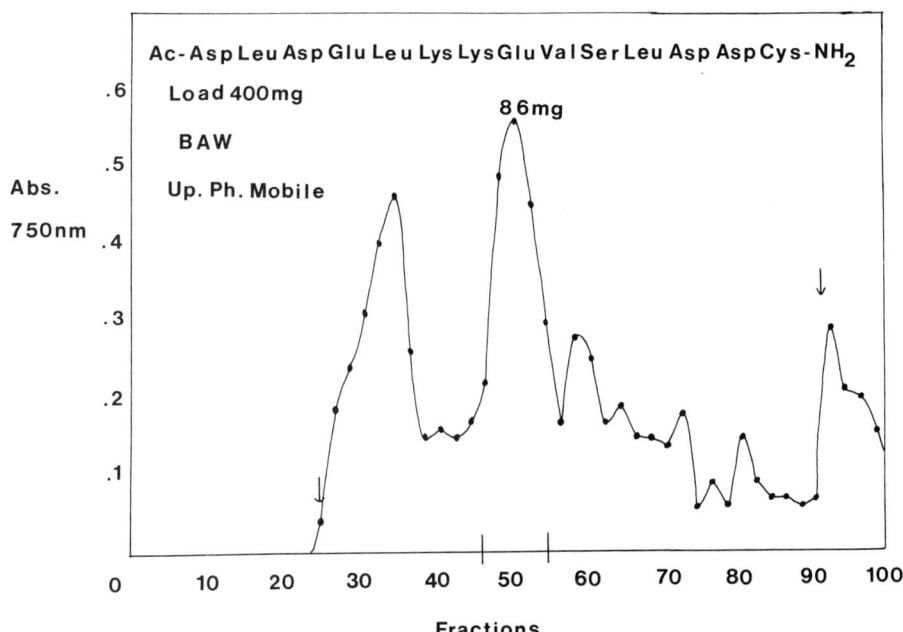

Fig. 9 Chromatography of a tetradecapeptide on the HFCPC in the conditions shown. The fractions were analyzed by the Folin Lowry reaction. Fractions 47-55 contained the purified peptide. The arrow at fraction 25 indicates the solvent front and at fraction 92 is indicated when the contents were pumped out and fractions collected. This is an example of an acidic peptide without an aromatic group.

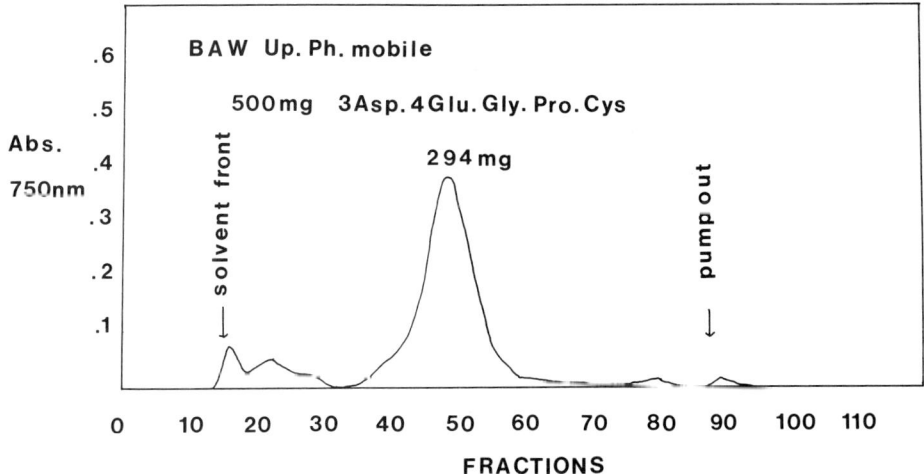

Fig. 10 Chromatography of an acidic decapeptide in the BAW solvent system with the upper phase mobile in the HFCPC in the usual conditions.

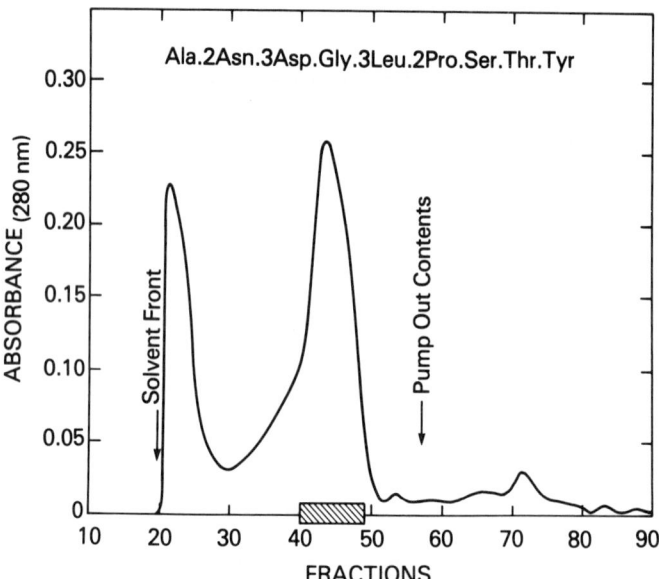

Fig. 11 Chromatography of an acidic pentadecapeptide with an aromatic group in the HFCPC in the BAW system with the upper phase mobile. The sample load was 1 g. The pure peptide was present in fractions 40-49.

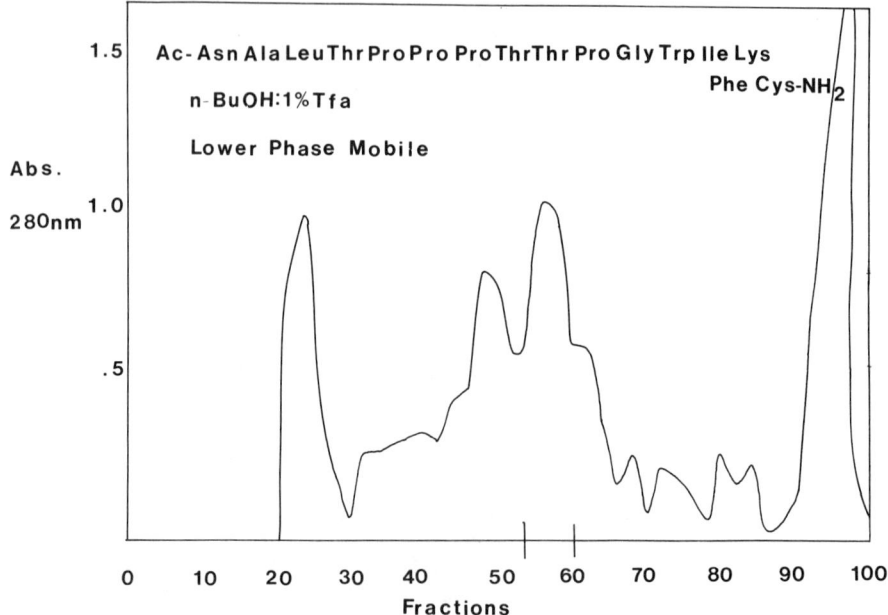

Fig. 12 Chromatography of a positively charged peptide with aromatic groups in the 1% Tfa/n-butanol system with the lower aqueous phase as the mobile phase. The purified peptide was located in fractions 54-60.

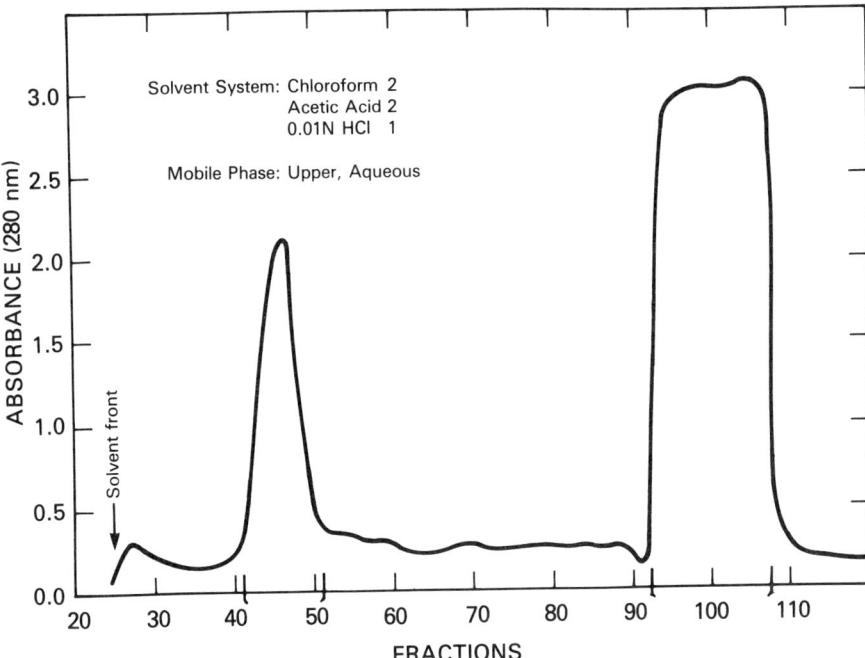

Fig. 13 Purification of 500 mg of a CCK derivative in the chloroform/acetic acid/0.01 N HCl solvent system in the HFCPC in the usual conditions. Fractions of 10 ml were collected and 130 mg of pure peptide was contained in fractions 41-51. The second peak contained impurities.

we have synthesized neuropeptides with modifications that usually make the compounds less water-soluble. Thus, loading these compounds on RP-HPLC is difficult. In CCC however, loading is not a problem in two-phase solvent systems. Nevertheless, alternate solvent systems had to be found for these peptides as they had high K values in the n-butanol systems. Thus, polar solvent systems used for DNP amino acid separations [12] were tried and found to be successful. Figure 13 is the chromatogram of a naphthoyl-CCK analog in chloroform/acetic acid/0.1 N HCl (2:2:1). The compounds were eluted with the aqueous upper phase. Later the solvent system was modified by removing the HCl. Many bombesin analogs were purified with this solvent system (Fig. 14) [15]. Currently, other strong solvent systems, or compositions using formic acid and toluene, have been proposed for the isolation of longer hydrophobic peptides [26]. Probably different solvent systems will appear in the future.

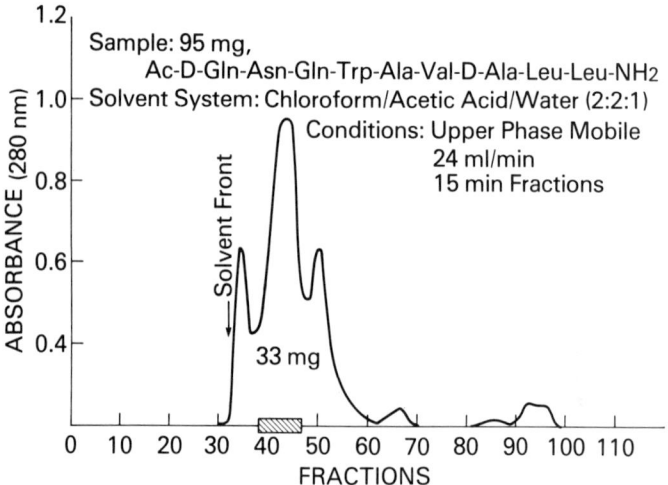

Fig. 14 An example of the purification of a bombesin analog by CCC in the HFCPC in the conditions as indicated. (Reprinted from Ref. 15.)

VI. MULTILAYER COIL PLANET CENTRIFUGE

A significant advance in the technology happened when Ito discovered unilateral hydrodynamic equilibrium occurring at high centrifugation rates in the multilayer coil [27]. The phases are unilaterally separated to either end in the coil by a strong force, against which it is necessary to pump the mobile phase at a faster speed. The stationary phase fraction is high, 80%, and so is the efficiency. Faster rotation and elution flow rate are possible in the smaller compact width of the multilayered coil of tubing compared to the original wider instrument with single-layered coils of tubing wrapped on long rods. Thus for certain solvent systems, rapid chromatography is possible in the multilayer coil planet centrifuge (MLCPC) pictured in Fig. 15. Conditions of rapid chromatography were tried with peptides but problems occurred due to the lack of stationary phase retention with the n-butanol solvent systems [28]. It was found that solvent systems with large density differences between the phases such as chloroform or hexane aqueous systems are held in the MLCPC at high speeds. However, butanol aqueous systems,

Fig. 15 The multilayer coil planet centrifuge with one concentric coil of 1.6-mm-i.d. tubing with a capacity of approximately 325 ml. The inflow and outflow is through the top. This commercially available unit rotates horizontally or sideways.

especially the BAW system, with a smaller density difference are not, because the butanol solvent systems have longer settling times after mixing [28]. Since settling time is reduced at higher temperatures, experiments were conducted at 45°C. Separations of synthetic bombesin and CCK peptides [29] were accomplished using 1-5% aqueous Tfa or DCA/n-butanol solvent systems. An example of the purification of the same peptide as shown in Fig. 7 is shown in Fig. 16 in another solvent system at elevated temperature [30]. Under these conditions, the peptide was purified in a short time. Only certain solvent compositions work in the MLCPC at the high speed of 800-1000 rpm and flow rates of 60-120 ml/hr [29]. The conditions are summarized in Table 2. For smaller molecules, the MLCPC is a high-performance instrument useful for preparative and analytical applications such as natural products isolation and purification of synthetic organic compounds. Nevertheless, for peptides which are best separated in the heavy alcohols, high-speed CCC conditions have not yet been completely developed. Development of heated instrumentation would be useful for this purpose.

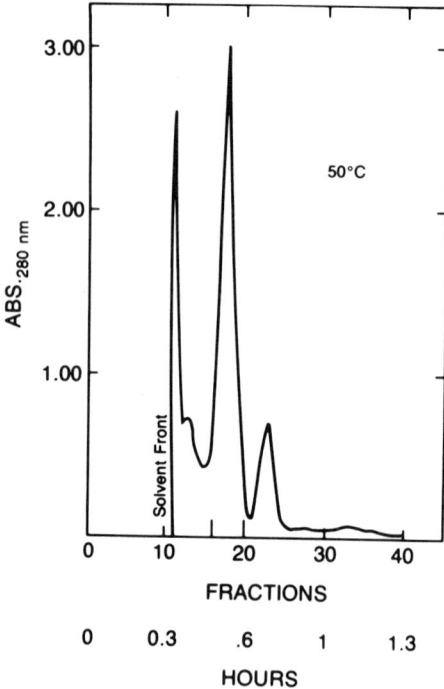

Fig. 16 Chromatography of Ac-Asp-Tyr-Met-Gly-Trp-Met-Asp-NH$_2$ in the MLCPC in 0.2 M NH$_4$ acetate, pH 9.5/n-butanol at 50°C. The lower phase was used as the mobile phase and the sample load was 15 mg. 10 mg was recovered from fractions 16-20. The flow was 150 ml/hr and centrifugation was 800 rpm. Fractions of 2 min or 6 ml were collected. (Reprinted from Ref. 30.)

VII. MULTICOIL COUNTERCURRENT CHROMATOGRAPH (MC-CCC)

More recently, Ito incorporated aspects of the former designs into an instrument with compact coils, mounted as in the original HFCPC [31]. Instead of one coil on a holder (Fig. 17) [23], a few coils in series are mounted on a column holder. In this instrument four coils are each mounted around two counterbalancing column holders (Fig. 18). Thus this coil planet centrifuge has eight coils connected in series. The total volume of the coils is more than that of the HFCPC.

The instrument is indicated in the flow diagram of the CCC technology (Fig. 2) and orientation J in Fig. 1 of the diagram of synchronous planetary motion. This assembly is only 18 cm wide,

Fig. 17 The multicoil countercurrent chromatograph or "eccentric" multilayer coil planet centrifuge equipped with eight coils in series. At left is the controller with digital rpm display. At the inlet is connected a pressure gauge and the outlet tubing emerges at right to connect to fraction collector.

$$\beta = \frac{r}{R}$$

Fig. 18 Cross section diagram of interconnected coils in the MC-CCC.

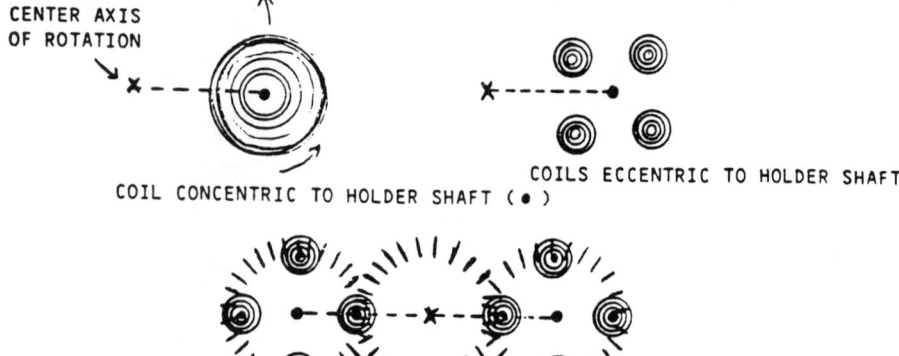

Fig. 19 Cross-section diagram of coils in the ML-CPC and the MC-CCC.

and the capacity is maintained by the series of compact multilayered coils instead of single-layered coils wrapped around long rods. The coils are equally spaced around the holder shaft but displaced parallel to the central axis of rotation. Ito calls this position "eccentric" (eccentric multilayer CPC) as distinguished from "concentric" in which the coil is on the holder shaft (Fig. 19) as for the MLCPC. All the coils are connected in series. The tubing enters the system through the central axis of revolution to a holder shaft and out to wind around a coil. From the outer layer the tubing goes to the center of the next coil and so on. The exit tubing passes from the holder shaft to inside the stationary pipe and out the center shaft through an opening on the other side. The present instrument built by Varex has heating capability to 50°C [23]. The rotary unit is encased in a steel box, which is lined with anodized aluminum, and has a top lid and front window. The name of the instrument has been simplified to the multicoil countercurrent chromatograph (MC-CCC).

A. Operation

The MC-CCC is presently the most versatile of all the Ito coil planet centrifuges with the capability of operating with all the solvent systems at room temperature. It cannot be operated at the very high speeds and the flow rates of the MLCPC but it does run at higher rpm and flow rates than the HFCPC: 500-800 rpm and a flow of 60 ml/hr compared to 400 rpm and a flow rate of 24 ml/hr. Thus the MC-CCC is a higher performance instrument than the HFCPC.

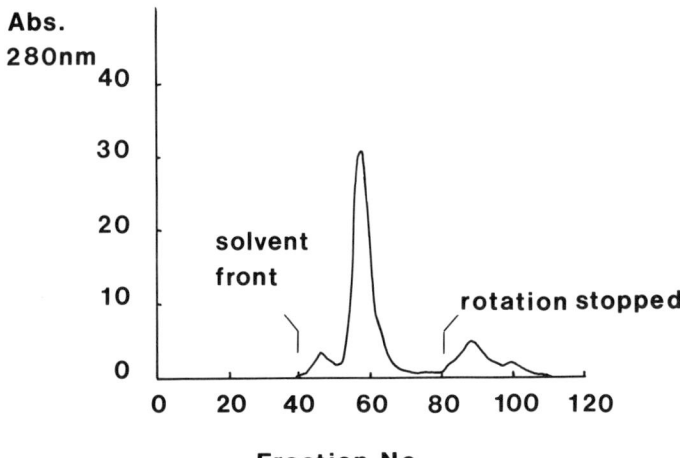

Fig. 20 Total absorbance of the fractions of a chromatography of 225 mg of Ac-Glu-Glu-Trp-Asp-Pro-Ser-Asp-Gln-Glu-Pro-Cys-NH$_2$ in a MC-CCC, NIH prototype of four coils in series. The sample was run in the lower phase of the BAW system at a flow of 60 ml/hr, 850 rpm. The K value was 1.6 and, as a result of analysis of the fractions shown in Fig. 21, the fractions were combined in three pools: 53-56, 27 mg; 57-59, 54 mg, purest material; and 60-67, 37 mg. (Reprinted from Ref. 32.)

In recent experiments the stationary phase fraction was found to average 40% with three commonly used solvent systems [23]. The phase fraction is slightly higher than that of the older instrument. A few peptides have been chromatographed in this instrument that were also chromatographed in the HFCPC. For these peptides the resolution was improved [32,23]. An undecapeptide was chromatographed with the lower phase mobile (Fig. 20) and the fractions across the peak analyzed by HPLC. As shown in Fig. 21, the center of the peak contained pure peptide.

The solvent systems used in CCC are chosen with the optimal K of less than 1. These conditions are comparable to isocratic elution, which affords maximum separation. Gradients in CCC have been used and would be suitable for speeding up elution of late components [33]. Another advantage of CCC for peptide purification is the capability to remove the large mass of impurities present in crude samples especially of longer peptides. The peptides after

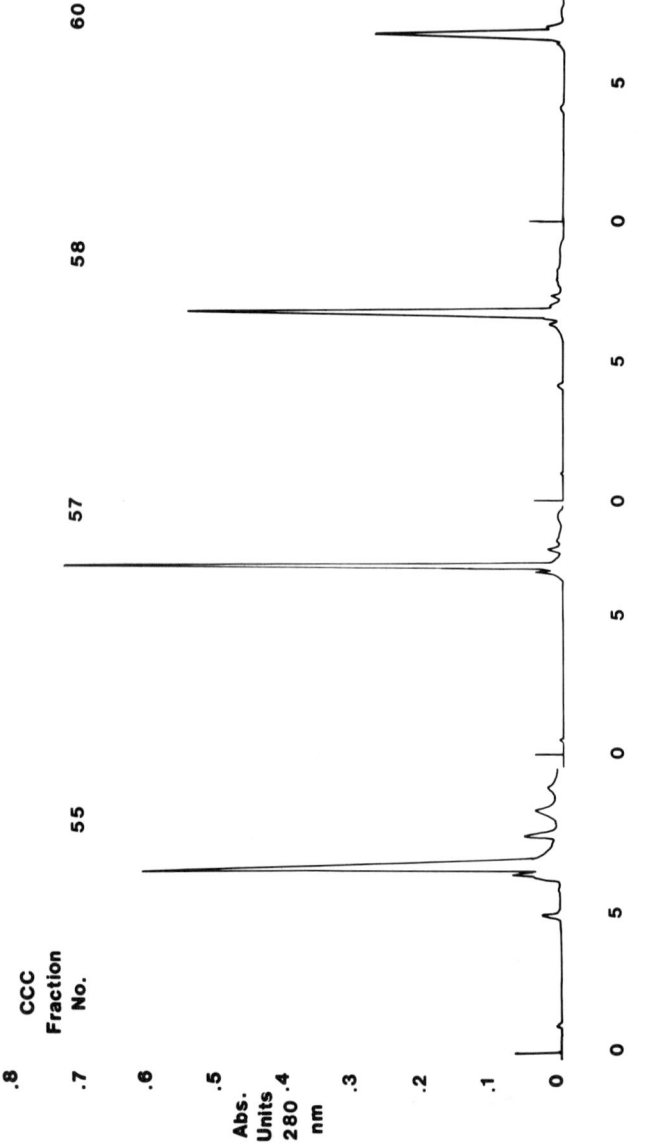

Fig. 21 Analytical HPLC of fractions of the separation in Fig. 20. Run on a μBondapak C_{18} column (15 × 0.4 cm i.d.) in 0.1% phosphoric acid with a gradient of 10-30% acetonitrile in 10 min, at a flow of 0.8 ml/min with detection at 280 nm. (Reprinted from Ref. 32.)

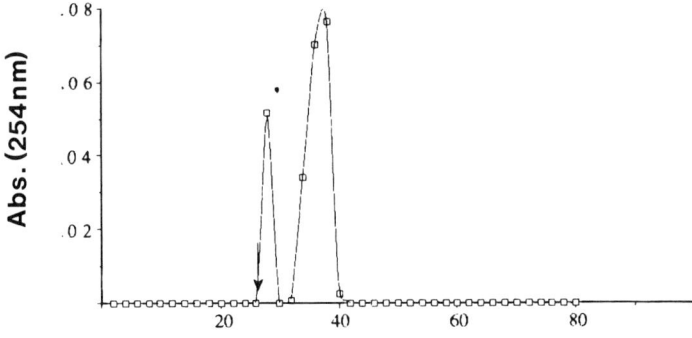

Fig. 22 Chromatography of myristoyl-Gly-Asn-Ile-Phe-Ala-Asn-Leu-Phe-Lys-Gly-Leu-Phe-Gly-Lys-Lys-Glu-NH_2 on the MC-CCC. A 26-mg previously purified sample was run in the BAW system with the upper phase mobile at 500 rpm, 60 ml/hr, and 15-ml fractions collected. The absorbance at 254 nm was determined. The solvent front emerged at fraction 27 and purified peptide was contained in fractions 33-37. (Reprinted from Ref. 18.)

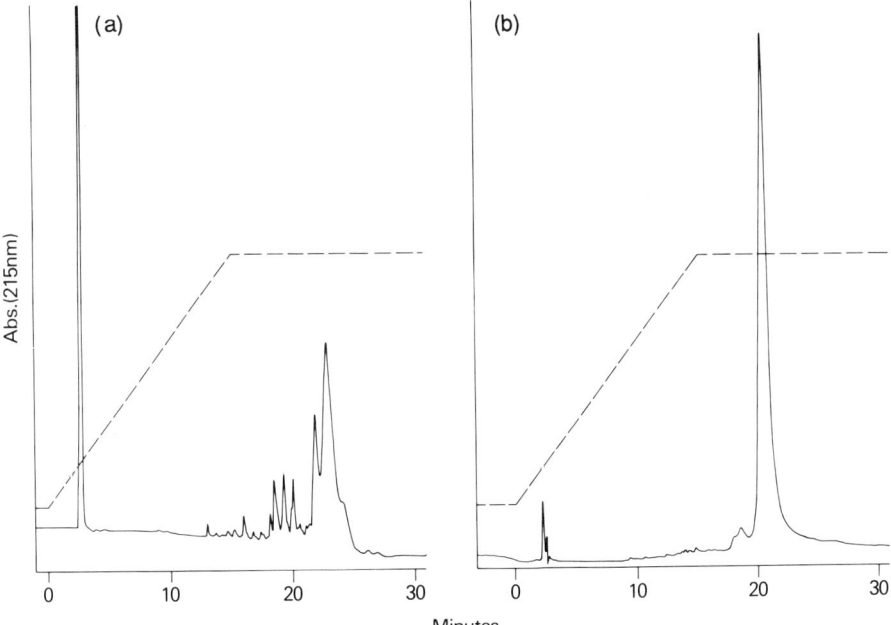

Fig. 23 (a) Analytical HPLC of the sample loaded in the CCC of Fig. 22 under conditions described in Fig. 21 except that the gradient of acetonitrile was from 20% to 60% in 15 min at a flow of 0.8 ml/min. (b) Analytical HPLC of the peptide recovered from the CCC in similar conditions except that the flow was 1 ml/min. (Reprinted from Ref. 18.)

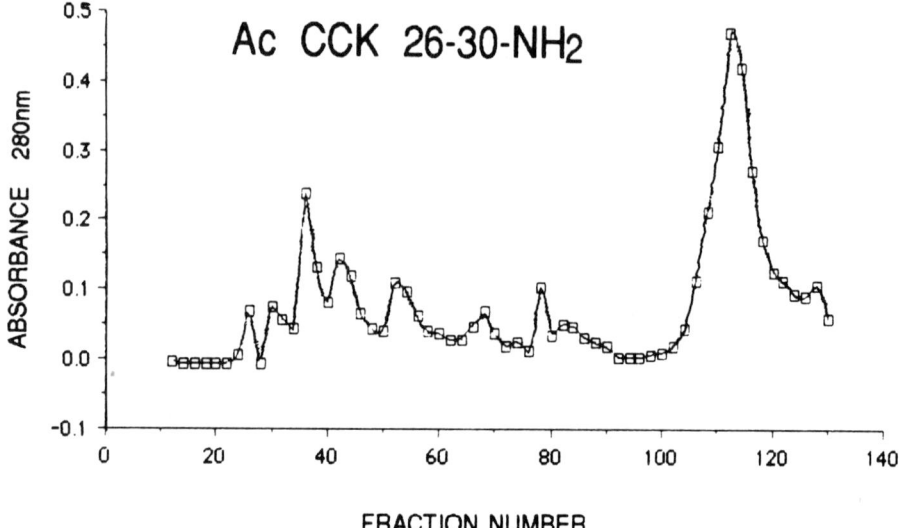

Fig. 24 Countercurrent chromatography of Ac-Asp-Tyr-Met-Gly-Trp-NH$_2$ in the MC-CCC in the BAW solvent system with the lower phase mobile at 500 rpm and 60 ml/hr. The sample load was 200 mg of unpurified peptide and 56 mg pure peptide resulted. (Reprinted from Ref. 23.)

solid phase synthesis have been treated with strong acids such as HF, traces of which can react with the silica support of packed columns. In the purification of a myristylated peptide, the mass of impurities removed after purification by preparative HPLC and CCC were comparable [18]. In Figs. 22 and 23 are the results of a CCC separation. The recovery of pure peptide was 10%. It is better to remove the side product material as it is in CCC than to have it be deposited on HPLC columns since the impurities shorten the useful lifetime of the columns.

The CCK fragment peptide Ac-CCK-(26-30)NH$_2$ was chromatographed by the MC-CCC using the BAW system with the lower phase mobile. The result of loading 200 mg is shown in Fig. 24 [23]. Fractions 107-117 contained 56 mg of pure peptide. The K was 0.157 and the N calculated from the run was 1600. Analytical HPLC of the crude and purified peptide is shown in Fig. 25. These chromatograms show the resolution of the method in removing impurities. This compound was also purified by preparative HPLC on a 1-in.-i.d. column and the results are shown in Fig. 26. The recovery of pure peptide from both methods was the same. Thus, for peptides with

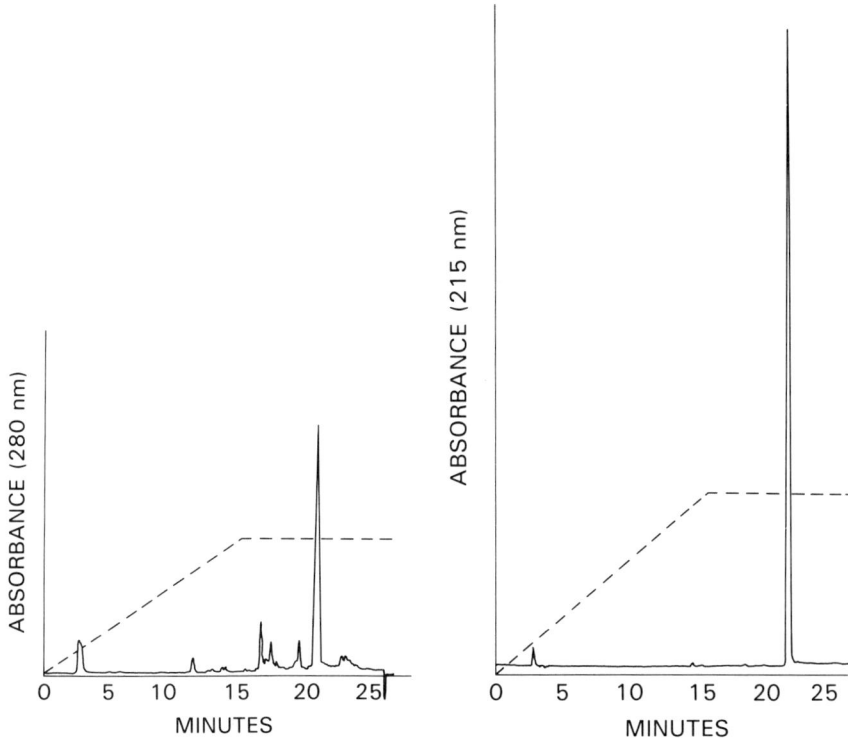

Fig. 25 Analytical HPLC of unpurified and purified acetyl-CCK (26-30) NH_2. Left is approximately 20-µg sample on a 5-µm spherical 200 A ODS, 0.4 × 15 cm, YMC Inc. column in 0.1% phosphoric acid and a gradient of acetonitrile from 10% to 30% in 15 min at a flow of 0.8 ml/min. Right is purified peptide in same conditions except that detection was at 215 nm.

a suitable partition coefficient, the capability for chromatography by CCC is similar to preparative HPLC in relatively high-resolution columns. A preparative purification of Ac-CCK-(26-29)NH_2 is shown in Fig. 27. A sample load of 300 mg was chromatographed with a yield of 278 mg of purified peptide [23].

B. Capability

The MC-CCC is the most versatile CCC apparatus for preparative chromatography of peptides, smaller organic molecules, and larger proteins. The resolution is high for preparative separations in

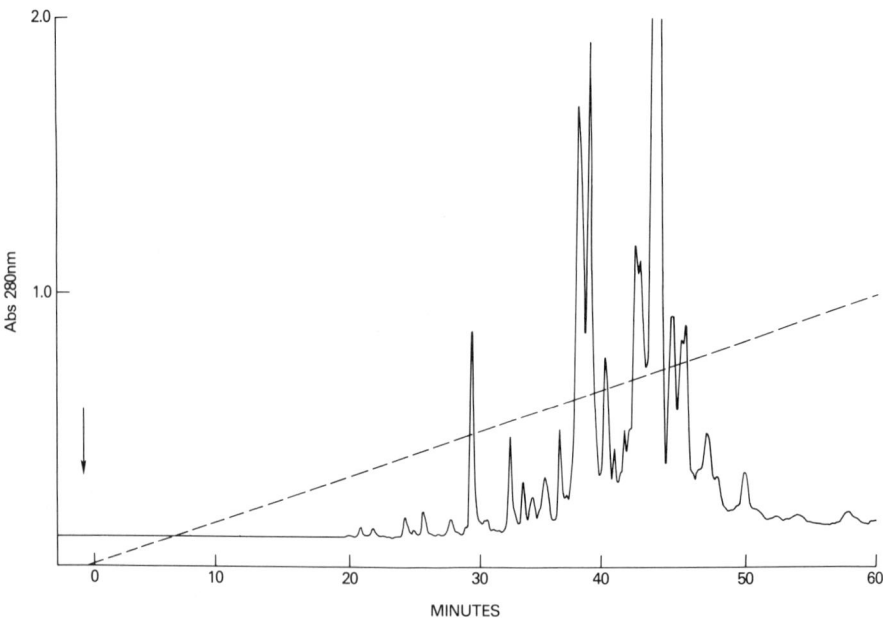

Fig. 26 Preparative HPLC of 100 mg CCK (26-30) NH_2 on a 20 × 250 mm column of 10-μm spherical ODS 120 A column, YMC Inc., in the solvent system of 0.1% aqueous Tfa and a gradient of 0.1% Tfa/acetonitrile, 0% to 30%, in 60 min at 15 ml/min using a Waters LC 3000 solvent delivery system. The arrow indicates start of the gradient after passing of the void volume. The major peak at 43-44 min contained pure peptide, approximately 25 mg.

solvent systems found suitable for the sample. For most peptides the solvent systems listed in Table 1 are applicable. With appropriate solvent systems, the present instrument will separate highly cationic compounds in high capacity, using partition chromatography combined with gradients and even ion exchange mechanisms. Protein separations by MC-CCC have only begun to be developed. Biologically active enzymes and receptors can be separated by the two-phase aqueous systems containing polyethylene glycol which can be run in this instrument [31]. The polyethylene glycol can be chemically modified to perform affinity chromatography [34].

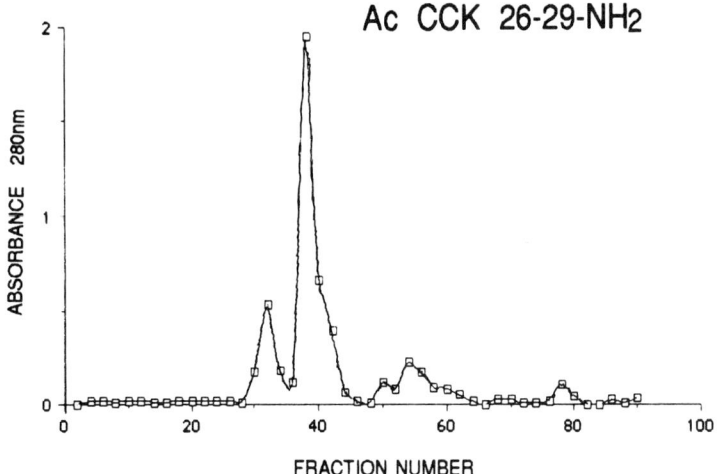

Fig. 27 Preparative chromatography of 300 mg Ac-Asp-Tyr-Met-Gly-NH_2 in the MC-CCC in the same conditions as in Fig. 25. The solvent front was at fraction 27 and 278 mg of pure peptide was recovered from fractions 37-40. The K value was 1.05 and N was calculated to be 1444 theoretical plates. (Reprinted from Ref. 23.)

VIII. POTENTIAL INSTRUMENTS FOR CCC OF PEPTIDES

Other CCC technologies have potential for high-resolution preparative chromatography of peptides. These instruments may have general or particular use in biotechnology processes. Ito noted that only part of the planetary schemes in Fig. 1 have been experimentally tested [35]. The nonsynchronous planetary instrument designs need to be studied. Future work on these instruments will be to assess the capabilities of the resulting hydrodynamic solvent behavior. Some instruments of the synchronous schemes are presently being researched and have noteworthy potential.

A. Cross-Axis Coil Planet Centrifuge (CPC)

Ito has been modifying the design of the cross-axis CPC to increase the capability of the highly efficient multilayer coil and increase the retention of the solvent systems with small density difference. The cross-axis CPC has the coil's axis of rotation perpendicular to the central axis of revolution as in the scheme X of Fig. 1. Along with a lateral placement of the coil on the column holder, the stationary phase is highly retained at high rpm and flow. A sample load of

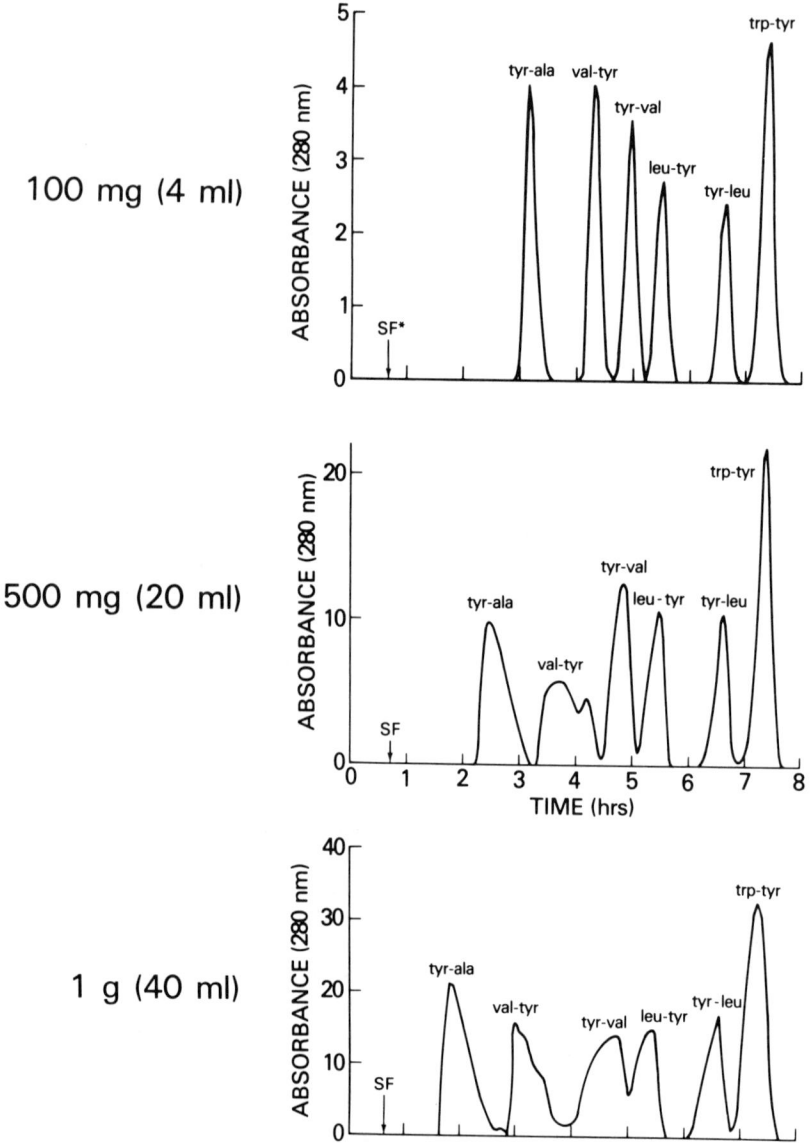

Fig. 28 Chromatography of dipeptides on the cross-axis coil planet centrifuge mounted with a multilayer coil of 2.6-mm-i.d. tubing with a total volume of 400 ml. The effect on retention and resolution is seen in the elution of increasing amounts of sample as indicated in the figure. The solvent system used was a gradient of the aqueous mobile phase of n-butanol/DCA/0.1 M NH_4 formate to n-butanol/0.1

1 g was separated in this instrument showing the potential for high-resolution preparative capability [36] (Fig. 28).

B. "Foam" Chromatography

Since the coil planet centrifuge (CPC) has a head and a tail, the less dense phase moves to one end due to the coil direction. Ito found that air bubbles migrate to one end during centrifugation [37]. Thus by arranging a sample feed in the center of the coil and liquid entry and gas feed at the opposite ends of the coil, particular components can be separated due to their surfactant interaction with the bubbles. These components are carried away to the foam end where they are collected in a concentrated state. The other hydrophilic components are eluted out the liquid end. Therefore, from large volumes of aqueous solutions, hydrophobic compounds can be separated and concentrated without extracting solvents, filters, or detergents, etc. [38]. In Fig. 29 the components of bacitracin are shown. The bacitracin was chromatographed in water in a CPC with N_2 bubbling through. The hydrophobic components were selectively isolated out the bubble or foam end.

The obvious application of this technology can be the isolation of products from fermentation reactors. In addition, proteins have been separated according to their surfactant capability [39]. By adding ionic surfactants, certain components are carried to the bubble end separating them from noncharged or similarly charged compounds. By use of a surfactant ligand (some of which can be peptides) this method can be used for affinity chromatography of membrane proteins and subcellular particles. Therefore this technology has significant possibilities for applications in biotechnology.

M NH_4 formate (1% DCA to 0%) with the n-butanol phase as the stationary phase. The sample mixture was dissolved in the upper stationary phase of the starting phase system. The apparatus was centrifuged at 800 rpm and the mobile phase was pumped at 120 ml/hr in the tail-to-head elution mode. (Reprinted from Ref. 36.)

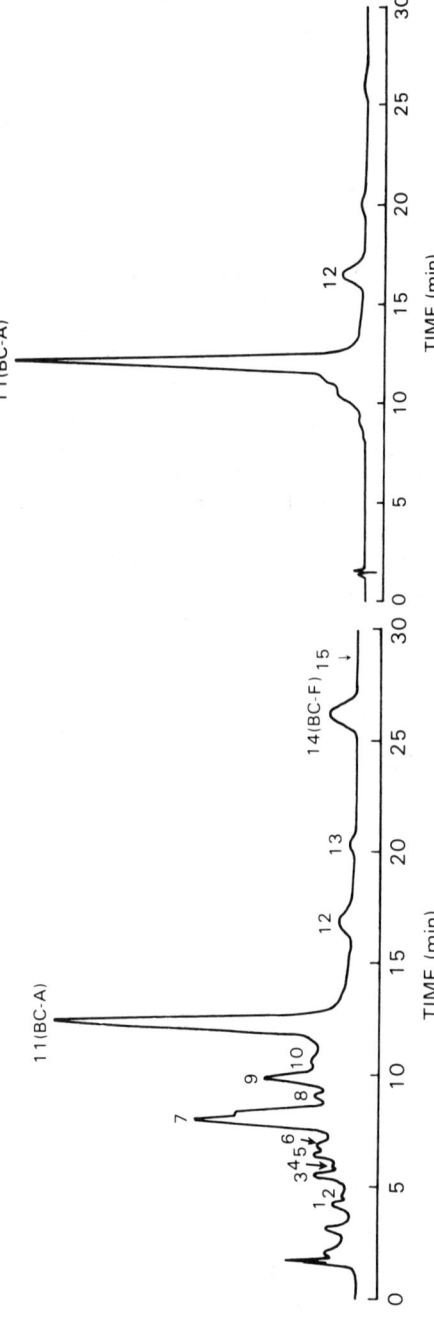

Fig. 29 The results of "foam" chromatography of a sample of bacitracin in water (0.5 ml of a 1% solution). A multilayer CPC was modified for sample loading at the center of the coil and N_2 entry at the head end and liquid entry at the tail. The coil was run at 500 rpm at a flow of 3.2 ml/min. Water elution continued at the tail-to-head direction and liquid fractions were collected and foam fractions collected at the other end. Above are shown the HPLC analysis of the sample loaded (left) and of a foam fraction (right). For HPLC a Shimadzu chromatograph with a Capcell Pak C_{18}, 0.46 × 15 cm column (Shiseido) is used with an isocratic mobile phase of 0.04 M Na_2HPO_4, pH 9.4, and 62% methanol at 1 ml/min and with detection at 234 nm. (Reprinted with permission from Ref. 38.)

Table 3 Types of Peptides Separated by CCC

Peptide	Conditions	Ref.
Aromatic dipeptides	HFCPC, MLCPC, MC-CCC, BAW, LP	12, 46, 32
	MLCPC, B/DCA/NH$_4$ formate gradient, LP	33
Acidic nonaromatic dipeptides	HFCPC, BAW, UP	45
Up to 10-mers:		
Tyr-Gly-Gly	HFCPC, BAW, UP	12
Cholecystokinin analogs	HFCPC, MC-CCC, BAW, Tfa/B	14, 23
Sulfated CCK and analogs	HFCPC, NH$_4$OAc/B, UP	14
D-Ala peptide T amide	HFCPC, 0.1% Tfa/B, UP	41
Fluorenylmethyloxycarbonyl CCK analog	MLCPC, Chl/A/W, UP	15
Naphthoyl-CCK peptides	HFCPC, Chl/A/W, UP	45
βLPH (61-69) analog	HFCPC, 0.1% Tfa/B, LP	13
Neutral aromatic peptides	HFCPC, BAW, LP	47
Longer than 10-mers:		
Acidic, basic, or neutral with aromatic groups	HFCPC, BAW, Tfa/B	15
	MC-CCC, BAW, LP	32
Acidic	HFCPC, BAW, Tfa/B	12, 15
Gramicidins	MLCPC, HFCPC, Chl/benzene/M/W	33
Bombesin	MLCPC, DCA/B, Tfa/B, LP	20, 29
Myristoyl peptides	HFCPC, MC-CCC, BAW, UP	45, 18
Insulin	HFCPC, DCA/sec-but., LP	29

Abbreviations used: B = *n*-butanol; A = acetic acid; W = water; Chl = chloroform; M = methanol; UP = upper phase mobile; LP = lower phase mobile. Others as defined in the text.

IX. SUMMARY

In Table 3 are listed the types of peptides separated by CCC, including the instruments used and the solvent conditions. In the last 20 years, intriguing discoveries about the hydrodynamic behavior of liquids in open-coiled tubing have introduced many possibilities for chromatographic separations. A fruitful area of endeavor certainly lies in further elucidation of the mechanisms. However, it is evident that CCC has the capability to separate all types of substances: organic compounds, proteins, membrane proteins, subcellular particles, as well as peptides, the intermediate-sized molecules. The potential for applications also lies at various levels of the scale such as analytical, laboratory, and, very importantly, industrial processes.

ACKNOWLEDGMENT

Peptide Technologies is grateful to the Small Business Innovation Research Program of the National Institutes of Health for grants NS22319, CA44399, and GM40833 providing support for the research reported here.

REFERENCES

1. A. J. P. Martin and R. L. M. Synge, Biochem. J. 35, 1358 (1941).
2. Y. Ito, M. A. Weinstein, I. Aoki, R. Harada, E. Kimura, and K. Nunogaki, Nature 212, 985 (1966).
3. T. Tamimura, J. J. Pisano, Y. Ito, and R. L. Bowman, Science 169, 54 (1970).
4. Y. Ito and R. L. Bowman, Science 167, 281 (1970).
5. L. C. Craig, W. Hausmann, E. H. Ahrens, Jr., and E. J. Harbenst, Anal. Chem. 23, 1236 (1951).
6. O. Post and L. C. Craig, Anal. Chem. 35, 641 (1963).
7. A. V. Schally, R. M. G. Nair, T. W. Redding, and A. Arimura, J. Biol. Chem. 246, 7230 (1971).
8. W.-Y. Huang, R. C. C. Chang, A. J. Kastin, D. H. Coy, and A. V. Schally, Proc. Natl. Acad. Sci. USA 76, 6177 (1979).
9. Y. Ito and R. L. Bowman, J. Chromatogr. Sci. 8, 315 (1970).
10. J. K. Snyder, K. Nakanishi, K. Hostettmann, and M. Hostettmann, J. Liquid Chromatogr. 7, 243 (1984).
11. Y. Ito, Anal. Biochem. 100, 271 (1979).
12. Y. Ito and G. J. Putterman, J. Chromatogr. 193, 37 (1980).
13. M. Knight and W. A. Klee, J. Biol. Chem. 254, 10426 (1979).
14. M. Knight, A. M. Kask, and C. A. Tamminga, J. Liquid Chromatogr. 7, 351 (1984).
15. M. Knight, J. D. Pineda, and T. R. Burke, Jr., J. Liquid Chromatogr. 11, 119 (1988).
16. W. D. Conway, in *Countercurrent Chromatography Theory and Practice* (N. B. Mandava and Y. Ito, eds.), Chromatogr. Sci. Series Vol. 44, Marcel Dekker, New York, 1988, p. 443.

17. Y. Ito and W. D. Conway, J. Liquid Chromatogr. 8, 2915 (1988).
18. M. Knight, S. Gluch, K. Takahashi, T. D. Dang, and R. A. Kahn, J. Chromatogr. 538, 141 (1991).
19. W. D. Conway and Y. Ito, J. Liquid Chromatogr. 7, 275 (1984).
20. M. Knight, Y. Ito, P. Peters, and C. diBello, J. Liquid Chromatogr. 8, 2281 (1985).
21. H. Oka and Y. Ito, J. Chromatogr. 475, 2229 (1989).
22. J. M. Stewart and J. D. Young, Solid Phase Peptide Synthesis, Pierce Chemical Co., Rockland, Ill., 1985.
23. M. Knight and S. Gluch, J. Liquid Chromatogr. 13, 2351 (1990).
24. M. Knight, A. M. Kask, and C. A. Tamminga, Peptides: Structure and Function (V. J. Hruby and D. Rich, eds.), Pierce Chemical Co., Rockland, Ill., 1984, p. 75.
25. M. Knight, Y. Ito, J. D. Gardner, C. A. Tamminga, and T. N. Chase, J. Chromatogr. 301, 277 (1984).
26. A. Foucalt and K. Nakanishi, J. Liquid Chromatogr. 11, 2455 (1988).
27. Y. Ito, J. Chromatogr. 301, 377 (1984).
28. Y. Ito and W. D. Conway, J. Chromatogr. 301, 405 (1984).
29. M. Knight, Y. Ito, J. L. Sandlin, and A. M. Kask, J. Liquid Chromatogr. 9, 791 (1986).
30. M. Knight, Y. Ito, A. M. Kask, C. A. Tamminga, and T. N. Chase, J. Liquid Chromatogr. 7, 2525 (1984).
31. Y. Ito and H. Oka, J. Chromatogr. 457, 393 (1988).
32. M. Knight and Y. Ito, J. Chromatogr. 484, 319 (1989).
33. Y. Ito, J. Sandlin, and W. G. Bowers, J. Chromatogr. 244, 247 (1982).
34. S. D. Flanagan, G. Johansson, B. Yost, Y. Ito, and I. A. Sutherland, J. Liquid Chromatogr. 7, 385 (1984).
35. Y. Ito, J. Chromatogr. 538, 3 (1991).
36. Y. Ito, Sep. Sci. Technol. 22, 1989 (1987).
37. Y. Ito, J. Liquid Chromatogr. 8, 2131 (1985).
38. H. Oka, K. Harada, M. Suzuki, H. Nakazawa, and Y. Ito, Anal. Chem. 61, 1998 (1989).
39. M. Bhatnagar and Y. Ito, J. Liquid Chromatogr. 11, 21 (1988).
40. A. T. James and R. L. M. Synge, Biochem. J. 50, 114 (1952).
41. T. R. Burke, Jr., and M. Knight, J. Chromatogr. 411, 431 (1987).
42. E. J. Harfenist and L. C. Craig, J. Am. Chem. Soc. 74, 3083 (1952).
43. H. Tsugita, Jikken Kagaku Koza 2, 324 (1956).
44. T. P. King and L. C. Craig, in Methods of Biochemical Analysis, Vol. 10 (D. Glick, ed.), Interscience, New York, 1962, p. 201.

45. M. Knight, in *Countercurrent Chromatography: Theory and Practice* (N. B. Mandava and Y. Ito, eds.), Chromatogr. Sci. Series Vol. 44, Marcel Dekker, New York, 1988, p. 583.
46. Y. Ito, in *Countercurrent Chromatography: Theory and Practice* (N. B. Mandava and Y. Ito, eds.), Chromatogr. Sci. Series Vol. 44, Marcel Dekker, New York, 1988, p. 407.
47. G. J. Putterman, M. B. Spear, and F. Perini, J. Liquid Chromatogr. 7, 341 (1984).
48. J. M. Stewart, J. W. Ryan, and A. H. Brady, J. Med. Chem. 17, 537 (1974).

5
Boronate Affinity Chromatography

Ram P. Singhal and S. Shyamali M. DeSilva *The Wichita State University, Wichita, Kansas*

I.	INTRODUCTION	294
II.	BORONATE COMPLEX FORMATION	295
	A. Chemistry of Interaction	295
	B. Mechanisms of Interaction	298
III.	CONDITIONS TO ENHANCE COMPLEX FORMATION	300
	A. Reaction Environment	300
	B. Nature of Boronate Ligand	301
	C. Nature of *cis*-Diols and Bond Distances	302
IV.	BORONATE LIGANDS USED FOR AFFINITY CHROMATOGRAPHY	303
V.	CHEMISTRY OF ACTIVATION AND COUPLING OF LIGANDS	308
	A. Carbodiimide Ligand Coupling Method	308
	B. Cyanogen Bromide Agarose Activation Method	310
	C. Periodate Oxidation Dextrose Activation Method	311
	D. Bisoxirane Method for Activation of Carbohydrate Matrices	311
VI.	AFFINITY MATRICES USED FOR IMMOBILIZATION OF DIFFERENT LIGANDS	312

VII.	METHODS FOR STUDYING COMPLEX FORMATION	314
	A. Ion Exchange Chromatography	315
	B. Affinity Chromatography Using Immobilized Boronates (Frontal Analysis)	316
	C. Absorption Spectroscopy	316
	D. Proton and ^{11}B-NMR	319
	E. Optical Rotation	321
VIII.	EXAMPLES OF APPLICATIONS OF BORONATE AFFINITY CHROMATOGRAPHY	321
	A. Small Biomolecules	321
	B. Large Biomolecules	321
IX.	CONCLUDING REMARKS	329
	REFERENCES	330

I. INTRODUCTION

Recent advances in affinity chromatography have created a wide variety of important applications for this powerful technique of purification of diverse biomolecules, such as antibodies, antigens, and enzymes. The instant acceptance of this technique has resulted from its design and adaptability for the sensitive and quantitative assay of many biomolecules. Though the interaction of boronic acid with vicinal alcohols (cis-diols) has been known for several decades, it has only been in the past 15-20 years that this reaction has been exploited for the purification of important biomolecules [1,2]. Applications of the phenylboronate matrices include the separation of ribonucleosides, nucleotides, and oligonucleotides from their deoxy derivatives in many situations [3-11]; the separation and assay of modified nucleosides from common nucleosides [12]; nucleotides from cAMP [13]; the assay of benzo(a)pyrene-DNA adducts in cells [14]; the isolation of the nucleotidyl peptides [15]; the assay of catechols [16]; catechol estrogens and other hormones [17]; the separation of different sugars [18]; the isolation of a specific transfer RNA (tRNA) from 19 other tRNAs [1,12,19,20]; the separation of capped from uncapped messenger RNA [21]; the separation of ADP-ribosyl-protein from common proteins [22]; the separation of γ-interferon and IgG [23]; the characterization of specific membrane glycoproteins [24]; the separation of glycosylated proteins [25]; and the purification of serine proteases from other enzymes [26].

The phenylboronate matrix has also been used in clinical analyses. For example, glycosylated hemoglobins are increasingly used for the assessment of glycemia [27,28]. Immobilized phenylboronates have been applied to measure the level of glycosylated hemoglobin [29-31]. The levels of nucleosides and other metabolites have been measured in the intestinal mucosa in healthy individuals and cancer patients; a correlation is claimed between enhanced amounts of specific metabolites and the presence of gastric cancer. These metabolites can be characterized and assayed by boronate chromatography. Similarly, hypoxanthine, uridine, and inosine have been linked to other disorders [32,33]. Substituted phenylboronates have even been shown to exhibit significant antimicrobial activity against common pathogens [34]. This brief survey of the current literature indicates the significance of boronate affinity chromatography and the possible impact of this technique on current research.

II. BORONATE COMPLEX FORMATION

A. Chemistry of Interaction

The interaction of boronates with cis-diols is the basis for all separation techniques involving boronates. The major structural requirements for complex formation between boronates and cis-diols are that the two hydroxyl groups should be on adjacent carbon atoms and be held in a coplanar configuration.

The strength of the interaction increases with the length of the carbon chain in simple polyalcohols in which the hydroxyl groups have the ability to rotate freely around the carbon-carbon bonds. However, the situation is somewhat different in closed-ring carbohydrates, where the configuration of the hydroxyls is stereochemically fixed. Carbohydrates, which normally do not have 1,2-cis-diol groups, can still react with borate because of ring flexing and mutarotation. For example, furanoses react more easily than pyranoses because of ring flexing [35,36]; and even though the two stereoisomers of glucose (α-D-glucopyranose and β-D-glucopyranose) cannot form complexes with borate, the open-chain mutarotation intermediate can [37,38]. In polysaccharides where the glycosidic linkages reduce the number of cis-diols, only the terminal sugar residues are available for complex formation. Compounds such as cis-inositol and triethanolamine make tridentate complexes with borate or boronate as shown in Fig. 1a [39]. Bisbidentate complexes are only possible with borate ions, not boronate, because of the presence of only three hydroxyl groups in the latter compound (Fig. 1b).

Catechols and their derivatives also yield strong complexes with both borate and boronate by a mechanism identical to that observed with other cis-diols [40]. Moreover, the boronate complex formed with aryl diols is more stable than that formed with sugar cis-diols [41a,b]. The complex formation observed with 4,5-dihydroxynaphthalene derivatives is an exception to the adjacent carbon atom rule [42].

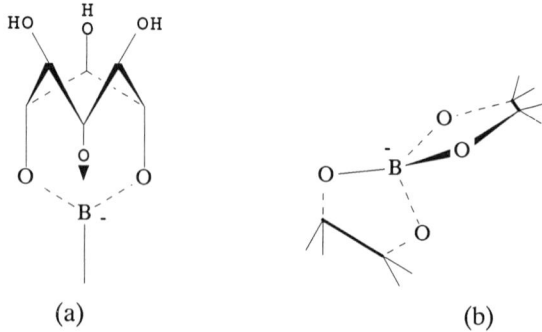

Fig. 1 Unusual interactions of borate anions with polyalcohols. (a) Formation of a tridentate complex between the borate anion and an inositol molecule. (b) Formation of a bisbidentate complex between a borate anion and two cis-diols. (Adapted from Ref. 39.)

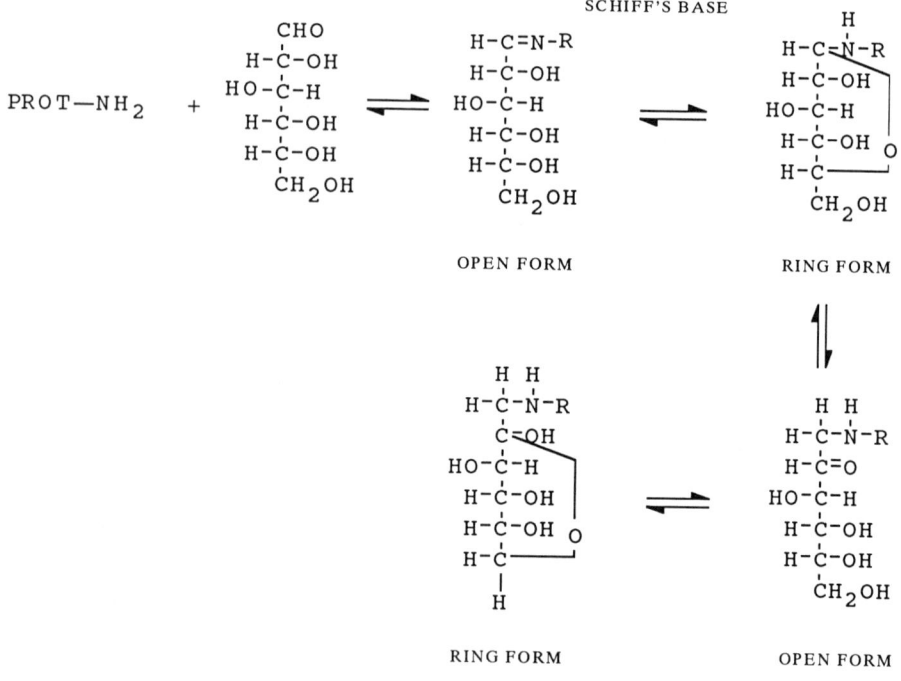

Fig. 2 Schematic representation of the reaction of glucose with the N-terminal valine residue of hemoglobin (β-chain) resulting in the so-called Hb A_{1c} species. (Adapted from Ref. 43.)

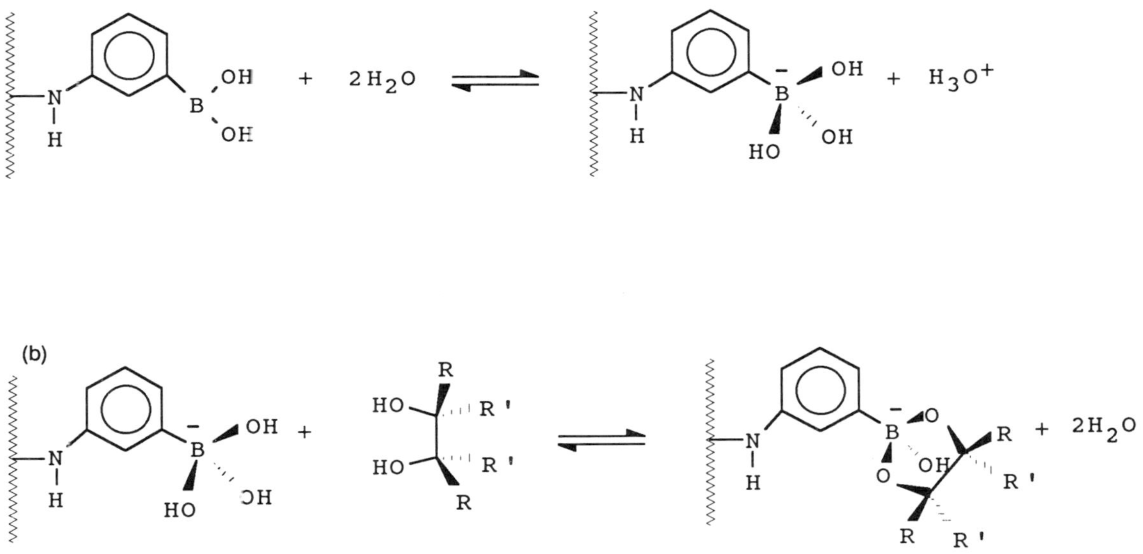

Fig. 3 Interaction of phenylboronic acid ligands with 1,2-cis-diol compounds. (a) Ionization of the coplanar, trigonal boronate to a tetrahedral structure. (b) Boronate diester formation between a boronate anion and a cis-diol. (Adapted from Ref. 46.)

The interaction of glycosylated hemoglobin with boronate is rather interesting. Conjugated glucose reacts with the N-terminal valine residues of the β chain of hemoglobin (Hb A_{1c}) to make a Schiff base adduct, which undergoes an Amadori rearrangement to form a 1-deoxyfructopyranose, as shown in Fig. 2 [43-45]. The latter compound exhibits the necessary cis-diols enabling it to react with boronate and yield a stable complex.

B. Mechanisms of Interaction

Two reaction mechanisms have been proposed for the interaction of boronate with cis-diols. In one case, the trigonal, coplanar boronic acid ionizes to form a tetrahedral boronate anion as shown in Fig. 3a. This anion can then react with a cis-diol to yield a cyclic boronate ester, as shown in Fig. 3b [46]. In the other proposed mechanism, a sequential nucleophilic attack of the diol oxygen atoms on boronic acid causes the formation of an anionic and a neutral species in equilibrium, as shown in Fig. 4 [47]. The basic esters with tetrahedrally coordinated boron atoms are found in alkaline solutions, e.g., pH > 8. However, only the neutral esters are formed at lower pH values, which hydrolyze in aqueous solutions. Therefore, the latter type of complexes can only be observed in organic solvents [48-50].

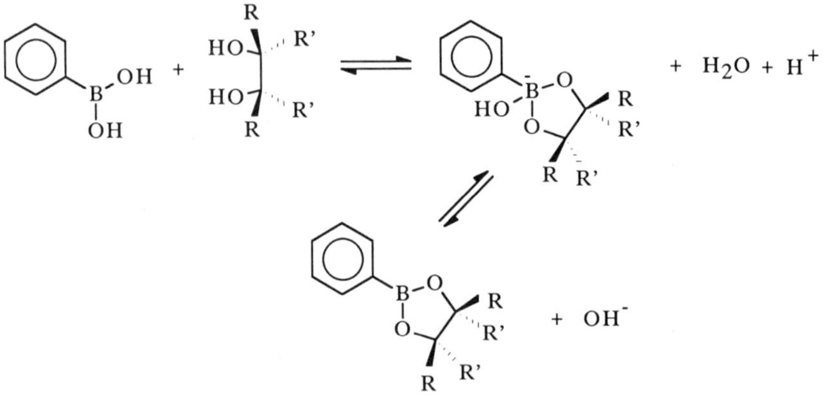

Fig. 4 Complex formation between neutral benzeneboronic acid and cis-diols. A sequential nucleophilic attack of the diol oxygen atoms on boronic acid causes the formation of an anionic and a neutral species in equilibrium. (From Ref. 47.)

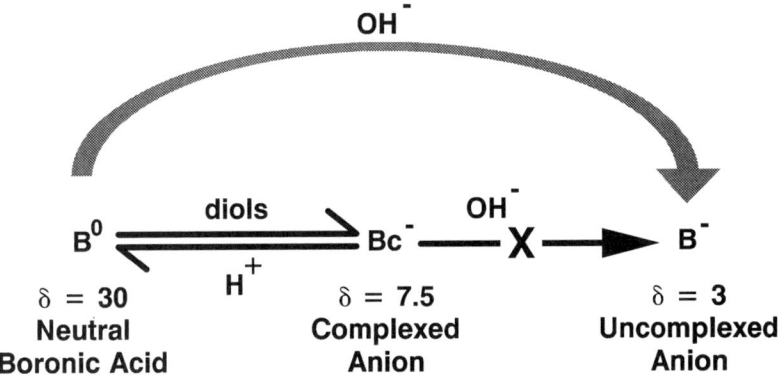

Fig. 5 Reaction mechanism proposed on the basis of ionization and complexation of ligands studied by ^{11}B-NMR and spectroscopic methods. Neutral boronic acid (B^0) undergoes a facile complex formation in the presence of cis-diols, yielding a complexed anion ($B\bar{c}$). This complex does not dissociate under alkaline conditions. The neutral boronic acid undergoes ionization under alkaline conditions. (Data from Refs. 41a and 69.)

A reaction mechanism proposed on the basis of ionization and complexation of ligands studied by ^{11}B-NMR and spectroscopic methods was described recently [41a]. These results, summarized in Fig. 5, indicate that the neutral phenylboronic acid (B^0) undergoes a facile complex formation in the presence of cis-diols, yielding a complexed anion ($B\bar{c}$). This complex forms approximately one pH unit below the ionization of the phenylboronic acid derivative and does not dissociate under alkaline conditions. That is, once the complex is formed, both the ligand and the affinity molecule are protected against the adverse alkaline condition of the medium. The neutral boronic acid undergoes ionization as usual under alkaline conditions [41a]. (See Sections VII.C and D for details.)

The mechanism of complex formation of a phenylboronate having an electron-withdrawing group is shown in Fig. 6. The presence of an electron-withdrawing group in the immediate vicinity of boron lowers the ionization constant of phenylboronic acid. For example, phenylboronic acid exhibits a pK_a value of 8.8, but the introduction of a nitro group in the meta or para positions of the boron atom lowers the pK_a value to 7.15 and 7.00, respectively. The nitro group in the ortho position causes an internal cyclization with vicinal boronic acid [41a].

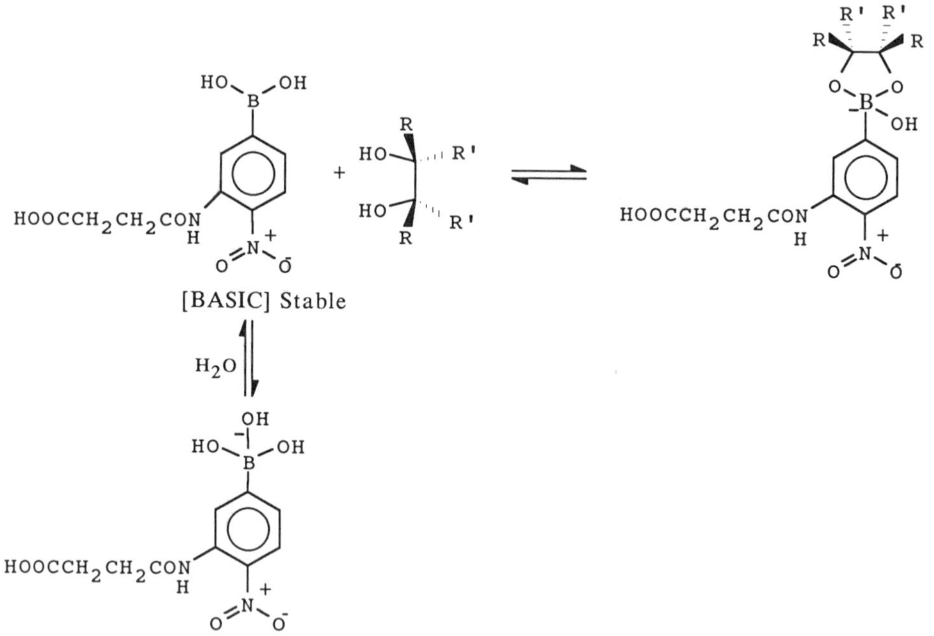

Fig. 6 Mechanism of complexation between N-(6-nitro-3-dihydroxyborylphenyl)succinamic acid and cis-diols. Note the electron-withdrawing effect on the boron atom. (Adapted from Ref. 41a.)

III. CONDITIONS TO ENHANCE COMPLEX FORMATION

A. Reaction Environment

There is very limited information available concerning the influence of the reaction environment (pH, buffer, solvents, metal ions, temperature, etc.) on the complex formation between borate/boronate and cis-diols. Using ^{11}B-NMR, van Duin et al. showed that esters of boric acid and borate in aqueous medium exhibit the highest stability at a pH value in which the sum of the charges of the free esterifying species is equal to the charge of the ester [51]. Accordingly, the borate diester of an α-hydroxycarboxylic acid exhibits the highest stability at a pH value which is equal to the pK_a of α-hydroxycarboxylic acid, where the latter neutral acid (L) and its anionic species (L$^-$) are present in equal amounts and the boric acid (B^0) is present in its neutral form: $B^0 + L + L^- = B^-L_2 + 3H_2O$. However, the boronate monoesters are preferentially present under pH conditions where $pK_a (L) < pH < pK_a (B^0)$, i.e., the carboxylic acid is ionized and the boric acid is present only in its neutral form: $B^0 + L^- \leftrightarrow B^-L + H_2O$. In a later study, these authors demonstrated the stability of borate esters with polyhydroxyls using ^{11}B chemical shifts [52]. The order of complex stability according to their results is tridentate > bidentate > monodentate.

McCutchan et al. studied the reaction conditions necessary to enhance the stability of the complex of phenylboronate (matrix) with tRNAs, which had unesterified cis-diols at their 3' ends, i.e., ribofuranose of residue A76 [20]. Chromatographic separations were examined at different temperatures and with different buffer conditions, e.g., pH, Mg^{2+} concentration, and ethanol. The authors note that higher concentrations of Mg^{2+} and Na^+ ions stabilize the complex between tRNA and the matrix-bound boronate, but tRNA goes out of solution under these conditions (see Section VIII.B below for details).

B. Nature of Boronate Ligand

A boronate ligand should possess several major characteristics in order to be considered useful in affinity chromatography for the purification of biomolecules. A major limitation of the use of phenylboronate ligands for affinity chromatography is a result of their high ionization constant. The most commonly used ligand, 3-aminophenylboronic acid (abbreviated here as 3aPBA), has a relatively lower ionization constant (pK_a = 8.75) because of the amine group in the meta position. Several attempts have been made to synthesize boronate ligands having lower pK_a values. An electron-withdrawing (nitro) group has been introduced into the phenyl ring to make boronate more acidic. For example, some of the ligands synthesized to achieve this effect are 2-nitro- and 4-nitro-3-succinamidobenzene boronic acids [41a, 53], 3-nitro-4-(6-aminohexylamido)phenyl boronate [54], {4-[(hexamethylenetetramine)methyl]phenyl} boronic acid (pK_a = 7.04) [55], and 4-(N-methyl)carboxamidobenzene boronic acid (pK_a = 7.86) [56]. In addition, 2-{[(4-boronphenyl)methyl]-ethylammonio}ethyl and 2-{[(4-boronphenyl)methyl]diethylammonio}-ethyl groups have been introduced to a dextran (Sephadex) matrix in order to create positively charged quaternary amine groups which are claimed to lower the pK_a value of boronic acid [36].

Some affinity ligands do not complex well with cis-diols in the absence of a spacer arm. In these cases the ligand must be held away from the matrix in order to provide free interaction and avoid steric hindrance. The boronate ligand is usually reacted with an acid anhydride (such as succinic anhydride) in order to introduce a small spacer arm and also to offer a functional group for easy coupling with the matrix.

C. Nature of cis-Diols and Bond Distances

The nature of the cis-diol is important in producing a stable complex with boronate, an important factor in their separation using boronate matrices. The chromatographic behavior of a number of carbohydrates has been investigated in this context [57]. D-Glucose is mainly present in aqueous solutions in the pyranose form with C1 conformation. It complexes with boron only through the 1,2-hydroxyls of the α anomer and the complex formed is less stable, whereas the complex formed with D-sorbitol is much more stable since D-sorbitol in its open-chain configuration can form a number of five- and six-membered cyclic complexes with tetrahedral boron. These open-chain sugar alcohols have the so-called acyclic "pure" cis-hydroxyls or 1,3-hydroxyls, which ensure the greatest conformational stability of these complexes. The retention volumes of D-glucose and D-ribose on anion exchangers are very similar since both are present in the pyranose configuration. But the slightly increased retention time observed for D-ribose may be due to the 1,3-diaxial arrangement of its hydroxyl groups, thus yielding a somewhat more stable six-membered cyclic complex. However, D-glucose has been found to behave differently in alkaline media (pH 8.4) where it binds rather strongly to the anion exchanger; it can only be eluted with a 0.1 M buffer solution of pH 6. The stability of this complex has been attributed to the possible interconversion of its pyranose form to a furanose form in alkaline media. This argument is supported by the fact that furanose structures are more stable for complex formation [37]. Because of this, there is an additional possibility of the formation of complexes utilizing the 5- and 6-hydroxyls of both α and β anomers of D-glucose. Thus, the stability constant for the glucose complex formed by participation of 5- and 6-hydroxyls can be compared to the characteristic stability constant of boronate complexes of acyclic sugar alcohols. It is also possible that the intermediate acyclic form of D-glucose can participate in the complex formation [58].

The chromatographic behavior of D-fructose on the anion exchange matrix is very similar to that observed for D-sorbitol in solutions of pH 6.5. This complex formation has also been credited to the open-chain configuration of these sugars in solution and the availability of acyclic hydroxyls because ketoses are found to be more decyclized than aldoses in aqueous solutions [59,60].

Both trigonal and tetrahedral boron in boronates have been reported to form strainless, six-membered cyclic esters utilizing 1,3-diols. Strainless, five-membered cyclic boronates are also formed, but only when the boron is tetrahedral. Thus, five-membered complexes (esters) formed between boronates and vicinal cis-diols of carbohydrates are generally strained, and they tend to strongly react with bases [61,62]. This frequently leads to easy hydrolysis of the five-membered complexes compared to the six-membered complexes, which are more stable and fail to undergo facile hydrolysis [63].

IV. BORONATE LIGANDS USED FOR AFFINITY CHROMATOGRAPHY

A large number of boronate ligands have been synthesized, but only 3aPBA (or its derivatives) has been extensively used for immobilization in affinity chromatography. The amine functionality in boronate ligands obviously has been very convenient for immobilization to the inert matrices. A number of coupling chemistries utilize the amino groups of boronate ligands either by introducing a carboxyl group via a spacer arm [4] or by coupling it directly to the matrix [12, 26,64]. In addition, the 3-amino substitution, as mentioned earlier, lowers the pK_a below that of simple phenylboronate or boric acid [36].

The 3aPBA ligand has been modified in various ways in order to improve its performance. Structures of several ligands derived from 3aPBA are shown in Fig. 7. The main criteria used in these modifications have been to lower the pK_a value of the boronate anion to a value more compatible with the stability of biomolecules and to reduce the secondary hydrophobic binding caused by the phenyl ring. Several strategies have been attempted, including the modifications of the amine group and the substitution of the phenyl ring with other electron-withdrawing groups. Mostly the amino group is modified to introduce a different functional group for coupling purposes and to add a spacer arm to prevent steric interactions with the matrix. For example, succinyl-3aPBA is coupled via carboxyl groups [4] or simply a spacer arm is introduced, as in the case of 6-aminocaproyl-3-aminophenylboronate [65] and 3-(N-succinimidoxycarbonyl)aminophenylboronate [11]. However, other modified ligands, e.g., p-(ω-aminoethyl)phenylboronate and p-vinylbenzeneboronate, have no effect on the pK_a value or the hydrophobicity of the phenyl ring [26,66].

The introduction of electron-withdrawing groups into the phenyl ring stabilizes the tetrahedral boronate anion required for the interaction with cis-diols, even at lower pH values [67]. For example, a nitro group introduced into the phenyl ring causes boronate to be more acidic [41a,53]. The ortho-nitro derivative, N-(4-nitro-3-dihydroxyborylphenyl)succinamic acid, is found to be the best among the different isomers [41a]. The boronate ligands that can perform at the needed acidic and neutral pH values were recently synthesized by Singhal et al. [41a]. Ortho-, para-, and meta-nitro derivatives of succinamidophenylboronic acid using nitronium trifluoromethanesulfonate were prepared. Preferential substitution of the

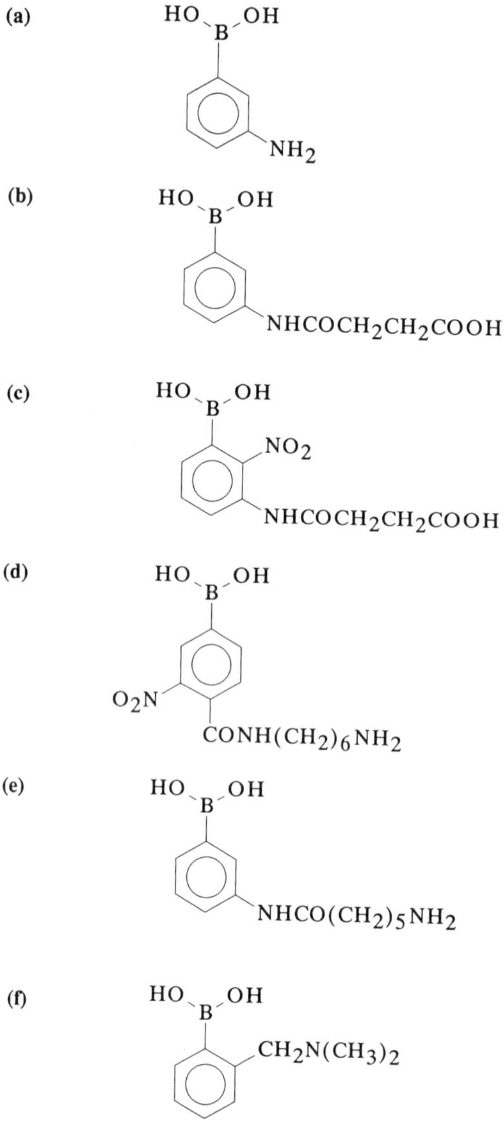

Fig. 7 Structures of various modifications of phenylboronic acid. (a) 3-Aminophenylboronic acid, (b) N-(3-dihydroxyborylphenyl)-succinamic acid, (c) N-(4-nitro-3-dihydroxyborylphenyl)succinamic acid, (d) 3-nitro-4-(6-aminohexylamido)phenylboronic acid, (e) 6-aminocaproyl-3-aminophenylboronic acid, (f) O-dimethylaminomethyl-benzeneboronic acid.

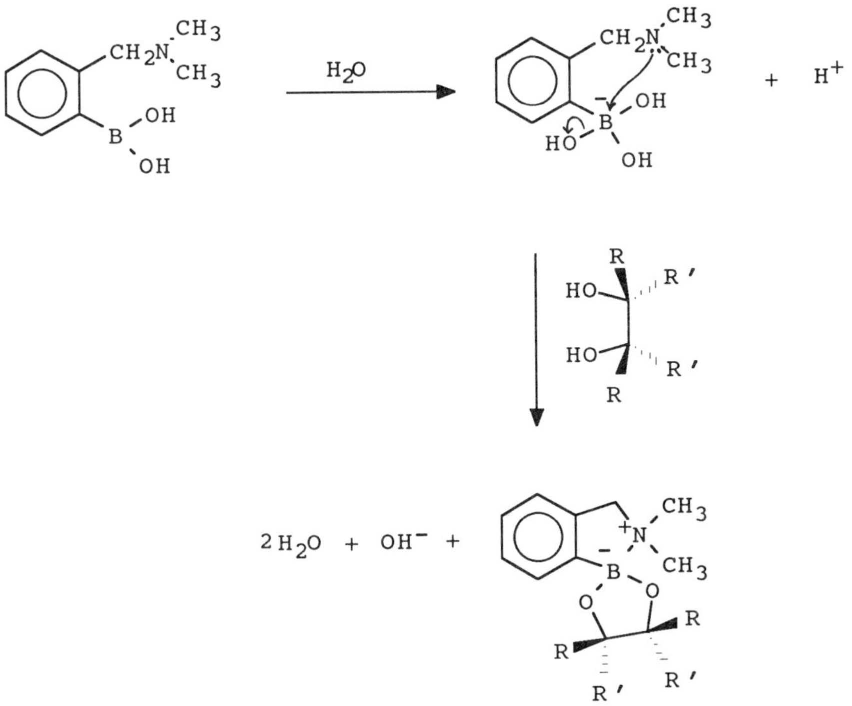

Fig. 8 Formation of a stable boronate complex involving intramolecular B-N interaction between dimethylaminomethyl group and boronate anion of the ligand, and with a cis-diol. (Adapted from Ref. 49.)

nitro functionality into the ortho position of the boronic acid was exploited by the selective use of acetic anhydride for the reaction medium. This method under selected reaction conditions yielded mostly an ortho nitro derivative (pK_a = 7.4). [Overall yield of nitro derivatives was approximately 25% (ortho- = 23%; para- = 2%) of the starting material.]

A simple variation of these nitro derivatives is 3-nitro-4-(6-aminohexylamido)phenylboronate in which the length of the side chain is increased and the functional group is an amine instead of a carboxyl group (Fig. 7) [54]. There is also an example of a ligand in which the tetrahedral boronate anion is stabilized by a dimethylaminomethyl group attached to the phenyl ring via an intramolecular B-N interaction in the binding group (Fig. 8) [49]. This 2-dimethylaminomethylphenylboronic acid [68] is highly specific for aliphatic diols, whereas little binding is observed for aromatic diols, amines, or monoalcohols [50].

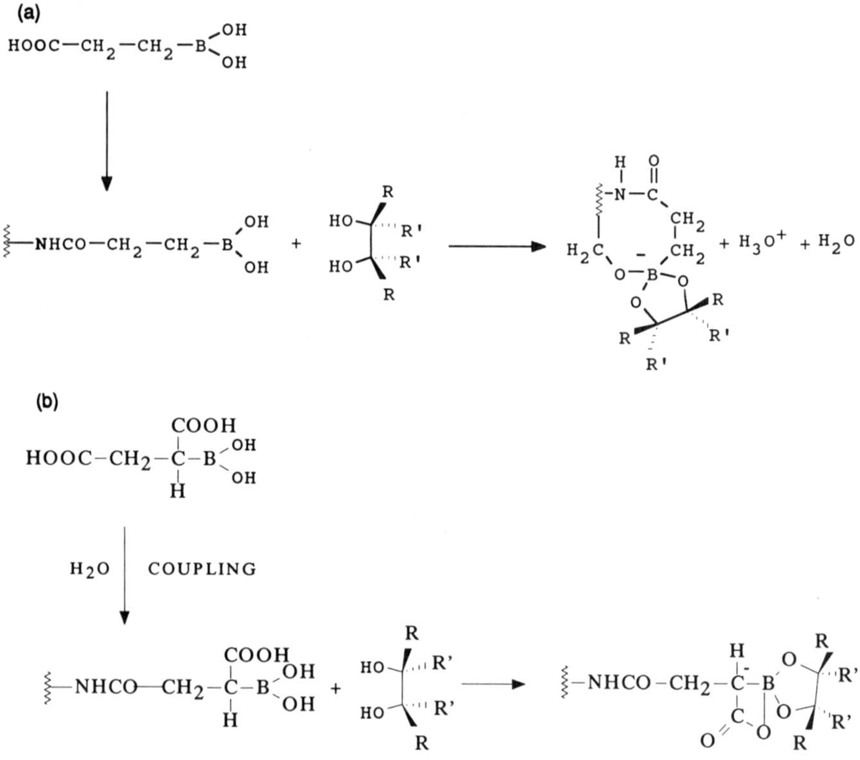

Fig. 9 Formation of complexes between (a) 2-carboxyethaneboronic acid or (b) 1,2-dicarboxyethaneboronic acid with cis-diols. Note that the complex formation in (a) greatly depends on the availability of an alcohol group from the matrix. In case (b), though the hydroxyl group is available within the ligand structure, the four-membered ring is highly strained, and therefore the complex is unstable. (From Ref. 41a.)

Another method of stabilizing this phenylboronate anion is by use of bipolar derivatives; for example, boronate ligands with quaternary ammonium groups. One of the strongest known phenylboronic acids is {4-[(hexamethylenetetramine)methyl]phenyl}boronic acid (pK_a = 7.04) [55]. Two other similar compounds are 2-{[(4-boronphenyl)methyl]ethylammonio}ethyl and -diethylammonio}ethyl derivatives [36]. There is a recent report of the use of 4-carboxybenzeneboronic acid, 4-N-methylcarboxamidobenzeneboronic acid, 4-(N-octyl)carboxamidobenzeneboronic acid, 3-nitro-4-carboxybenzeneboronic acid, and 2-nitro-4-carboxybenzeneboronic acid in

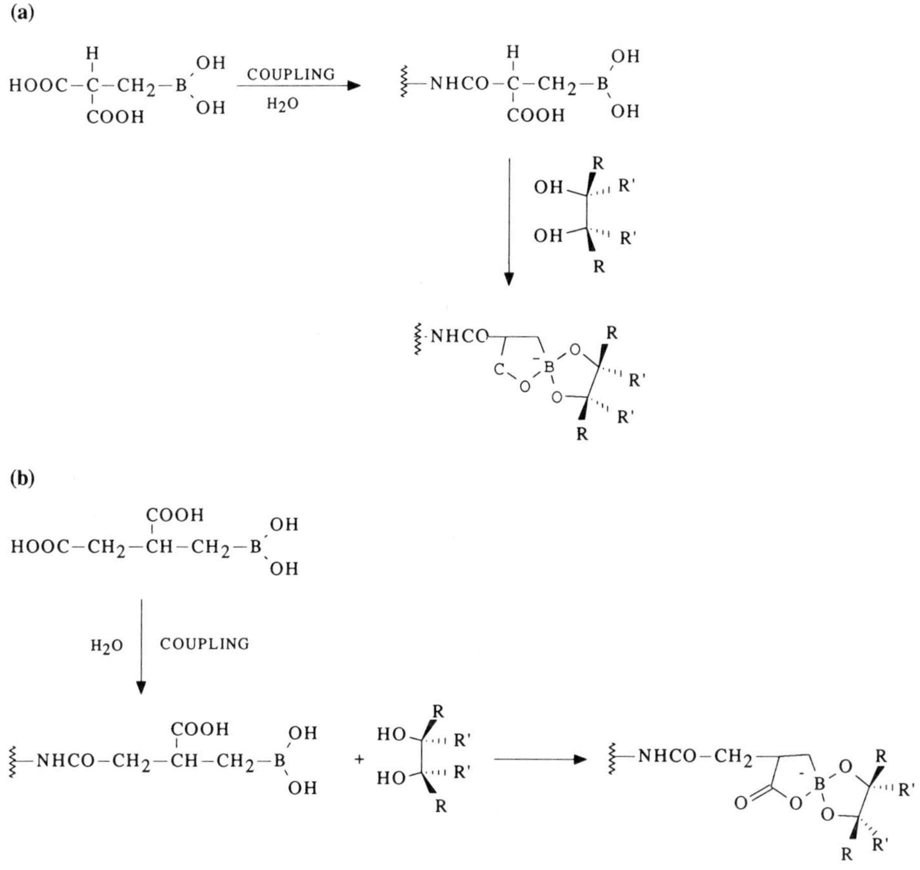

Fig. 10 A five-membered stable complex formation between cis-diols and (a) β,β'-dicarboxyethaneboronic acid, and (b) β,γ-dicarboxypropaneboronic acid. (From Ref. 41b.)

boronate affinity chromatography [56]. The authors suggest that the 4-carboxamidophenylboronic acid functionality should provide sufficient acidity, diol-binding capacity, and hydrolytic stability, thus making it an excellent ligand for the preparation of affinity matrices.

Because of the hydrophobic nature of the phenylboronates, it is difficult to use them for the separation of macromolecules such as proteins, glycoproteins, and polynucleotides; they allow nonspecific binding of undesired biomolecules by hydrophobic interactions.

Therefore, synthesis of boronate ligands without the phenyl ring
but possessing other characteristics required for stable complex
formation has been attempted [69]. Of several compounds that have
been designed, 2-carboxyethaneboronic acid and 1,2-dicarboxyethane-
boronic acid form complexes with cis-diols as shown in Fig. 9a and
b, respectively. Although these ligands were expected to complex
very well with cis-diols, the stability of the complexes formed are
in question because of the strained four-membered ring formed with
1,2-dicarboxyethaneboronic acid and the eight-membered (variable-
size) ring formed involving a hydroxyl group of the matrix in 2-
carboxyethaneboronic acid (Fig. 9).

The strained nature of the boronate-cis-diol complex can be
avoided by not having the carboxyl group on the α carbon to the
boronic acid (Fig. 9b). Moreover, synthesis of the latter compound
is extremely difficult. We are currently involved in the synthesis
of β,β'-dicarboxyethaneboronate (Fig. 10a) and β,γ-dicarboxypropane-
boronate (Fig. 10b) ligands. Each ligand forms a stable five-mem-
bered complex and offers a free carboxylic functionality for coupling
to the matrix. The complex formation is expected to be greatly enhanced by the presence of a vicinal hydroxyl group in the ligand.

V. CHEMISTRY OF ACTIVATION AND COUPLING OF LIGANDS

Several coupling reactions have been used to couple boronate ligands
to inert solid supports involving different chemistries. These reactions include activation of the matrix by water-soluble carbodiimide,
cyanogen bromide, periodate oxidation, and bisoxirane.

A. Carbodiimide Ligand Coupling Method

The carbodiimide coupling reaction has been widely used to couple
boronate ligands to various solid supports. The fact that the coupling can be carried out in aqueous solutions makes it very convenient to use with most solid supports. The carbodiimide activates
carboxyl groups, thus allowing them to react with amino groups to
form amide linkages as shown in Fig. 11. Weith et al. used this
coupling reaction in order to link 3aPBA to an aminoethylcellulose
matrix [4]. They first reacted 3aPBA with succinic anhydride to
introduce a carboxyl functionality and a spacer arm, which was then
reacted with the amino groups of the aminoethylcellulose matrix.
This method has been used extensively by several investigators for
coupling ligands to cellulose [12,16,19], acrylamide [13,53], and
agarose [10,26]. Alternatively, the matrix could be converted to
the hydrazide form and acylated with a carboxylic anhydride to
provide a spacer arm with a carboxyl group which can then be

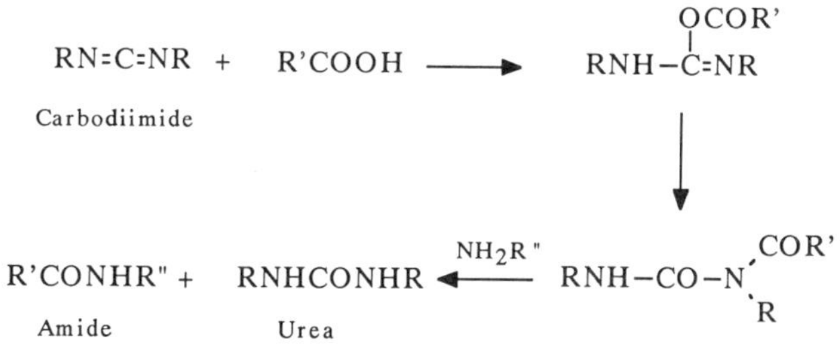

$$RN{=}C{=}NR + R'COOH \longrightarrow RNH-\overset{OCOR'}{\underset{}{C}}{=}NR$$

Carbodiimide

$$R'CONHR'' + RNHCONHR \xleftarrow{NH_2R''} RNH-CO-N\overset{COR'}{\underset{R}{\diagdown}}$$

Amide   Urea

Fig. 11 Immobilization of boronate ligands involving activation of carbonyl groups (matrix) with a water-soluble carbodiimide to couple with an amino group (ligand) and form stable amide linkage. (See text for details.)

activated by carbodiimide to react with an amine functionality as shown in Fig. 12. Several matrices have been prepared by this method using acrylamide-based solid supports [7,70].

A slightly different approach was used by Olsson, who diazotized the hydrazide with nitrous acid and then reacted it with 3aBPA directly without reacting it with succinic anhydride, which involves the carbodiimide activation reaction [64].

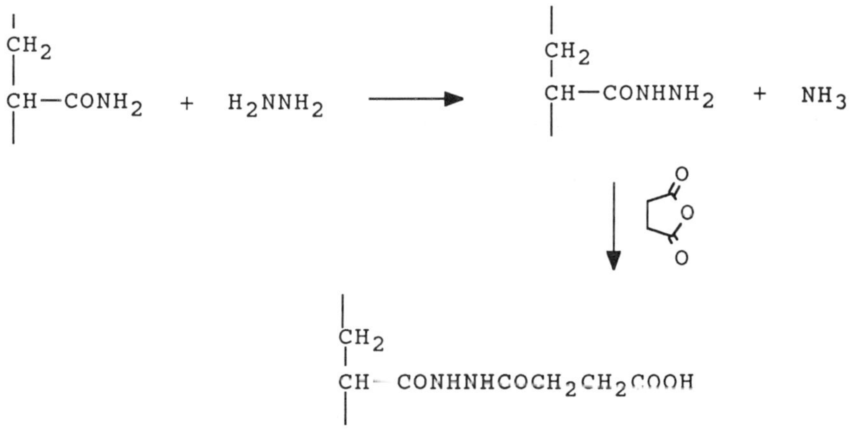

$$\begin{array}{c}|\\CH_2\\|\\CH-CONH_2\\|\end{array} + H_2NNH_2 \longrightarrow \begin{array}{c}|\\CH_2\\|\\CH-CONHNH_2\\|\end{array} + NH_3$$

$$\downarrow \text{(succinic anhydride)}$$

$$\begin{array}{c}|\\CH_2\\|\\CH-CONHNHCOCH_2CH_2COOH\\|\end{array}$$

Fig. 12 The acyl hydrazide derivative of polyacrylamide, prepared by reacting the matrix with hydrazine, can be reacted with a carboxylic acid anhydride to yield a reactive carbonyl group for ligand coupling. (See text for details.)

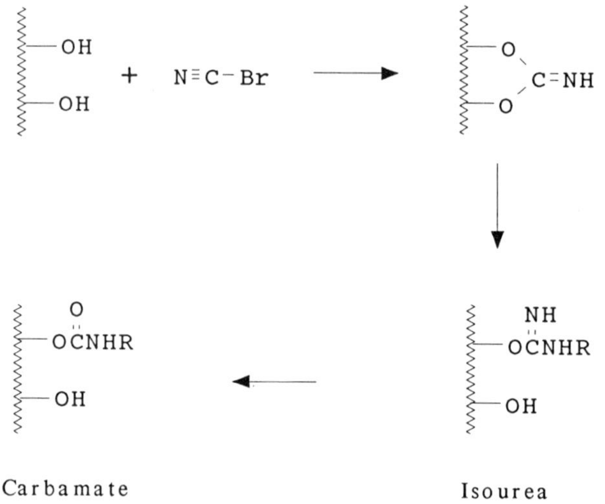

```
~OH                        ~O
        + N≡C-Br    ⟶          C=NH
~OH                        ~O
```

```
     O                          NH
~OCNHR                     ~OCNHR
                    ⟵
~OH                        ~OH
```

Carbamate   Isourea

Fig. 13 Activation of polysaccharide matrix with cyanogen bromide makes the matrix reactive with the amine groups of the ligand yielding carbamate or isourea derivatives. (See text for details.)

B. Cyanogen Bromide Agarose Activation Method

The hydroxyl groups of polysaccharide matrices are activated with cyanogen bromide (CNBr). This makes them reactive with the amino groups of the ligand and form carbamate or isourea derivatives (Fig. 13). The activation of matrix hydroxyl groups by CNBr has been used extensively for the immobilization of proteins to agarose beads. This procedure has found only limited application for the boronate ligands, since the ligand coupling to the activated matrix must be carried out under fairly basic conditions, e.g., pH > 10. Only 5% of CNBr yields the desired product:

$$\text{Matrix-OH} + \text{CNBr} + \text{OH}^- \rightarrow \text{matrix-O-C}\equiv\text{N} + \text{Br}^- + \text{H}_2\text{O}$$

The remainder of CNBr (95%) is readily hydrolyzed:

$$\text{CNBr} + 2\text{OH}^- \rightarrow (\text{O-C}\equiv\text{N})^- + \text{Br}^- + \text{H}_2\text{O}$$

Previously, cyanate esters were regarded as short-lived intermediates and coupling was assumed to proceed via imidocarbonates [matrix-O-C(=N$^+$H$_2$)-NH-ligand] at pH 7 (pK$_a$ = 9.5), thus contributing to unwanted anion exchange characteristics. The isourea derivative is susceptible to nucleophilic attack. This activation procedure also results in significant crosslinking of the matrix, depending on the amount of CNBr used for the activation reaction.

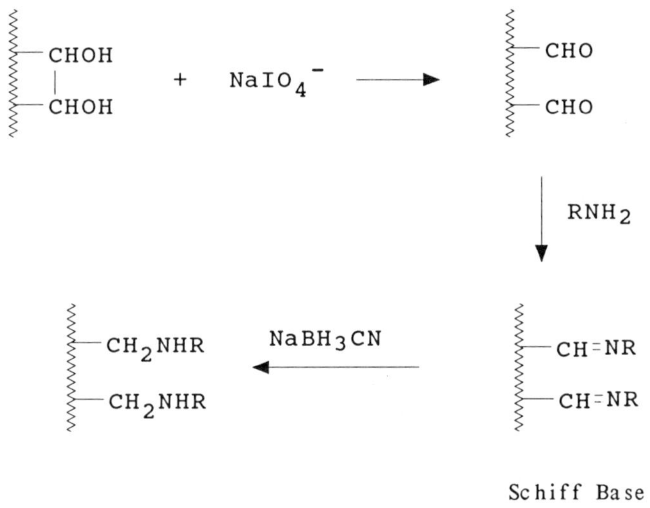

```
⌇—CHOH                              ⌇—CHO
  |       + NaIO₄⁻    ⟶
⌇—CHOH                              ⌇—CHO
                                        |
                                        | RNH₂
                                        ▼
⌇—CH₂NHR      NaBH₃CN               ⌇—CH=NR
              ◀─────────
⌇—CH₂NHR                            ⌇—CH=NR

                                    Schiff Base
```

Fig. 14 The periodate oxidation of vicinal diols of polysaccharide matrix is used to couple the ligand via amino group followed by reduction of the Schiff base. (See text for details.)

Cyanogen bromide has been used to immobilize 6-aminocaproyl-3-aminophenylboronate to crosslinked agarose beads, e.g., Sepharose CL-6B [65].

C. Periodate Oxidation Dextrose Activation Method

Periodate oxidation of the vicinal hydroxyl groups of polysaccharide matrices results in reactive dialdehydes which enables them to form Schiff bases with ligand amino groups (Fig. 14). This method can be used in place of CNBr activation. The *meta*-periodate ($NaIO_4$)-activated gel, after washing to remove the iodate, can be stored at 4°C for several days without appreciable loss of activity. The dialdehydes easily form Schiff base with the ligand amine groups between pH values of 4 and 6. The adduct is stabilized by reducing it with a borohydride ($NaBH_4$ or $NaBH_3CN$) to form stable secondary amines.

D. Bisoxirane Method for Activation of Carbohydrate Matrices

The reaction of bisoxirane with the hydroxyl groups of a carbohydrate matrix is another method of coupling. Bisoxirane, e.g., 1,4-butanediol diglycidoxy ether, reacts readily with hydroxy- or amine-

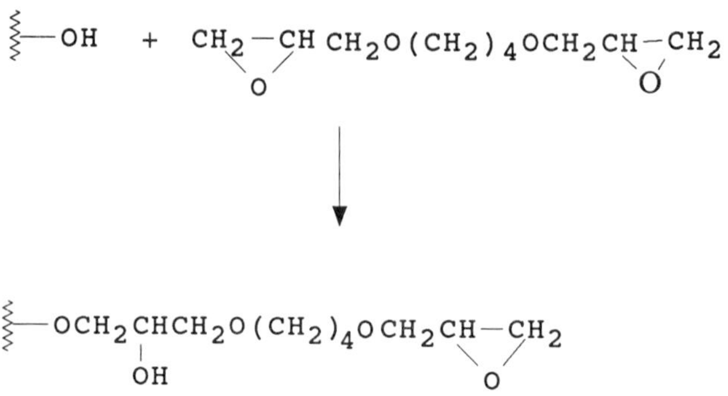

$$\text{\textasciitilde}\text{—OH} + CH_2\text{—}CH\,CH_2O(CH_2)_4OCH_2CH\text{—}CH_2$$
$$\underset{O}{\diagdown\diagup} \qquad\qquad\qquad\qquad \underset{O}{\diagdown\diagup}$$

$$\downarrow$$

$$\text{\textasciitilde}\text{—OCH}_2\underset{OH}{CH}CH_2O(CH_2)_4OCH_2CH\text{—}CH_2$$
$$\qquad\qquad\qquad\qquad\qquad\qquad \underset{O}{\diagdown\diagup}$$

Fig. 15 Activation of polysaccharide matrix with bisoxirane (epoxy) results in a free epoxy group ready to react with nucleophilic (amino, hydroxyl, and sulfhydryl) groups of the ligand. (See text for details.)

containing gels at an alkaline pH yielding gel derivatives with a long hydrophilic and an epoxy-activated matrix with a free epoxy group to react with nucleophilic amine, hydroxyl, and sulfhydryl groups of the ligand (Fig. 15). The oxirane-coupled ligands provide a long-chain hydrophilic spacer arm. This method has been used to couple 3aPBA to agarose beads, such as Sepharose 6B [71].

VI. AFFINITY MATRICES USED FOR IMMOBILIZATION OF DIFFERENT LIGANDS

A number of boron-containing affinity matrices have been synthesized using different inert solid supports like cellulose, acrylamide, silica, porous glass, agarose, and dextrans. Some of these matrices are commercially available under various trade names. The first boronate matrix was synthesized by coupling 3aPBA to aminoethyl- and carboxymethylcellulose materials using carbodiimide coupling reaction [4]. Satisfactory affinity chromatography results have been obtained for the separation of sugars and nucleic acid components using these matrices. Several other investigators have similarly synthesized cellulose-based matrices [16,19,20]. A commercially available acetylated cellulose-based matrix, DBAE-cellulose (Collaborative Research), has been used for the separation of small and large biomolecules [10,15,72,73]. A cellulose-based matrix has also been coupled to 4-bromophenylboronate to yield a different kind of matrix with quaternary amine groups. The positive charge generated by these groups has been claimed to lower the pK_a of the boronate or at least allow stable complex formation with diols at a lower pH value [36]. In spite of all these successful investigations, there are a number of drawbacks in using cellulose-based matrices for affinity chromatography, such as poor flow characteristics, low binding capacity, and high residual charges [36]. The unwanted charges have been removed by derivatizing the residual groups, i.e., acetylation in DBAE-cellulose.

In order to overcome these problems, attempts have been made to find alternate solid supports, e.g., with better flow characteristics. Acrylamide-based matrices were then developed and several methods used for the immobilization of different ligands. In the first method, 3aPBA was treated with methacryloyl chloride to form a polymerizable derivative which was then crosslinked using tetramethylene dimethacrylate by radical polymerization [74]. Although this method resulted in very high ligand concentrations and excellent binding capacity for nucleotides [75], it could not be adapted to obtain spherical beads [76]. Another polyacrylamide-based matrix was prepared by modifying the method of Weith et al. [4] and using it for the separation of a number of biomolecules containing cis-diol groups [13,53,77]. The other method was to use hydrazine activation to couple polyacrylhydrazide to succinic anhydride and then use carbodiimide activation to couple it to 3aPBA [7]. Boronate gels made by this method have been used for the chromatographic separation and quantitation of nucleosides [7,64,70,78] and purification of modified nuclear proteins [22]. Olsson has obtained a slightly different matrix by modifying the step after hydrazide activation.

The acrylamide-based boronate gels, boric acid gel (Aldrich) and Affi Gel 601 (BioRad), also have been used for the isolation of catechol compounds, nucleosides, etc. [9,17,79-82]. p-Vinylbenzene boronate as a monomer is copolymerized with styrene, acrylamide, or used preferentially in an interstitial homopolymerization on porous polystyrene beads [66] and used for the purification of DOPA. The interstitial homopolymerization afforded the most useful matrix with sufficient capacity and sorption and desorption characteristics. Although the acrylamide-based matrices are more resistant to degradation against heat and microorganisms compared to the cellulose-based matrices, they exhibit a strong tendency to swell or shrink with changes in pH or ionic strength, are not stable in alkaline conditions, and yield very poor ligand concentrations, especially those matrices made for the separation of macromolecules.

Dextran-based matrices also were made by introducing quaternary ammonium groups as described earlier (see cellulose-based matrices) [36]. They give satisfactory results apart from the fact that these matrices also exhibit enhanced swelling and shrinking with changes in pH and ionic strength of the medium.

Deactivated silica [26,83-86] and porous glass [12,19] are among the inorganic solid supports that have been examined in an attempt to increase the pressure-handling capacity of the matrices for HPLC applications. These matrices have been used for the separation of carbohydrates and nucleic acid components. However, the amount of ligand immobilized in these matrices was relatively very small. A nine-carbon aliphatic phenylboronate has been synthesized and coated in chloroform on microparticulate poly(chlorotrifluoroethylene) beads [12]. This "reversed-phase boronate" matrix has been successfully used in aqueous solutions for the separation of aminoacyl-tRNAs from unesterified free tRNAs and ribonucleotides from deoxyribonucleotides.

The many advantages that crosslinked agarose matrices exhibit over other affinity matrices, such as high molecular weight exclusion limit, low nonspecific adsorption, and excellent flow characteristics, prompted investigators to use agarose-based solid supports to immobilize different boronate ligands. Carboxylated agarose matrices have been prepared using carbodiimide activation [10,12,26] and used very satisfactorily for the affinity chromatography of biomolecules [21,65]. 4-(ω-Aminoethyl)phenylboronate has been coupled to carboxylated agarose and used to purify serine proteases [26]. The relatively low ligand concentrations and the presence of varying amounts of residual carboxyl groups were among the disadvantages of this material.

A novel coupling technique was employed some years ago for the preparation of an agarose-based boronate gel, Matrex Gel PBA (Amicon) [76]. This gel was claimed to have as much as 100 μmol of boron per ml of the gel; the ligand, 3aPBA, was coupled covalently to a crosslinked agarose support. This boronate gel was reported to be stable under a wide range of operating conditions, and the method was claimed not to introduce any residual charges or hydrophobic groups into the matrix. Separation of enzymes, membrane proteins, and most importantly, glycosylated hemoglobin has been accomplished using Matrex Gel PBA, to name a few applications [23-25,34,87]. Glycogel B (Pierce) is another commercially available 6% crosslinked agarose-based boronate gel that has been successfully employed for the separation and quantitation of glycosylated hemoglobin in diabetic patients [45,88-90] (see Section VIII.B for details).

VII. METHODS FOR STUDYING COMPLEX FORMATION

There are several methods available to study the complex formation of boronates with *cis*-diols, such as electrochemical and calorimetric titrimetry, conductivity, optical rotation, spectrophotometry, nuclear magnetic resonance, differential solubility or chemical reactivity, chromatography, electrophoresis, enzyme inhibition, fluorescent labeling, X-ray diffraction, and absorption spectroscopy. Some of these techniques used frequently are discussed below in detail; they can be used for accurate determination of binding capacities and also binding constants of the matrix.

A. Ion Exchange Chromatography

Both anion and cation exchange matrices have been applied for making complexes of sodium borate with hydroxyl groups of different types of sugars. The anion exchange method most commonly used involves binding of borate anions to the anion exchange matrix and then washing of unexchanged borate ions [38.91]. Only the solutes having affinity for the borate anion bind to this *borate column* while the others elute freely from the column. In order to maintain the boric acid in its anionic form, the operation must be carried out under alkaline conditions (pH 9.0-9.5). The molecules bound to the borate matrix are eluted by lowering the pH or by increasing the salt concentration of the eluent. Alternatively, a borate-alcohol complex is formed in the mobile phase, such as by incorporating a small amount of sodium tetraborate in the eluent and then applying a mixture of sugars to this column. The borate-diol complex so formed exhibits an anionic charge and therefore it readily exchanges with the anion exchange column. The complexed anions are retained more strongly by the anion exchanger than those molecules which fail to form an anionic complex. Historically, 3' nucleotides were separated from 5' nucleotides by this technique; only the latter nucleotides form a boronate complex. The ability of different sugars to exchange with the matrix differs widely because of the orientation of the hydroxyl groups involved in making the complex and the overall confirmation of the molecule. The difference in the structure is the basis of their distinctions in retention by the borate column. Weak anion exchangers have also been used in the same way [92].

When cation exchange resins are used, the sugars are applied at an acidic pH and the retained molecules are desorbed by a borate buffer of an alkaline pH [8,93]. In this case, the sugars are apparently not retained by the exchange characteristics of the cation exchange resin since both are charged negatively. After the addition of the borate buffer, the sugar perhaps forms a complex with the borate and this induced negative charge allows them to be *excluded* from the matrix. Boronate molecules bound to anion exchange resins could also be eluted using competitive affinity molecules which have *cis*-diol groups. An example is the use of sugar alcohols like sorbitol to elute bound molecules from ion exchange as well as immobilized boronate matrices.

B. Affinity Chromatography Using Immobilized Boronates (Frontal Analysis)

This method is used to determine the binding capacity and binding constant of an affinity matrix. A buffer containing the affinity molecule (e.g., adenosine) which can bind to the boronate matrix at a given pH (pH 8.5) is applied continuously to a column packed with the affinity matrix until saturation is reached and the effluent concentration of the respective molecule is equal to the concentration of the feed solution. Then the column is washed until all the bound molecules are eluted and nothing is coming out of the column, and a buffer containing the same molecule (at the same concentration) but at a different pH (pH 5.5) where it entirely fails to bind with the column matrix is passed through the column [94]. The elution profile for this kind of analysis reveals important information about the interaction of the ligand and the affinity molecules as shown in Fig. 16. The frontal volume (V_f) is obtained as the difference in the elution volumes between the 50% saturation point for the binding and control experiments. The binding capacity is calculated as follows:

$$\text{Binding capacity} = \frac{V_f \times [T]}{V_t}$$

where V_f is the frontal volume, V_t is the total column volume, and [T] is the concentration of the feed solution. The apparent dissociation constant can be calculated from the following equation [95]:

$$K_{diss.} \text{ (apparent)} = \frac{V_t[L] - V_f[T]}{V_f}$$

where [L] is the ligand concentration in the matrix.

C. Absorption Spectroscopy

The possibility of studying the complex formation between boronates and cis-diols using absorption spectroscopy has been reported by several investigators. The method is based on the observation that free boronic acid and boronate anion exhibit differences in their absorption spectra [41a]. A change in the absorption spectrum of the boronate ligand before (boronic acid) and after (boronate anion) complexing with cis-diols is expected in this method. 3-Aminophenylboronic acid and several other ligands (the model ligands) and cis-diols of carefully selected (exhibiting no overlapping) absorption spectra are used in order to monitor characteristic changes in the spectrum after complex formation [41a]. In this method, two sets of

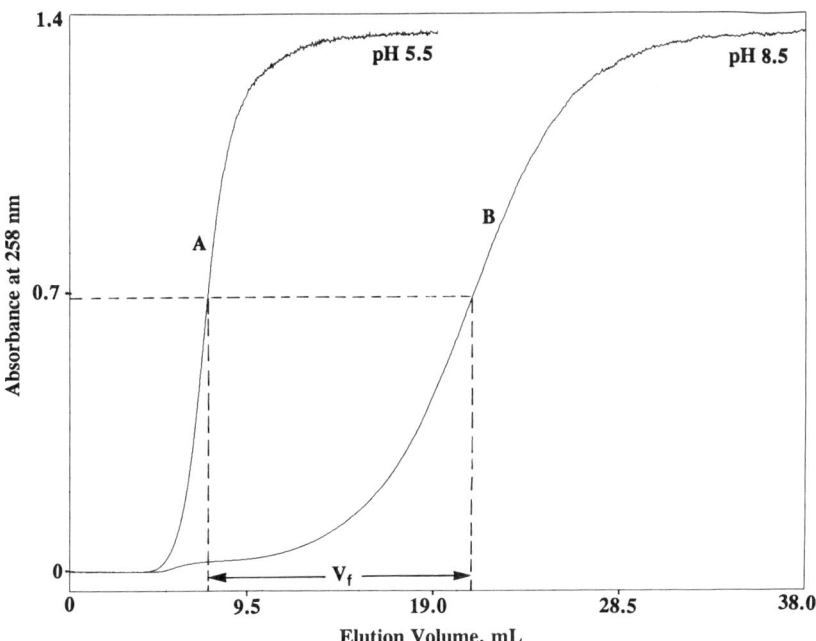

Fig. 16 Binding capacity of a boronate affinity column determined by frontal uptake of the affinity molecule (adenosine) by the matrix. Curve A was obtained by passing 0.22 mM solution of adenosine in a 50 mM phosphate buffer, pH 5.5 (a noncomplexing environment). Curve B was similarly obtained using adenosine under complexing conditions (pH 8.5). The frontal volume (V_f), 13.7 ml, was derived from the volume difference between the 50% saturation points (midpoints) of the two curves. (Based on data from Refs. 41a and 112.)

experiments are carried out, one to examine the ionization of the acid and the other to study the ionization of boronic acid in the presence of a complexing diol. Identical amounts of boronic acid are delivered into buffered solutions of different pH values. One solution pH is selected where boronic acid is present only in the nonionized form (an acidic pH) and the other in the ionized form (very basic pH). Other solution pHs are selected one pH unit below and above the pK_a of the boronic acid with 0.1 or 0.2 pH unit increments. These solutions of identical amounts of boronic acid (and a diol), but at different pH values, are read at two selected wavelengths. For example, 3aPBA showed absorption maxima at 280.5 and 295 nm in acidic and basic solutions, respectively.

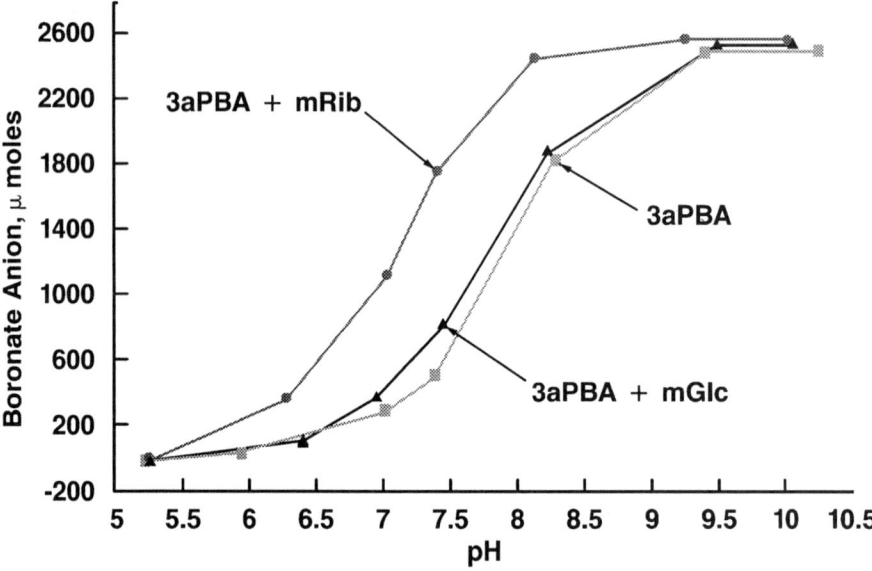

Fig. 17 Study of interaction by absorption spectroscopy, without involving immobilization, between a boronate ligand (3aPBA, 3-aminophenylboronic acid) and polyol affinity molecules (mRib, β-methylribofuranoside; mGlc, α-methylglucopyranoside). (See text for details.) (From Ref. 41a.)

Therefore, the absorbance of this compound was monitored at those two wavelengths while gradually raising the pH of the solution from pH 5.3 to 10.25 (Fig. 17). The amount of boronate anion (B^-) was then determined at different pH values using the following equation:

$$[B^-] = \frac{(A_{\lambda 1} - A_{\lambda 2}) - \{[B^0](\varepsilon B^0_{\lambda 1} - \varepsilon B^0_{\lambda 2})\}}{\varepsilon B^-_{\lambda 1} - \varepsilon B^-_{\lambda 2}}$$

where $A_{\lambda 1}$ = absorbance at wavelength λ_1 nm,

$A_{\lambda 2}$ = absorbance at wavelength λ_2 nm,

$[B^-]$ = molar concentration of boronate anion,

$[B^0]$ = molar concentration of boronic acid (uncharged molecule),

$\varepsilon B^0_{\lambda 1}$ = extinction coefficient of neutral boronic acid at λ_1 nm,

$\varepsilon B^0_{\lambda 2}$ = extinction coefficient of neutral boronic acid at λ_2 nm,

$\varepsilon B^-_{\lambda 1}$ = extinction coefficient of boronate anion at λ_1 nm, and

$\varepsilon B^-_{\lambda 2}$ = extinction coefficient of boronate anion at λ_2 nm.

Previously, the binding constants of boric acid [96] and boronic acid [42] to chromatographic ligands were determined. In these cases the ionization constants and the extinction coefficients for free and bound ligands were determined by potentiometric titration and by applying extrapolation methods to the absorption spectra, respectively. More recently, Soundararajan et al. [56] used the method described by Gorman and Darnall [97] in order to obtain both the pK_a and the diol formation constants of four arylboronic acid derivatives. For pK_a determination of an arylboronic acid, two matched spectrophotometric cells, a reference $[B]_R$ and a sample $[B]_S$, are filled with solutions of equal concentrations of boronic acid. The pH of the reference solution is maintained at a value at least two pH units less than the expected pK_a of boronic acid, where it is essentially in the nonionized form, $[B]_R$. The sample solution is maintained at a pH somewhat higher than the pK_a so that boronic acid is present in both ionized and nonionized forms. A series of "difference spectra" is taken for an experiment in which the total concentration of boronic acid is varied in each cell but in such a way that $[B]_S = [B]_R$, while the pH in each cell is maintained at the selected values. Another set of difference spectra is taken in an identical manner except that a different constant pH value is chosen for the boronic acid solution in the sample cell. By determining the difference in absorbance at two selected wavelengths, i.e., δA_1 and δA_2, respectively, and by knowing the values of the total concentration of boronic acid, i.e., $[B_T]$ at each wavelength, the concentration of the complexed boronate anion can be determined from equation 25 in [56]. Furthermore, the value of the binding constant for a cis-diol, e.g., fructose, can be determined by substituting the values of the complexed boronate anion, the hydrogen ion concentration, total boronate concentration, and appropriate value for K_a in equation 23 of [56].

D. Proton and ^{11}B-NMR

Proton and ^{11}B-NMR have been used to detect and measure complex formation between borate or boronate with cis-diols [35,59]. When tetraborate, $(B_4O_7)^{2-}$, is added to a solution of D-glucose, a complex is formed which can be observed by monitoring the chemical shifts in ^{1}H- or ^{11}B-NMR spectroscopy. The ^{1}H-NMR spectral analysis has shown that the signal for H-1αp disappears and a low field signal for H-1αf appears gradually as D-glucose is titrated with tetraborate indicating a pyranose-to-furanose transformation [35]. These results have suggested that the complex formed is a D-glucofuranose-1,2-borate.

Chemical shifts of ^{11}B-NMR of anionic, neutral, and complexed-anionic species of boronates are expected to change after complexing

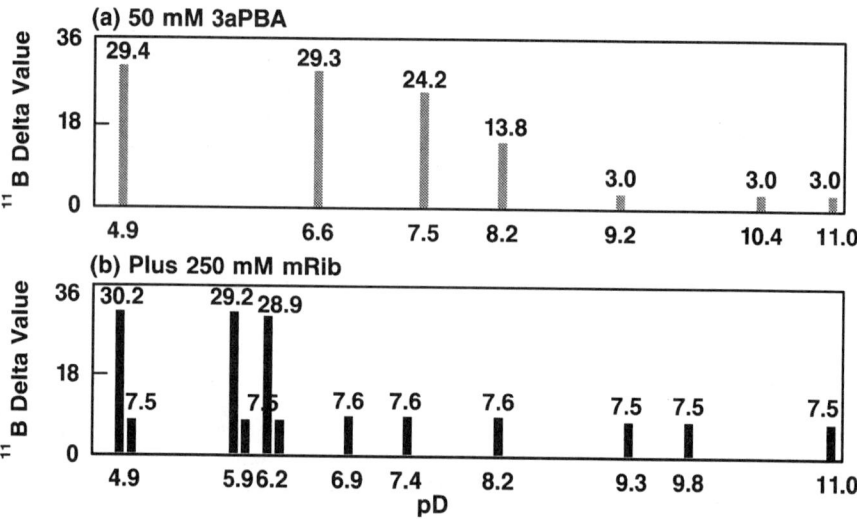

Fig. 18 Study of interaction by ^{11}B-NMR spectroscopy, without involving immobilization, between (a) a boronate ligand (3aPBA, β-methylribofuranoside) and (b) an affinity molecule (mRib, β-methylribofuranoside). (See text for details.) (From Ref. 41a.)

with cis-diols because of differential electronic shielding of the ^{11}B atom in the species. The ^{11}B resonance frequency of phenylboronate ligands in solution gradually shifts from δ30 to δ3 with an increase in the pH (Fig. 18). The observed signal has been assigned to the equilibrium state [$B^0 \leftrightarrow B^-$] between the B^0 and B^- species [51]. Since the conversion of neutral to anionic species occurs at a very fast rate, it escapes detection on the ^{11}B-NMR time scale.

The addition of a cis-diol compound to the boronate ligand solution results in complex formation and therefore specific signals can be distinguished for the complexed-boronate anions (B_c^-) and the neutral (B_0) molecules. For example, the complexed species exhibits a characteristic chemical shift at δ7.5, whereas the neutral molecules exhibit a chemical shift at δ30 (see Fig. 18). The absence of signals at δ3 indicates completion of the forward reaction, and thus it is possible to conclude that the favored reaction is the forward reaction resulting in the complex formation. Moreover, the boronate cis-diol complex (B_c^-) is apparently stable and fails to break down to the free boronate anion even at higher pH values. In fact, compounds that are usually unstable in alkaline solutions exhibit their original spectra upon lowering of the solution pH, e.g., catechols (complexed with boronate).

E. Optical Rotation

Optical rotation methods rely on the change in the optical activity of carbohydrates upon formation of complexes with borate [35,47,98]. This phenomenon is caused by changes in the equilibrium of the carbohydrates when they complex with borates, i.e., the equilibria between the aldo and keto isomers can be displaced if one isomer reacts to a greater extent than the other with a complexing reagent. For example, the presence of borate in the reaction of D-glucose with alkali increases the yield of D-fructose. Wavelengths in the range of 200-400 nm are commonly used. Shifts in the wavelength maxima of rotation are observed when the borate-to-carbohydrate ratio is varied; Aronoff et al. used this property for a polarimetric titration of D-glucose by tetraborate [35]. They were able to demonstrate the existence of only two species, Glc and $GlcB^-$, by this technique. The species Glc_2B^- was very minor, unless indistinguishable both in dispersion (λ) and intensity (α).

VIII. EXAMPLES OF APPLICATIONS OF BORONATE AFFINITY CHROMATOGRAPHY

A. Small Biomolecules

Separation and quantitation of a vast number of small biomolecules having cis-diol groups has been accomplished using boronate affinity chromatography. Historically, 5'-nucleoside monophosphates were separated from their 3' counterparts by borate complexing [90]. Among the first molecules to be separated on the immobilized boronate matrices were ribonucleosides [4,7,9,64,70,75,78], ribonucleotides [13,81], and catechols [16,17,79,80]. There is a more recent report of the separation of 2-hydroxycarboxylic acids using an immobilized boronate matrix [82].

B. Large Biomolecules

Among the macrobiomolecules separated by boronate affinity chromatography are aminoacyl-tRNA (AA-tRNA) from free (nonacylated) tRNAs [12,19,20,72], oligoribonucleotides from deoxyoligoribonucleotides [10,73], specialized transfer RNAs [12,21], enzymes, proteins, modified peptides [15,22-24,26,54,65,77,86,87], and glycosylated hemoglobin from nonglycosylated hemoglobin [25,71,89,100,101].

Aminoacyl Transfer RNA

The basis for the separation of AA-tRNA from nonacylated tRNA is the interaction of the 2',3'-cis-hydroxyl groups of the terminal adenosines of nonacylated tRNAs with the boronate affinity matrix to form stable complexes [19,75]. The binding of uncharged tRNAs is enhanced by the use of an alkaline buffer. However, alkaline

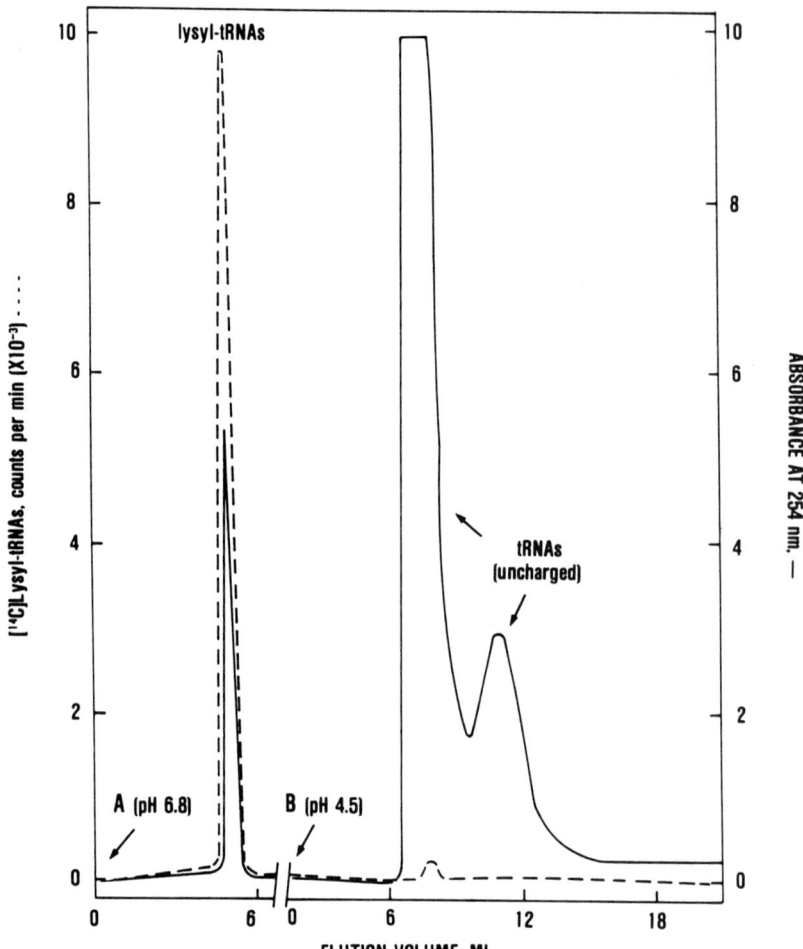

Fig. 19 Separation of an aminoacyl-tRNA from uncharged tRNAs by reversed phase boronate chromatography at neutral pH. About 1 mg of *E. coli* tRNA containing [^{3}H]lysyl-tRNAs (50,000 cpm) in a buffer A (pH 6.8) was applied and then eluted with buffer B (pH 4.5). About 4 nmol of lysyl-tRNA was recovered from other (uncharged) tRNAs. (From Ref. 1. Reproduced with permission of Elsevier Sci. Publ. B. V., Amsterdam.)

conditions cause hydrolysis of the AA-tRNA ester bond, resulting in uncharged tRNAs which form complex with the boronate matrix (similar to the other uncharged tRNAs) [102]. To achieve satisfactory complex formation at the *least* alkaline or neutral pH, modification of the complexing solvent is recommended, e.g., high salt contents, Mg^{2+}, ethanol [20,103]. To remove nonspecific anionic binding to the affinity matrix, acetylation of the boronate matrix after ligation has been carried out, such as N-[N-(m-dihydroxyborylphenyl)succinamyl]aminoethylcellulose (AEB-cellulose).

For example, the separation of lysyl-tRNA from other uncharged tRNAs has been accomplished by using an N-(m-dihydroxylborylphenyl)carbamylmethylcellulose (CMB-cellulose) column [12]. On a novel reversed phase boronate (RPB) matrix, both uncharged and AA-tRNAs were retained by a combination of the hydrophobic interactions with the ligand (alkylboronate) and the boronate affinity functionality [12]. To weaken interactions between AA-tRNAs and the matrix, relatively high concentrations of chloride and magnesium ions were required. However, these ionic conditions were insufficient to break the boronate complex formed between the matrix and the uncharged tRNAs. Strongly held uncharged tRNAs required significantly higher concentrations of chloride ions, but only neutral or mildly acidic conditions (a low eluent pH), thus preserving the desired ester bond in AA-tRNA. Uncharged tRNAs form the complex with the decanoylboronate ligand on reversed phase boronate (RPB) columns at least one unit below the pK_a value of the phenylboronic acid. The desired AA-tRNA, which fails to complex with the boronate ligand, is eluted in the void volume of the column. An example of such separations involving the isolation of liver lysyl-tRNA from other tRNAs (complexed with the matrix) is shown in Fig. 19. The presence of a 1-nm-long tail and the absence of negatively charged vicinal groups in the matrix were apparently responsible for complexing at a lower (pH 6.8) than the usual alkaline pH needed for cellulose- or polyacrylamide-linked phenylboronic acid matrices [12].

The recovery of tRNAs from CMB-cellulose and RPB column matrices was satisfactory (92-97%) but incomplete from the AEB-cellulose column (65-70%). Chromatography on other matrices (benzeneboronate linked to porous glass, agarose, or Sepharose beads via >1.0-nm-long spacer arm) gave little or no separation of AA-tRNA from the uncharged tRNAs, and recovery of the material was dismally low (about 30%). The low recovery of tRNAs is presumably caused by strong hydrophobic interactions between purine and pyrimidine bases of the polynucleotide and the long spacer arm of the matrix. However, such strong interactions are not exhibited by CMB-cellulose and RPB column matrices, which each contain only approximately 1-nm-long spacer arms [12].

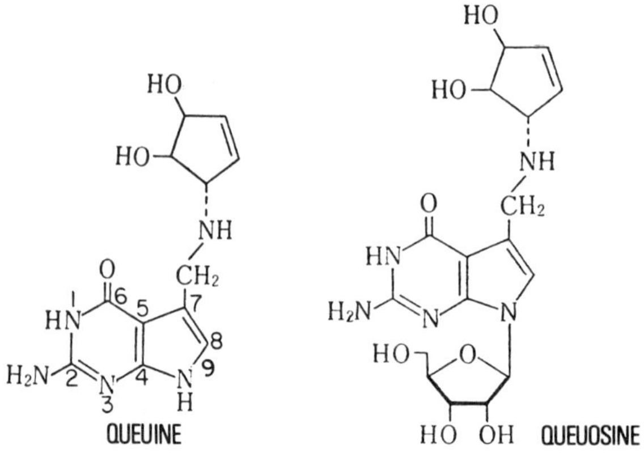

Fig. 20 Structure of a modified purine found in the "wobble position" (position 34) of specific tRNAs. The base, queuine, having a dihydroxycyclopentene ring, tends to form a stable complex with boronate ligands. The nucleoside, queuosine, can additionally form complex via cis-diols of the ribose. Unmodified queuosine is found in four specific tRNAs and eukaryotic aspargine and histidine tRNAs. (Data from Refs. 1 and 104a.)

Fig. 21 Structure of hexosyl-queuosine of eukaryotic tRNAs. Mannosyl-queuosine is present in liver aspartate tRNA and galactosyl-queuosine in liver tryosine tRNA. Since cyclopentene cis-diols are substituted in hexosyl, they are less effectively retained on a boronate column than those tRNAs containing unmodified Q residues. The hexosyl-Q tRNAs can be easily isolated by lectin affinity chromatography. (Data from Refs. 1 and 104b.)

Specialized Transfer RNAs

The "wobble" base, located in position 34, is modified in 61% of eukaryotic and 47% prokaryotic tRNAs [104]. Similarly, out of 49 mammalian tRNAs, 31 contain a modified residue in this position (R. Singhal, unpublished results). A replacement of guanine (G) by a modification of G residue, called queuine (Q), occurs especially at the wobble position in asparagine, histidine, aspartate, and tyrosine tRNAs. These four specific prokaryotic tRNAs have a simple Q residue which contains a *cis*-diol, whereas mammalian tRNAs contain either an unmodified Q, e.g., in tRNA^{Asn} and tRNA^{His} (Fig. 20), or a modified Q. In the latter case, a hexose esterified via one hydroxyl group of the diol present in the Q structure, e.g., mannose-Q in tRNA^{Asp} and galactose-Q in tRNA^{Tyr} (Fig. 21). The unsubstituted Q, containing *cis*-diols, reacts with the boronate ligand and forms an anionic complex at a slightly alkaline pH—similar to the reaction of *cis*-diols of ribose at the 3' end of the tRNAs with a boronate ligand. The bacterial tRNA contains only unsubstituted Q and therefore each one of the four Q-containing tRNAs complex with the boronate matrix. However, only two mammalian Q-containing tRNAs having unsubstituted Q (tRNA^{Asn} and tRNA^{His}) form complex with the boronate matrix. (The other two mammalian tRNAs with sugar-substituted Q residues are resolved on lectin affinity matrices specific for the hexose [1].)

Simple Q-containing tRNAs tend to complex with the boronate matrix more strongly than those lacking the Q residue in their structure, i.e., complexation via ribose *cis*-diols. A group separation of Q-tRNAs from other tRNAs on a CMB-cellulose affinity matrix has been achieved. This simple separation can be scaled up to process large amounts of tRNAs. For example, Q-tRNAs from mammalian and bacterial sources were resolved from unfractionated, bulk tRNA samples on CMB-cellulose columns. Typically, a 0.5-g sample of bovine liver tRNA, as shown in Fig. 22, was applied to a CMB column. The column was first eluted with a basic buffer (pH 8.7) and then with an acidic buffer (pH 4.5). Material from each peak was pooled and precipitated with ethanol. Levels of different Q-tRNAs in the two peaks were determined by aminoacylation. Analysis showed that only a very small proportion of hexose-substituted Q-tRNAs, but mostly all of the Q-tRNAs, were retained by the column [12]. Elution conditions can be examined in order to reduce the interaction between Q diols and the matrix while maintaining the complex formation between the ribose diols and the matrix.

Glycosylated Hemoglobin and Plasma Proteins

The association between glycosylated hemoglobins and diabetes mellitus has been known for more than 20 years [105]. Since then the

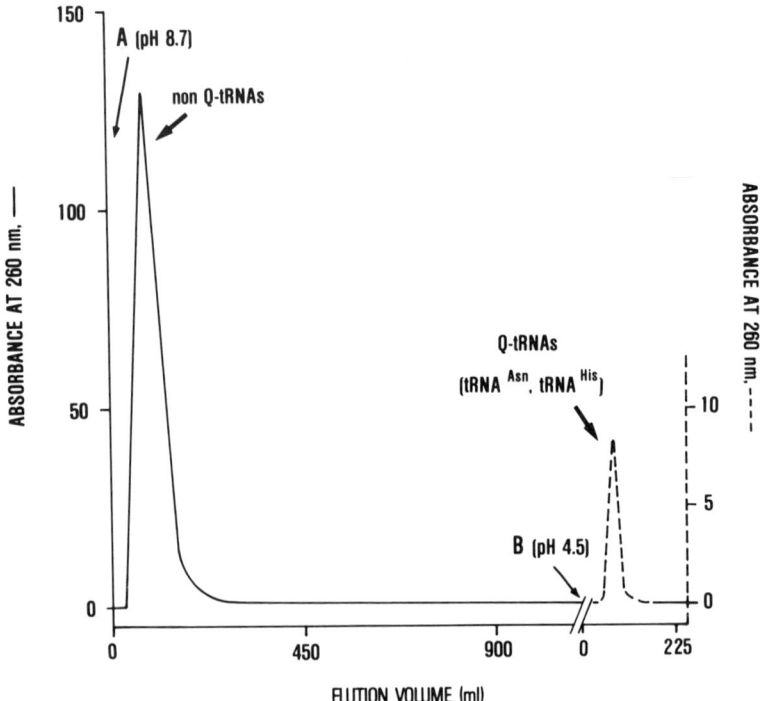

Fig. 22 Group separation of unsubstituted-Q-containing tRNAs (tRNA^{Asn} and tRNA^{His}) from other tRNAs by boronate affinity chromatography (see legend to Fig. 19 for details). (From Ref. 1. Reproduced with permission of Elsevier Science Publ. B.V., Amsterdam.)

determination of glycosylated hemoglobins has been a valuable tool for the assessment of long-term glycemia in diabetic patients [27, 28,106,107]. Glycosylated hemoglobins are formed by nonenzymatic attachment of glucose, glucose-6-phosphate, and fructose-1,6-biphosphate to hemoglobin. The glucose-hemoglobin adducts are the clinically important molecules since they are formed and accumulated in the erythrocytes in proportion to the prevailing blood glucose concentration. Glucose attachment to the hemoglobin can occur at several sites of this macromolecule. For example, glucose can couple to N-terminal amino groups of both α and β chains and also to certain ε-amino group of the lysine residues [92]. The major glycosylated hemoglobin fraction, Hb A_{1c}, is formed by the reaction of glucose and the N-terminal valine residue of the β chain, whereas

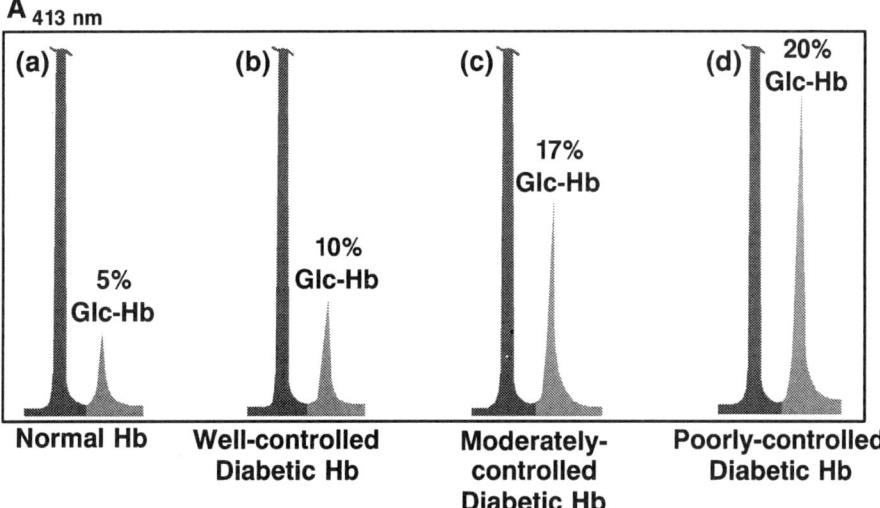

Fig. 23 High-performance separation of hemoglobin from glycosylated hemoglobin (Gly-Hb) on a hydrophilic vinyl-based boronate matrix. A sample of diluted whole blood (10 μl) is applied to the boronate column and eluted with a buffer of pH 8.2 followed by a buffer containing sorbitol. The effluent was monitored at 413 nm. The entire analysis was completed in less than 10 min. The Gly-Hb contents increase with the severity of the diabetic condition. (From Ref. 112.)

other fractions—Hb A_{1a_1}, Hb A_{1a_2}, and Hb$_{A1b}$—are derived from the reaction of a phosphorylated hexose (i.e., glucose-6-phosphate or fructose-1,6-biphosphate) again with the N-terminal residue of the β chain [108].

Hemoglobins glycosylated at the N-terminal α chains or ε-amino group of lysine residues elute at the leading edge of the main hemoglobin peak [92,109]. This portion of glycosylated hemoglobin represents approximately 50% of the total glycosylated hemoglobin. Fluckiger et al. demonstrated that the extent of glycosylation at the β-N terminus and at all other sites is similar, and that the total glycohemoglobin is quantitatively retained on a boronate agarose affinity support [89]. However, approximately 50% of the HB A_{1a+b} fraction and less than 10% of Hb A_{1c} are not retained [110].

Boronate affinity chromatography of hemoglobin is based on the complex formation between boronate and the cis-diol groups of the carbohydrate moiety of glycosylated hemoglobin (for reaction mechanism, see Section II). The bound glycosylated hemoglobin is

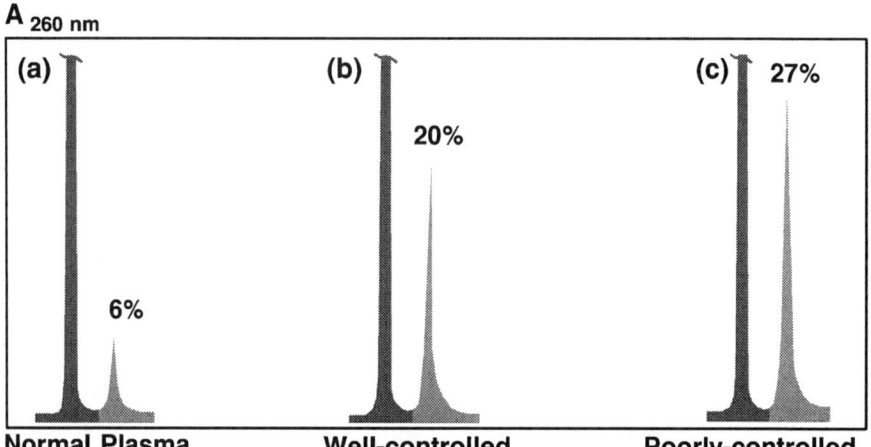

Fig. 24 High-performance separation of plasma proteins from glycosylated plasma proteins on a vinyl-based boronate column. The effluent was monitored at 260 nm. The glycosylated plasma protein contents increase with severity of diabetic condition (for details, see legend to Fig. 23). (From Ref. 112.)

generally desorbed by the inclusion of a competitive diol, e.g., sorbitol in the eluent. The results are expressed as a percentage of the boronate-bound hemoglobin (Hb) and determined from the following relation between bound and unbound peak areas (A_{413}) monitored at 413 nm:

$$\text{Percent of bound Hb} = \frac{A_{413} \text{ bound Hb} \times 100}{A_{413} \text{ bound} + 2.5 \times A_{413} \text{ unbound}}$$

So far these affinity separations have been mostly carried out in a stepwise manner such as using Glycogel of Pierce (Rockford, Ill.). However, boronate ligands attached to crosslinked agarose or to a hydrophilic vinyl polymer matrix can be used in a high-performance chromatography setup while monitoring the absorbance at 413 nm for hemoglobin and at 280 for all glycosylated plasma proteins. An example of the separation of hemoglobin from glycosylated hemoglobin, carried out under high-performance conditions, are shown in Fig. 23 [112]. These separations were achieved on a hydrophilic vinyl-based boronate matrix while using a competing polyalcohol for elution. The hemoglobin elutes after the void volume, but glycosylated hemoglobin (Gly-Hb) is retained and can be eluted when sorbitol is added to the eluent. The relative amount of the latter peak increases with the severity of the diabetic condition. Similarly, plasma proteins can be resolved from glycosylated plasma proteins, as shown in Fig. 24 [112]. These separations are carried out in a way similar to hemoglobin separations, but the effluent is monitored at 260 nm. Again, the glycosylated plasma proteins increase with the severity of the diabetic condition.

IX. CONCLUDING REMARKS

The boronate affinity chromatography method offers promise for instant identification and simple purification of numerous metabolites of interest, however, the current technique suffers from lack of satisfactory boronate ligands and ready-made column matrices. Only one boronate derivative, 3-aminophenylboronic acid (3aPBA), is commercially available for immobilization, and among the ready-made matrices available, each contains a derivative of 3aPBA (or simply a boric acid gel). This ligand (3aPBA), besides being unstable under alkaline conditions, paradoxically can function satisfactorily only under alkaline conditions for a large number of biomolecules. That is, in addition to the ligand, several metabolites and macromolecules of interest are also unstable under required alkaline conditions of chromatography. Moreover, matrices currently used for immobilization and the techniques employed for immobilization require the use of undesirable alkaline reaction conditions, thus yielding a boronate matrix with only a limited number of functional groups. Most biomolecules intended for separation by this technique are highly hydrophilic, but may also contain one or more hydrophobic groups (for example, nucleosides and polynucleotides, glycosylated hemoglobins, nervous system amines). In spite of these drawbacks, 3aPBA has been employed and often linked to a given solid support mostly because of availability of the compound and the matrix. The currently available matrix for separation of glycosylated hemoglobin, Glyco-Gel from Pierce Chemical, Rockford, Illinois, is a 3aPBA ligated to 6% crosslinked agarose beads. The matrix is highly unstable and breaks down easily (the ligand leaches out) upon repeated use. (Though the manufacturer recommends no more than 10 runs on these reusable columns, in our and others' [111] experiences the matrix can only be used for two to three applications.)

Ideally, boronate affinity ligands are needed that can function (form a stable complex) under neutral or acidic conditions and can be coupled to stable solid supports. To overcome the difficulties associated with 3aAPB, novel ligands and coupling chemistries have been proposed which hold promise for satisfactory matrices, i.e., an affinity matrix that can yield a stable complex under desired reaction conditions.

NMR, absorption spectroscopic, and polarimeteric techniques have assisted in understanding the mechanism of boronate complex formation with different biomolecules. This novel approach of examining the interaction (in solution) between the ligand and the affinity molecules (as opposed to using the chromatography column) offers a convenient means for studying the association and dissociation between the boronate ligand and the affinity molecule. The results can help design the most effective boronate ligand for a given affinity molecule. Moreover, the study of complexation in solution also assists in predicting the *feasibility of interaction* without requiring prior immobilization of the ligand to the matrix. The availability of new, carefully designed boronate ligands can be expected to yield highly desirable affinity matrices for the isolation of a great number of biological substances whose purification in molecular biology as well as in clinical chemistry is in increasing demand.

ACKNOWLEDGMENTS

This research was supported by a grant from Wesley Foundation, Wichita, Kansas (grant T8707011).

REFERENCES

1. R. P. Singhal, J. Chromatogr. *266*, 359 (1983).
2. A. Bergold and W. M. Scouten, Boronate Chromatography, in Chem. Anal. *66*, 149 (1983).
3. J. K. Inman and H. M. Dintzis, Biochemistry *8*, 4074 (1969).
4. H. L. Weith, J. L. Weibers, and P. T. Gilham, Biochemistry *9*, 4396 (1970).
5. M. Rosenberg, J. L. Wiebers, and P. T. Gilham, Biochem. *11*, 3623 (1972).
6. E. A. Ivanova, I. E. Kolodkina, and A. M. Yurkevich, J. Gen. Chem. U.S.S.R. *44*, 409 (1974).
7. M. Uziel, L. H. Smith, and S. A. Taylor, Clin. Chem. *22*, 1451 (1976).
8. R. Moran and W. Werkheiser, Anal. Biochem. *88*, 688 (1978).
9. E. H. Pfadenhauer and S.-D. Tong, J. Chromatogr. *162*, 585 (1979).
10. B. Pace and N. R. Pace, Anal. Biochem. *107*, 128 (1980).
11. N. W. Y. Ho, R. E. Duncan, and P. T. Gilham, Biochemistry *20*, 64 (1981).
12. R. P. Singhal, R. K. Bajaj, C. M. Buess, D. B. Smoll, and V. N. Vakharia, Anal. Biochem. *109*, 1 (1980).

13. J. H. Hageman and G. D. Kuehn, Anal. Biochem. 80, 547 (1977).
14. D. Pruess-Schwartz, S. M. Sebti, P. T. Gilham, and W. M. Baird, Cancer Res. 44, 4104 (1984).
15. A. E. Annamalai, P. K. Pal, and R. F. Colman, Anal. Biochem. 99, 85 (1979).
16. M. Sugumaran and H. Lipke, Anal. Biochem. 121, 251 (1982).
17. S. Higa, T. Suzuki, A. Hayashi, I. Tsuge, and Y. Yamamura, Anal. Biochem. 77, 18 (1977).
18. A. Gasion, T. Wood, and L. Chiltemerere, Anal. Biochem. 118, 4 (1981).
19. R. E. Duncan and P. T. Gilham, Anal. Biochem. 66, 532 (1975).
20. T. F. McCutchan, P. T. Gilham, and D. Soll, Nucl. Acid Res. 2, 853 (1975).
21. H.-E. Wilk, N. Kecskemethy, and K. P. Schafer, Nucl. Acid Res. 10, 7621 (1982).
22. H. Okayama, K. Ueda, and O. Hayashi, Proc. Natl. Acad. Sci. USA 75, 1111 (1978).
23. G. T. Williams, A. P. Johnstone, and P. D. G. Dean, Biochem. J. 205, 167 (1982).
24. G. T. Williams, A. P. Johnstone, V. Bouriotis, and P. D. G. Dean, Biochem. Soc. Trans. 9, 137 (1981).
25. F. A. Middle, A. Bannister, A. J. Bellingham, and P. D. G. Dean, Biochem. J. 209, 771 (1983).
26. V. H. Akparov and V. M. Stepanov, J. Chromatogr. 155, 329 (1978).
27. H. F. Bunn, Am. J. Med. 70, 325 (1981).
28. H. F. Bunn, Diabetes 30, 613 (1981).
29. D. C. Klenk, G. T. Hermanson, R. I. Krohn, E. K. Fujimoto, A. K. Malia, P. K. Smith, J. D. England, H. M. Wiedmeyer, R. R. Little, and D. E. Goldstein, Clin. Chem. 28, 2088 (1982).
30. B. J. Gould, P. M. Hall, and G. H. Cook, Clin. Chim. Acta 125, 41 (1982).
31. R. Fluckiger, T. Woodtli, and W. Berger, Diabetes 33, 73 (1984).
32. K. Nakano, K. Shindo, T. Yaraka, and H. Yamamoto, J. Chromatogr. 332, 21 (1956a).
33. K. Nakano, K. Shindo, T. Yaraka, and H. Yamamoto, J. Chromatogr. 332, 127 (1956b).
34. S. Mubarak, J. B. Stanford, and J. K. Sugden, Drug Dev. Ind. Pharm. 10, 1131 (1984).
35. S. Aronoff, T. Chen, and M. Cheveldayoff, Carbohydr. Res. 40, 299 (1975).
36. A. M. Yurkevich, I. I. Kolodkina, E. A. Ivanova, and E. I. Pichuzhkina, Carbohydr. Res. 43, 215 (1975).
37. J. Boeseken, Adv. Carbohydr. Chem. 4, 189 (1949).
38. P. Jandera and J. Churacek, J. Chromatogr. 98, 55 (1974).

39. S. J. Angyal, D. Greeves, and V. A. Pickles, Carbohydr. Res. 35, 165 (1974).
40. H. Steinberg, in *Organoboron Chemistry*, Vols. 1-3, John Wiley and Sons, New York, 1974.
41. (a) R. P. Singhal, B. Ramamurthy, N. Govindaraj, and Y. Sarwar, J. Chromatogr. 543, 17-38 (1991).
    (b) R. P. Singhal, S. S. M. DeSilva, and Y. Sarwar (submitted).
42. P. Sienkiewicz and D. Roberts, J. Inorg. Nucl. Chem. 42, 1559 (1980).
43. R. J. Koenig, S. H. Blobstein, and A. Cerami, J. Biol. Chem. 252, 2992 (1977).
44. R. W. Fischer and K. H. Winterhalter, FEBS Lett. 135, 145 (1981).
45. R. Ducrocq, A. Cahour, S. Berriche, S. Intrator, and J. Elion, Protides Biol. Fluids 33, 651 (1985).
46. J. P. Lorand and J. O. Edwards, J. Org. Chem. 24, 769 (1959).
47. S. A. Barker, A. K. Chopra, B. W. Hatt, and P. J. Somers, Carbohydr. Res. 26, 33 (1973).
48. G. Wulff and W. Vesper, J. Chromatogr. 167, 171 (1978).
49. G. Wulff, J. Pure and Appl. Chem. 54, 2093 (1982).
50. G. Wulff, W. Dederichs, R. Grotstollen, and C. Jupe, in *Affinity Chromatography and Related Techniques* (T. C. J. Grinbau, J. Visser, and R. J. F. Nivard, eds.), Elsevier, Amsterdam, 1982, p. 207.
51. M. Van Duin, J. A. Peters, A. P. G. Kieboom, and H. Van Bekkum, Tetrahedron 40, 2901 (1984).
52. M. Van Duin, J. A. Peters, A. P. G. Kieboom, and H. Van Bekkum, Tetrahedron 41, 3411 (1985).
53. B. J. B. Johnson, Biochemistry 21, 6103 (1981).
54. T. A. Myohanen, V. Bouriotis, and P. D. G. Dean, Biochem. J. 197, 683 (1981).
55. E. I. Pichuzhkina, I. I. Kolodkina, and A. M. Yurkevich, Zh. Obshch. Khim. 43, 2275 (1973).
56. S. Soundararajan, M. Badawi, C. M. Kohlrust, and J. H. Hageman, Anal. Biochem. 178, 125 (1989).
57. E. Ivanova, S. Panchenko, I. Kolodkina, and A. Yurkevich, Z. Obshchei Khimii 45, 208 (1975).
58. H. Carminatti, S. Passeron, M. Dankert, and E. Recondo, J. Chromatogr. 18, 342 (1965).
59. J. M. Connor, J. Inorg. Nucl. Chem. 32, 3545 (1970).
60. P. J. Antikainen, Suomen Kemistilehti 31B, 255 (1958).
61. A. Finch and J. C. Lockhart, J. Chem. Soc., 3723 (1962).
62. A. Finch, P. J. Gardner, P. M. McNamara, and G. R. Wellum, J. Chem. Soc., 3339 (1970).
63. R. A. Bowie and O. C. Musgrave, J. Chem. Soc., 3945 (1963).

64. R. A. Olsson, J. Chromatogr. *176*, 239 (1979).
65. V. Bouriotis, I. J. Galpin, and P. D. G. Dean, J. Chromatogr. *210*, 267 (1981).
66. C. A. Elliger, B. G. Chan, and W. L. Stanley, J. Chromatogr. *104*, 57 (1975).
67. R. J. Ferrier, Adv. Carbohydr. Chem. and Biochem. *35*, 31 (1978).
68. R. T. Hawkins and H. R. Snyder, J. Amer. Chem. Soc. *82*, 3863 (1960).
69. R. P. Singhal, N. Govindaraj, and Y. Sarwar, *New Ligands for Boronate Chromatography*, International Bioanalytical Workshop, Lawrence, Kansas, May 21-24, 1989.
70. C. W. Gehrke, K. C. Kuo, G. E. Davis, R. D. Suits, T. P. Waalkes, and E. Borek, J. Chromatogr. *150*, 455 (1978).
71. P. D. G. Dean, P. J. Brown, and V. Bouriotis, U.S. Patent 4,269,605 (1981).
72. B. A. Roe, A. F. Stankiewicz, and C. Y. Chen, Nucl. Acid Res. *4*, 2191 (1977).
73. S. Ackerman, B. Cool, and J. J. Furth, Anal. Biochem. *100*, 174 (1979).
74. H. Schott, Angew. Chem. Int. Ed. *11*, 824 (1972).
75. H. Schott, E. Rudolff, P. Schmidt, R. Roychoudhury, and H. Kossel, Biochemistry *12*, 932 (1973).
76. S. Fulton, in *Boronate Ligands in Biochemical Separations*, Amicon, Danvers, Mass., 1981.
77. R. R. Maestas, J. R. Prieto, G. D. Kuehn, and J. H. Hageman, J. Chromatogr. *189*, 225 (1980).
78. G. E. Davis, R. D. Suits, K. C. Kuo, C. W. Gehrke, T. P. Waalkes, and E. Borek, Clin. Chem. *23*, 1427 (1977).
79. C. Hansson, G. Agrup, H. Rorsman, A.-M. Rosengren, and E. Rosengren, J. Chromatogr. *161*, 352 (1978).
80. C. F. Gelijkens, A. P. DeLeenheer, J. Chromatogr. *183*, 78 (1980).
81. R. Alvarez-Gonzalez, H. Juarez-Salinas, E. L. Jacobson, and M. K. Jacobson, Anal. Biochem. *135*, 69 (1983).
82. S. Higa and S. Kishimoto, Anal. Biochem. *154*, 71 (1986).
83. M. Glad, S. Ohlson, L. Hansson, M.-O. Mansson, and K. Mosbach, J. Chromatogr. *200*, 254 (1980).
84. M. Glad, L. Hansson, and C. Hansson, in *Affinity Chromatography and Biological Recognition*, Academic Press, New York, 1983.
85. M. Akashi, T. Tokiyoshi, M. Miyauchi, and K. Mosbach, Nucleic Acids Symp. Ser. *16*, 41 (1985).
86. F. B. Anspach, H. J. Wirth, K. K. Unger, P. Stanton, J. R. Davies, and M. T. Hearn, Anal. Biochem. *179*, 171 (1989).

87. R. De Cristofaro, R. Landolfi, B. Bizzi, and M. Castagnola, J. Chromatogr. 426, 376 (1988).
88. P. M. Hall, G. M. Cawdell, J. G. H. Cook, and B. J. Gould, Diabetologia 25, 477 (1983).
89. R. Fluckiger, T. Woodtli, and W. Berger, Diabetes 33, 73 (1984).
90. W. G. John, E. C. Albutt, G. Handley, and R. W. Richardson, Clin. Chim. Acta 136, 257 (1984).
91. J. Stepper and C. Steuart, Anal. Biochem. 34, 123 (1970).
92. R. Shapiro, M. J. McManus, C. Zalut, and H. F. Bunn, J. Biol. Chem. 255, 3120 (1980).
93. G. Mattok and D. Wilson, Anal. Biochem. 11, 575 (1965).
94. C. Lowe and P. Dean, in *Affinity Chromatography*, John Wiley and Sons, New York, 1974.
95. L. Nichol, A. Ogsten, D. Winzor, and W. Sawyer, Biochem. J. 143 (1974).
96. D. W. Tanner and T. C. Bruice, J. Am. Chem. Soc. 89, 6954 (1967).
97. E. G. Gorman and D. W. Darnall, Biochemistry 20, 38 (1981).
98. C. Zittle, Adv. Enzymol. 12, 493 (1951).
99. J. X. Khym and W. E. J. Cohn, Amer. Chem. Soc. 75, 1153 (1975).
100. A. K. Mallia, G. T. Hermanson, R. I. Krohn, E. K. Fujimoto, and P. K. Smith, Anal. Lett. 14, 649 (1981).
101. D. G. Willey, M. A. Rosenthal, and S. Caldwell, Diabetologia 27, 56 (1984).
102. R. A. Kopper and R. P. Singhal, Int. J. Biol. Macromolecules 1, 65 (1979).
103. G. Vogeli, T. S. Stewart, T. McCutchan, and S. Soll, J. Biol. Chem. 252, 2311 (1975).
104. (a) R. P. Singhal and P. A. M. Fallis, Progr. Nucl. Acids Res. Molec. Biol. 23, 227 (1979);
    (b) R. P. Singhal, Progr. Nucl. Acids Res. Molec. Biol. 28, 75 (1983).
105. S. Rhabar, Clin. Chim. Acta 22, 296 (1968).
106. D. E. Goldstein, R. R. Little, H.-M. Wiedmeyer, J. D. England, and E. M. McKenzie, Clin. Chem. 32, B64 (1986).
107. L. Kennedy and T. J. Lyons, Br. Med. Bull. 45, 174 (1989).
108. M. J. McDonald, R. Shapiro, M. Bleichman, J. Solway, and H. F. Bunn, J. Biol. Chem. 2327 (1978).
109. H. F. Bunn, R. Shapiro, M. McManus, L. Garrick, M. J. McDonald, P. M. Gallop, and K. H. Gabbay, J. Biol. Chem. 254, 3892 (1979).
110. R. Fluckiger and H. B. Mortensen, J. Chromatogr. 429, 279 (1988).

111. D. O. Sobel and K. M. Shakir, Diab. Metab. *13*, 575 (1987).
112. R. P. Singhal, S. S. M. DeSilva, and Y. Sarwar, Boronate Affinity Chromatography, talk presented at HPLC-90 Conference, Boston, May 20-25, 1990.

6
Chromatographic Methods for Determining Carcinogenic Benz(c)acridine

Noboru Motohashi *Meiji College of Pharmacy, Tokyo, Japan*
Kunihiro Kamata *Tokyo Metropolitan Research Laboratory of Public Health, Tokyo, Japan*
Roger Meyer *Herbert Laboratories, Irvine, California*

I.	INTRODUCTION	337
II.	COLUMN CHROMATOGRAPHY	341
III.	PAPER CHROMATOGRAPHY	344
IV.	THIN-LAYER CHROMATOGRAPHY	350
V.	HIGH-PERFORMANCE LIQUID CHROMATOGRAPHY	360
VI.	GAS-LIQUID CHROMATOGRAPHY	365
	REFERENCES	374

I. INTRODUCTION

Recently, benz[c]acridine and its alkylated derivatives (BAc) have been extensively studied because of their toxic and carcinogenic properties [1-19] (Table 1).

BAc have been found as trace pollutants in urban air particulates [20-29], tobacco smoke [30,31], automobile exhaust [32,33], marine sediments [34], lake sediments [35], and many air pollution source effluents [36-38]. BAc are also present in high-boiling petroleum distillates [39-41]. shale oil [42,43], coal tar [44-46], and coal liquefaction products [47-49].

Table 1 Carcinogenic Activity of Benz[c]acridines

Compound	Paintings				Injections				
	No. of mice	A C	B D	Epithelioma index (%)	No. of mice	Quantity injected (mg)	A B	B D	Sarcoma index (%)
Benz[c]acridine	12			0					
7-Methylbenz[c]acridine	10	8T 89	9 141	63					
8-Methylbenz[c]acridine					10	1 × 0.5			0
9-Methylbenz[c]acridine	10			0					
10-Methylbenz[c]acridine	10			?	20	1 × 0.5 1 × 0.1			0 0
7-n-Propylbenz[c]acridine	10			0					
9-Phenylbenz[c]acridine					20	1 × 0.2			0
5,7-Dimethylbenz[c]acridine	10			0	10	1 × 0.5			0
7,9-Dimethylbenz[c]acridine	10	7T 87.5	8 108	81	10	2 × 2.5	5T 71	7 175	41
7,10-Dimethylbenz[c]acridine	10	5T 62.5	8 113	56	10	2 × 2.5	6T 75	8 117	64
7,11-Dimethylbenz[c]acridine	10	2T 28	7 207	14	10	2 × 2.5	4T 50	8 117	42
10,11-Dimethylbenz[c]acridine	10			0	10	1 × 0.5			0
5,7,11-Trimethylbenz[c]acridine	10	2T 33	6 169	20	10	1 × 0.5			0

Compound							
7,8,11-Trimethylbenz[c]acridine	10	3T 75	4 172	43	10	1 × 0.5		0
7,9,10-Trimethylbenz[c]acridine	10	5T 82	6 174	48	10	1 × 0.5		0
7,9,11-Trimethylbenz[c]acridine	10	4T 57	7 194	29	10	2 × 2.5	4T 57	7 281	20
7,8,9,11-Tetramethylbenz[c]acridine	10	9T 100	9 201	50	10	2 × 2.5	1T 25	4 123	20
7-Methyl-9-ethylbenz[c]acridine	10	4T 57	7 156	37	10	3 × 0.5		0
7-Ethyl-9-methylbenz[c]acridine	10	2T 100	2 204	49	10	2 × 0.25		?
7-Ethyl-11-methylbenz[c]acridine	10			1 papiloma			
7-Methyl-9-phenylbenz[c]acridine	10	2T 50	4 240	21			
9,10-Tetramethylenebenz[c]acridine					20	3 × 1		0
8,9-Tetramethylene-11-ethylbenz[c]acridine					10	3 × 1		0
7,11-Dimethyl-10-chlorobenz[c]acridine	30	Impossible to maintain animals alive		About 25–30	20	1 × 0.5		0
7-Ethyl-10-chloro-11-methylbenz[c]acridine	10			0	20	1 × 0.5 1 × 0.25	1T 11	9 190	6
7-Methyl-9-fluorobenz[c]acridine	10	3T 100	3 145	69	20	3 × 0.5	4T 57	7 150	38
7-Methyl-11-fluorobenz[c]acridine					10	3 × 0.5	2T 20	10 223	9
7-Ethyl-9-fluorobenz[c]acridine	20			0	10	3 × 0.5		0

Table 1 (Continued)

Compound	Paintings				Injections			
	No. of mice	A C	B D	Epithelioma index (%)	No. of mice	Quantity injected (mg)	A C	B D	Sarcoma index (%)
7-n-Propyl-9-fluorobenz[c]-acridine					20	3 × 0.5			0
7-n-Amyl-9-fluorobenz[c]-acridine	10			0	20	3 × 0.5			0
7-Benzyl-9-fluorobenz[c]-acridine					20	3 × 0.5			0

A = number of animals with tumor (T); B = number of mice alive when first tumor appears; C = percentage of tumors (A/B × 100); D = average latent period; tumor index = (C/D × 100).

Source: Data from Ref. 1.

The identification of BAc and related chemicals in widely diversified samples presents some notable difficulties. These difficulties are principally due to the relatively small quantity of BAc contained in the samples and to the presence of many other materials in a wide variety of matrices. Many different methods have been suggested for separating and determining BAc. The more common techniques used are based on a variety of chromatographic separations—paper chromatography (PC), column chromatography (CC), thin-layer chromatography (TLC), high-performance liquid chromatography (HPLC), gas-liquid chromatography (GC)—and the determinations usually are by means of ultraviolet or fluorescence spectrophotometry. Very recently, BAc have also been studied using gas chromatography/mass spectrometry (GC/MS) and nuclear magnetic resonance (NMR).

As far as we know, the analytical chemistry of BAc has not yet been surveyed. This chapter will attempt to cover all relevant chromatographic works pertaining to the analysis of BAc. The period of this review is from late 1961 through 1988.

II. COLUMN CHROMATOGRAPHY

Column chromatography (CC) is especially used as a clean-up procedure for the elimination of interfering materials. Although CC is probably one of the most exacting chromatographic techniques to perform, it has the capacity to handle relatively large samples. The use of fraction collectors and automatic effluent monitoring devices helps reduce the time and effort involved in CC. To identify fractions in which the separated components are concentrated, one must laboriously examine all fractions with UV spectrometry (UV), fluorescence spectrometry, HPLC, or GC.

One successful column chromatographic system for the separation of azaarenes from coal tar pitch and air pollution source particulates was achieved on an alumina column by Sawicki et al. [36, 37,50-52] using pentane solutions containing ether and acetone as the developing solvent.

Adsorbents other than alumina have also been used for the separation of azaarenes. Engel and Sawicki [53] used a silica gel column to determine BAc in coal tar pitch. Snook [54,55] also recommends the use of a silicic acid column for the separation of BAc in azaarenes. Samples are eluted consecutively with petroleum ether, 25% benzene-petroleum ether, 50% benzene-petroleum ether, 50% benzene-ethyl ether, and ethyl ether. Each fraction was evaporated and analyzed by GC (Table 2). Snook [54,55] reported a two-step procedure for separating azaarenes in tobacco smoke by a method involving several solvent partitioning steps, silicic acid

Table 2 Silicic Acid Chromatography of Azaarenes

Compound	Percentage distribution of eluting solvent[a]					
	PE	B/PE (1:3)	B/PE (1:1)	B	B/E (1:1)	E
Naphthalene	—	100	—	—	—	—
Anthracene	—	100	—	—	—	—
Benzo[a]pyrene	—	100	—	—	—	—
Quinoline	—	—	—	—	79	21
6-Methylquinoline	—	—	—	—	83	17
Acridine	—	—	—	—	100	—
Benz[c]acridine	—	—	—	69	31	—
Phenanthridine	—	—	—	—	100	—
1-Azapyrene	—	—	—	—	100	—

[a] PE = petroleum ether; B = benzene; E = ethyl ether.
Source: Data from Ref. 55.

chromatography, and gel chromatography on Bio-Beads SX-12. The effluents were identified by UV, GC/MS, and HPLC.

Mckay et al. [39] discussed in detail the analysis of basic nitrogen compounds in high-boiling petroleum products. Extracted samples were dissolved in cyclohexane and passed through a column containing Cellex-P cation exchange cellulose. The eluate from the Cellex-P column was passed through an acidic alumina column and eluted first with methylene chloride to remove the weakly held bases and then with absolute ethanol to remove the strongly held bases. Solvent was evaporated from each fraction and the fractions were redissolved in cyclohexane. Each fraction was then placed on a basic alumina column and the column was eluted with 90% cyclohexane-methylene chloride, methylene chloride, and absolute ethanol. The structures of the separated compounds were examined in detail using fluorescence, mass spectrometry (MS), and infrared spectrometry (IR).

Kershaw [49] used the method developed by Mckay et al. [39] to separate nitrogen bases from coal-derived liquids.

Amberlyst-15 cation exchange chromatography has been used by Burchill et al. [44] to separate basic nitrogen compounds from coal tars. Cation exchange resins are used in the hydrogen form for petrochemical applications. Proton transfer to a basic nitrogen atom forms a positively charged ammonium ion which is strongly held by the sulfonate anion on the resin. The stationary phase was applied

Table 3 Elution Volumes into Arenes and Azaarenes by Chromatography on Strong Acid (I,II) or Strong Basic (III,IV) Ion Exchangers

Compound	Elution volumes (ml)			
	I	II	III	IV
Anthracene	9.2		16.6	17.7
Chrysene	9.2		17.2	
Benzo[a]pyrene	9.2	9.3	18.3	
Coronene	9.4		18.6	
N-Methylindole	—			
Quinoline	—	74.1		
Isoquinoline	—		17.0	
4-Azafluorene	—	74.2	17.2	
1-Azapyrene	—	78.4	17.0	17.5
7,8-Benzoquinoline	—	80.4	16.3	
Acridine	—	83.0	15.7	17.5
Benz[c]acridine	—	134.2	17.2	18.2
Benz[a]acridine	—	136.5	17.3	18.0
6,9-Dimethylbenz[c]acridine	—	198.1	17.2	
Dibenzo[a,h]acridine	—	266.4		
Dibenzo[a,j]acridine	—	267.8	18.0	18.3
Indole	9.2		—	74.8
Carbazole	9.2	9.4	—	98.3
β-Carboline(9H-pyrido[3,4-b]indole, 2-carboline)	9.1	9.3	—	99.4
Aribine(1-methyl-β-carboline)	9.1			91.4
Carbazinc(9,10-dihydroacridine, ms-dihydroacridine)	9.3			
2,3-Benzcarbazole	9.4		—	112.4
3,4,5,6-Dibenzcarbazole	9.4		—	124.8
1,2,7,8-Dibenzcarbazole	9.4	9.5	—	127.3

Solvent system: I = SP-Sephadex C-25, strong acid ion exchanger (methanol); II = SP-Sephadex C-25, strong acid ion exchanger (methanol-1% acetic acid); III = QAE-Sephadex A-25, strong basic ion exchanger (cyclohexane); IV = QAE-Sephadex A-25, strong basic ion exchanger (cyclohexane-1% triethanolamine); — = not enough to detect.

to the column as a suspension in tetrahydrofuran (THF) and desorption carried out with a mobile phase of propylamine-THF (2:1). Further, they examined the separation of nitrogen compounds from coal tars using an o-phthalonitrile (OPN) stationary phase chemically bonded to a Porasil C support (45).

Schultz et al. [56] also used a cation exchange resin to separate azaarenes in coal liquefaction products by benzene-MeOH-NH_3.

Klimisch and Beiss [57] examined the chromatographic behavior of the mixtures of the azaarenes—acridine- and carbazole type—and arenes on strong acidic or strong basic ion exchangers of Sephadex gels. By using liquid-liquid distribution procedures, a concentrate of azaarenes and arenes is produced from cigarette smoke condensate. This concentrate is separated by CC on Sephadex ion exchangers into three fractions which should contain carbazole-type, acridine-type, and arenes (Table 3).

The liver microsomal metabolites of the 7-methyl-BAc were applied to a Sephadex LH-20 column by both Boux and Holder [58] and Jacob et al. [59]. The column was eluted at 0.5 ml/min with a linear gradient from 20% to 100% methanol with isopropanol, respectively.

Grimmer et al. [40,60] separated various azaarenes (such as nonsubstituted-BAc and 7-methyl-BAc) from petroleum, petroleum products, and high-protein foods by using Sephadex LH-20 or SP-Sephadex C25 CC. They have examined the separation of arenes, azaarenes, and aromatic amines in the particle and vapor phases of mainstream and sidestream cigarette smoke by using Sephadex LH-20 and S-Sepharose CC [31].

Wakeham [35] discussed in detail the separation of azaarenes in recent lake sediments by a method involving several solvent partitioning steps on Sephadex LH-20, acidic silica gel, and alumina columns. Mixed azaarenes isolated from various samples were analyzed by glass capillary gas chromatography.

III. PAPER CHROMATOGRAPHY

Paper chromatography (PC) separations are still of interest because the technique is simple, small sample sizes can be used, and the R_f values are more reproducible than in other separations. This last point is particularly important since it permits some identification of the separated components. Table 4 lists the PC R_f values of azaarenes reported in the literature.

Sawicki et al. [50,61] examined the chromatographic behavior of azaarenes on PC. Whatman No. 1 paper was used in the chromatography. The solvent systems used were formamide-water (35:65 v/v) and dimethylformamide-water (35:65 v/v). In addition, the same solvent systems were applied to circular paper chromatography with satisfactory results. Spots obtained on the paper chromatograms were analyzed directly in the spectrophosphorimeter.

Table 4 Paper Chromatographic R_f Values of Benz[c]acridines

Compound	Stationary phase	Mobile phase	R_f value	Ref.
Benz[c]acridine	Whatman No. 1	(1) Formamide-water (36:65 v/v)	0.44	61
		(2) Dimethyl formamide-water (35:65 v/v)	0.78	61
	Whatman No. 3MM	(1) 1% acetic acid	0.10	63
		(2) 6 N HCl	0.41	63
		(3) Conc. ammonia	0.15	63
		(4) 10% aqueous pyridine	0.31	63
		(5) 0.1 N H_2SO_4	0.35	64
		(6) 3.5 N H_8SO_4	0.25	64
		(7) 0.17 N $CH_2Cl-COOH$	0.31	64
		(8) 0.17 N $CHCl_2COOH$	0.29	64
		(9) 0.17 N CCl_3-COOH	0.23	64
		(10) 0.1 N $(COOH)_2$	0.33	64
		(11) 0.1 N $CH_2(COOH)_2$	0.26	64
		(12) 0.1 N $(CH_2)_2(COOH)_2$	0.16	64
		(13) 0.17 N CH_3COOH	0.10	64
		(14) Aqueous acetate buffers (pH 0.48)	0.22	64
		(15) Aqueous acetate buffers (pH 3.95)	0.00	64
7-Methylbenz[c]acridine	Whatman No. 1	(1) Formamide-water (35:65 v/v)	0.34	61
		(2) Dimethyl formamide-water (35:65 v/v)	0.73	61
10-Methylbenz[c]acridine	Whatman No. 1	(1) 1% acetic acid	0.10	62
		(2) 6 N HCl	0.40	62
		(3) Conc. ammonia	0.13	62
		(4) 10% aqueous pyridine	0.39	62
7,9-Dimethylbenz[c]acridine	Whatman No. 1	(1) Formamide-water (35:65 v/v)	0.29	61
		(2) Dimethyl formamide water (35:65 v/v)	0.66	61

Table 4 (Continued)

Compound	Stationary phase	Mobile phase	R_f value	Ref.
7,9-Dimethylbenz[c]acridine	Whatman No. 3MM	(1) 0.1 N H_2SO_4	0.10	64
		(2) 3.0 N H_2SO_4	0.05	64
		(3) 0.17 N $CH_2Cl-COOH$	0.14	64
		(4) 0.17 N $CHCl_2COOH$	0.11	64
		(5) 0.17 N CCl_3-COOH	0.00	64
		(6) 0.1 N $(COOH)_2$	0.11	64
		(7) 0.1 N $CH_2(COOH)_2$	0.10	64
		(8) 0.1 N $(CH_2)_2(COOH)_2$	0.07	64
		(9) 0.17 N CH_3-COOH	0.10	64
		(10) Aqueous acetate buffer (pH 0.48)	0.06	64
		(11) Aqueous acetate buffer (pH 3.95)	0.03	64
7,10-Dimethylbenz[c]acridine	Whatman No. 1	(1) Formamide-water (35:65 v/v)	0.29	61
		(2) Dimethyl formamide-water (35:65 v/v)	0.66	61
		(3) 0.1 N H_2SO_4	0.11	64
		(4) 3.0 N H_2SO_4	0.05	64
		(5) 0.17 N $CH_2ClCOOH$	0.15	64
		(6) 0.17 N $CHCl_2-COOH$	0.11	64
		(7) 0.17 N CCl_3-COOH	0.00	64
		(8) 0.1 N $(COOH)_2$	0.13	64
		(9) 0.1 N $CH_2(COOH)_2$	0.10	64
		(10) 0.1 N $(CH_2)_2(COOH)_2$	0.07	64
		(11) 0.17 N CH_3-COOH	0.06	64
		(12) Aqueous acetate buffer (pH 0.48)	0.07	64
		(13) Aqueous acetate buffer (pH 3.95)	0.03	64
1,10-Dimethylbenz[c]acridine	Whatman No. 1	(1) 1% acetic acid	0.12	62
		(2) 6 N HCl	0.43	62
		(3) Conc. ammonia	0.04	62
		(4) 10% aqueous pyridine	0.27	62

Compound	Paper	Solvent	Rf	Ref
2,10-Dimethyl-benz[c]acridine	Whatman No. 1	(1) 1% acetic acid	0.10	62
		(2) 6 N HCl	0.28	62
		(3) Conc. ammonia	0.06	62
		(4) 10% aqueous pyridine	0.31	62
9,11-Dimethylbenz[c]acridine	Whatman No. 3MM	(1) 1% acetic acid	0.04	63
		(2) 6 N HCl	0.20	63
		(3) Conc. ammonia	0.04	63
		(4) 10% aqueous pyridine	0.18	63
		(5) Methanol-acetic acid-water (99:1:100)	0.30	63
		(6) Methanol-acetic acid-water (9:1:90)	0.08	63
		(7) Methanol-acetic acid-water (50:10:40)	0.79	63
		(8) Methanol-6 N HCl (40:60)	0.56	63
		(9) Methanol-2 N HCl (40:60)	0.43	63
		(10) Methanol-conc. ammonia (40:60)	0.50	63
		(11) Ethanol-water (40:60)	0.42	63
		(12) Methanol-10% pyridine (aq.) (40:60)	0.50	63
		(13) Isopropanol-acetic acid-water (50:10:40)	1.00	63
		(14) 1-Butanol-acetic acid-water (80:20:100)	0.60	63
		(15) 1-Butanol-pyridine-water (100:10:90)	0.04	63
		(16) 1-Pentanol-10% acetic acid (100:100)	0.94	63
		(17) Formamide-water (10:90)	0.13	63
		(18) Formamide-water (40:60)	0.39	63
		(19) Formamide-acetic acid-water (9:1:90)	0.06	63
		(20) Pyridine-water (20:80)	0.72	63
		(21) Pyridine-water (30:70)	1.00	63
		(22) 1-Pentanol-acetic acid-water (99.5:1:99.5)	0.93	63

Table 4 (Continued)

Compound	Stationary phase	Mobile phase	R_f value	Ref.
8,10-Dimethylbenz[c]acridine	Whatman No. 3MM	(1) 1% acetic acid	0.04	63
		(2) 6 N HCl	0.18	63
		(3) Conc. ammonia	0.05	63
		(4) 10% aqueous pyridine	0.17	63
		(5) Methanol-acetic acid-water (99:1:100)	comet	63
		(6) Methanol-acetic acid-water (9:1:90)	0.09	63
		(7) Methanol-acetic acid-water (50:10:40)	0.79	56
		(8) Methanol-6 N HCl (40:60)	0.57	63
		(9) Methanol-2 N HCl (40:60)	0.45	63
		(10) Methanol-conc. ammonia (40:60)	0.52	63
		(11) Ethanol-water (40:60)	0.42	63
		(12) Methanol-10% pyridine (aq.) (40:60)	0.53	63
		(13) Isopropanol-acetic acid-water (50:10:40)	1.00	63
		(14) 1-Butanol-acetic acid-water (80:20:100)	0.57	63
		(15) 1-Butanol-pyridine-water (100:10:90)	0.05	63
		(16) 1-Pentanol-10% acetic acid (100:100)	0.90	63
		(17) Formamide-water (20:80)	0.15	63
		(18) Formamide-water (40:60)	0.26	63
		(19) Formamide-acetic acid-water (9:1:90)	0.03	63
		(20) Pyridine-water (20:80)	0.66	63
		(21) Pyridine-water (30:70)	1.00	63
		(22) 0.1 N H_2SO_4	0.13	64

Determining Carcinogenic Benz[c]acridine / 349

	(23) 3.0 N H_2SO_4	0.04	64
	(24) 0.17 N $CH_2ClCOOH$	0.07	64
	(25) 0.17 N $CHCl_2-COOH$	0.09	64
	(26) 0.17 N CCl_3-COOH	0.00	64
	(27) 0.1 N $(COOH)_2$	0.11	64
	(28) 0.1 N $CH_2(COOH)_2$	0.10	64
	(29) 0.1 N $(CH_2)_2(COOH)_2$	0.07	64
	(30) 0.17 N CH_3-COOH	0.07	64
	(31) Aqueous acetate buffer (pH 0.48)	0.07	64
	(32) Aqueous acetate buffer (pH 3.95)	0.03	64
8,11-Dimethylbenz[c]acridine Whatman No. 3MM	(1) 1% acetic acid	0.04	63
	(2) 6 N HCl	0.21	63
	(3) Conc. ammonia	0.05	63
	(4) 10% aqueous pyridine	0.23	63
	(5) Methanol-acetic acid-water (99:1:100)	0.33	63
	(6) Methanol-acetic acid-water (9:1:90)	0.09	63
	(7) Methanol-acetic acid-water (50:10:40)	0.81	63
	(8) Methanol-6 N HCl (40:60)	0.60	63
	(9) Methanol-2 N HCl (40:60)	0.46	63
	(10) Methanol-conc. ammonia (40:60)	0.52	63
	(11) Ethanol-water (40:60)	0.43	63
	(12) Methanol-10% pyridine (aq.) (40:60)	0.52	63
	(13) Isopropanol-acetic acid-water (50:10:40)	1.00	63
	(14) 1-Butanol-acetic acid-water (80:20:100)	0.59	63
	(15) 1-Butanol-pyridine-water (100:10:90)	0.06	63
	(16) 1-Pentanol-10% acetic acid (100:100)	0.90	63
	(17) Formamide-water (20:90)	0.19	63
	(18) Formamide-water (40:60)	0.43	63
	(19) Formamide-acetic acid-water (9:1:90)	0.04	63
	(20) Pyridine-water (20:80)	0.69	63
	(21) Pyridine-water (30:70)	1.00	63

The work of Luly and Sakodynsky [62] was carried out to obtain data on the adsorption properties of azaarenes on cellulose (Whatman No. 1 or 3MM paper) and cellulose derivatives (carboxymethylcellulose What CM 50 paper, cellulose citrate Whatman CT 30 paper, and Amberlite WA-2 paper), as well as for other analytical purposes. By a suitable choice of simple aqueous solvents, numerous azaarenes could be separated by the chromatography on ordinary cellulose paper (Whatman No. 1).

Lederer and Roch [63] in a detailed survey on the analysis of azaarenes was able to reproduce Luly and Sakodynsky's work [62]. Additionally, they extended the work on PC to numerous solvents and attempted to separate from one another at least all the pure compounds at their disposal.

Caroli and Lederer [64] reported R_f values of azaarenes on Whatman 3MM paper which were developed with various concentrations of aqueous sulfuric acid, aqueous organic acids, and aqueous acetate buffers.

Because of the number of azaarene compounds, one is not able to separate them all satisfactorily in one step. The analyst is therefore required to consider the use of multiple solvent systems to achieve the necessary specificity. The use of PC therefore becomes too lengthy a procedure for routine use. Consequently, more recent workers have employed TLC.

IV. THIN-LAYER CHROMATOGRAPHY

Thin-layer chromatography (TLC) is an inexpensive and simple method for the separation of complex mixtures of azaarenes preliminary to their spectral characterization. The paper chromatographic methods are valuable in that these techniques are simple and are readily performed by untrained personnel. The disadvantages are the length of time needed for separation and the difficulty of separating some of the azaarenes. TLC is a much more rapid technique than PC and allows more groups of adsorbent-solvent systems with radically different separation patterns to be used. However, the reproducibility of R_f values is generally poorer than in PC. The reproducibility can be improved if adequate attention is applied to all of the experimental variables. Tables 5-7 summarize the more important reports of the use of TLC in the literature.

The work of Sawicki et al. [20,36,38,50,52,61,65-69] presents a comprehensive scheme of TLC methods for the systematic identification of azaarenes from each other, from arenes, and other types of aromatic compounds. Six diverse TLC systems that have proven highly valuable in the separation of azaarenes are (a) alumina-ether/pentane (1:19 v/v), (b) alumina-nitrobenzene/pentane (1:9 v/v),

(c) alumina-2-nitropropane/triethanolamine/pentane (1:0.001:9 v/v), (d) cellulose-water/dimethyl formamide (65:35 v/v), (e) cellulose-water/methanol (70:30 v/v), and (f) silica gel-ether/pentane (1:9 v/v). Used in conjunction, these systems give even better separations. Additional work reported two-dimensional TLC procedures for separation of azaarenes and its application to urban atmospheres, air pollution source effluents, and coal tar pitch [21,53,70]. The greatest value of two-dimensional TLC with mixed adsorbents is that it allows quick separation, characterization, and analysis of complicated molecules. The plates were coated with either alumina-cellulose (2:1) or silica gel-cellulose (2:1) and developed with pentane-ether followed by dimethyl formamide (DMF)-water. The identification was based mainly on the R_f values of each constituent and on the fluorescence color produced by various spray reagents. In addition, spectroscopic procedures (UV, quench fluorometry, and spectrophotofluorometry) have been described for eventual application for compound characterization. These procedures are used for quantitatively analyzing azaarenes in coal tar pitch, air pollution source effluents, and urban atmospheres.

Klemm et al. [71] reported their studies on the chromatographic absorption of azaarenes using benzene-chloroform on alumina oxide G. Simultaneous investigations of some azaarenes on alumina were reported by Sawicki et al. [65,68]. Klemm et al. [71], however, do not appear to be discussing a steric effect of interfering methyl groups in the BAc system studied by Sawicki et al. [65], who find 8,10-dimethyl-BAc = 7-methyl-BAc = BAc > 7.10-dimethyl-BAc = 7,9-dimethyl BAc on the R_f values.

Caroli and Lederer [64] examined the chromatographic behavior of azaarenes on polyamide thin-layer plates with acetone-water and methanol-water as the developing solvent. The order of movements is mainly based on the molecular weight and is essentially similar to those on cellulose paper.

Wang and Brockhaus [72] in a survey on the analysis of BAc in airborne particulates were able to reproduce Sawicki's work and, in addition, applied the same solvent systems to TLC with satisfactory results.

The important work of Shultz et al. [56] was also based on that of Sawicki but a different solvent system was employed. The basic fraction of coal tar pitch was eluted from a cation exchange resin column and evaporated to dryness. The residue was dissolved in benzene and spotted on the TLC plate. Development of the alumina G plate was made with isooctane-benzene.

A silanized silica gel plate, developed with benzene-hexane-pyridine (4:5:1), was used by Brocco et al. [26] to separate azaarenes in dust. In this way, azaarenes are separated from other components in dust, such as arenes and paraffins. Stanley et al. [25,

Table 5 TLC Separation of Benz[c]acridines

Compound	Stationary phase	Mobile phase
Benz[c]acridine	Cellulose	DMF^b-H_2O (35:65)
	MN-cellulose powder 300G	Acetic acid-H_2O (3:7)
		DMF-H_2O (35:65)
	Aluminum oxide G	Pentane-ether (9:1)
		Benzene-chloroform (1:1)
	Alumina	Pentane-ether (19:1)
		Pentane-nitrobenzene (9:1)
		Pentane-2-nitropropane-triethylamine (9:1:0.01)
		Pentane-ether (9:1)
		Chloroform-benzene (1:1)
	Silica gel G	Pentane-ether (7:3)
	Silica gel	Pentane-ether (7:3)
		Benzene-hexane-pyridine (4:5:1)
	Polyamide	Acetone-H_2O (6:4)
		Methanol-H_2O (8:2)
	Magnesium hydroxide	Benzene
	Alumina-cellulose (2:1)	Cyclohexane-ethyl acetate (19:1) and DMF-H_2O (35:65) Pentane and DMF-H_2O (35:65)
	Silica gel-cellulose (2:1)	Pentane-ether (9:1) and DMF-H_2O (35:65)
	Aluminum oxide G-cellulose D-O (95:5)	Hexane-ether (19:1) and Methanol-ether-H_2O (4:4:1)
	Alumina oxide G-silica gel G (1:1)	Pentane-ether (19:1)

Fluorescence color[a] (treated) and comments	R_f	Ref.
B (untreated)	0.26	20, 52 65, 66, 69
Quenched (3-methylthioaniline sprayed)		52
BG → G (NO_2-TFA)[c]		52
B → BG (TFA fumes)[d]	0.55	65
B → YG (TFA fumes)	0.26	50, 36
	0.53	36
OBr (untreated), activated 23-26 hr at 100-110°C	0.68	71
B → BG (TFA fumes)	0.53	38, 65 67
Quenched (wet plate), G (TFA fumes)	0.92	68
Quenched (wet plate), B (dry plate), G (TFA) fumes)	0.92	68
Activated 23 hr at 110°C	0.52	76
Activated 23 hr at 110°C	0.86	76
BG (untreated)	0.90	53
Activated at 110°C for 10 min.	0.73	76
	0.55	26
	0.23	64
	0.15	64
B (untreated), unactivated, development temperature (25°C)	0.30	76
Activated 20 hr at 105°C, development temperature (25°C)	0.50	76
Unactivated, development temperature (35°C)	0.06	76
Two-dimensional TLC		21
Two-dimensional TLC		21
YG (TFA fumes), two-dimensional TLC		53
Two-dimensional TLC		74
One-dimensional TLC		25

Table 5 (Continued)

Compound	Stationary phase	Mobile phase
Benz[c]acridine-7-carboxylic acid	Kieselguhr GF_{254}	Methanol-chloroform (3:7) Acetic acid-ethanol-ethyl acetate (1:12:19) Ethanol-chloroform (2:3)
7-Methylbenz[c]-acridine	Cellulose	DMF-H_2O (35:65) Acetic acid-H_2O (3:7)
	Alumina	Pentane-ether (19:1)
	Silica gel G	Pentane-ether (7:3)
	Kieselguhr G-Florisil	n-Pentane-benzene (19:1)
7-Hydroxymethyl-benz[c]acridine	Kieselguhr GF_{254}	Methanol-toluene (1:9) Ethyl acetate-hexane (1:3) Acetone-hexane (7:13)
10-Methylbenz[c]-acridine	Silica gel	Benzene-hexane-pyridine (4:5:1)
7,9-Dimethylbenz-[c]acridine	Cellulose	DMF-H_2O (35:65) Acetic acid-H_2O (3:7)
	Alumina	Pentane-ether (19:1)
	Silica gel	Pentane-ether (7:3)
	Polyamide	Acetone-H_2O (6:4) Methanol-H_2O (8:2)
	Alumina G + kieselguhr G (2:1)-acetylated cellulose	
	Kieselguhr G-Florisil	n-Pentane-benzene (19:1)
7,10-Dimethylbenz-[c]acridine	Cellulose	DMF-H_2O (35:65) Acetic acid-H_2O (3:7)
	Alumina	Pentane-ether (19:1)
	Silica gel G	Pentane-ether (7:3)
	Polyamide	Acetone-H_2O (6:4) Methanol-H_2O (8:2)
	Alumina G + kieselguhr G (2:1)-acetylated cellulose	
	Kieselguhr G-Florisil	n-Pentane-benzene (19:1)
8,10-Dimethylbenz-[c]acridine	Cellulose	DMF-H_2O (35:65) Acetic acid-H_2O (3:7)

Fluorescence color[a] (treated) and comments	R_f	Ref.
	0.31	79
	0.50	79
	0.40	79
LB → LYG (TFA fumes)	0.20	50, 65
	0.61	65
B → BG (TFA fumes)	0.53	65
BG (untreated)	0.85	53
One-dimensional dual-band TLC	0.03	75
	0.42	79
	0.23	79
	0.42	79
	0.55	26
LB → LYG (TFA fumes)	0.10	50, 65
	0.46	65
B → BG (TFA fumes)	0.45	65
BG (untreated)	0.87	53
	0.13	64
	0.08	64
Two-dimensional dual-band TLC		33
One-dimensional dual-band TLC	0.02	75
	0.10	65
	0.53	65
B → BG (TFA fumes)	0.46	65
BG (untreated)	0.87	53
	0.16	64
	0.12	64
Two-dimensional dual-band TLC		33
One-dimensional dual-band TLC	0.01	75
	0.13	65
	0.40	65

Table 5 (Continued)

Compound	Stationary phase	Mobile phase
8,10-Dimethylbenz-[c]acridine	Alumina	Pentane-ether (19:1)
	Polyamide	Acetone-H_2O (6:4) Methanol-H_2O (8:4)
7,11-Dimethylbenz-[c]acridine	Kieselguhr G-Florisil	n-Pentane-benzene (19:1)

[a]Fluorescence colors of spots = B, blue; G, green; L, light; Y, yellow.
[b]DMF = dimethyl formamide.
[c]NO_2-TFA = nitrogen dioxide-trifluoroacetic acid.
[d]TFA = trifluoroacetic acid.

73] also developed an analytical scheme for the identification of BAc in airborne particulates using an alumina oxide G-silica gel G (1:1) plate.

Another analytical separatory scheme based on two-dimensional TLC was reported by Matsushita et al. [74] where aliquots of the sample solution were applied on a 4 × 20 cm plate of Al_2O_3 G-cellulose D-O (95:5 w/w) and developed with n-hexane ether (19:1 v/v) for 15 cm in 35 min and then two-dimensionally developed on a 26% acetylated cellulose plate, 16 × 20 cm, with methanol-ether-water (4:4:1 v/v) for 10 cm in 60 min. Spots scraped from the plates were extracted with benzene and the purity of the sample was easily determined by fluorescence and excitation spectra.

Handa et al. [33] used a two-dimensional dual-band thin-layer plate (coated with alumina G + kieselgel G, 2:1 w/w% acetylated cellulose) for identification and determination of BAc in the basic tar of diesel and gasoline engine vehicles.

Yamauchi and Handa [75] attempted to preseparate the azaarene fraction in urban atmospheric particulate matter by dual-band TLC (Florisil and Kieselgel) with n-pentane-benzene (19:1 v/v).

Fluorescence color[a] (treated) and comments	R_f	Ref.
TB → B (TFA fumes)	0.54	65
	0.12	64
		64
One-dimensional dual-band TLC	0.08	75

Keefer and Johnson [76] concluded that magnesium hydroxide could be used as a thin-layer chromatographic adsorbent for the separation of azaarenes. There are, however, certain interesting differences between this adsorbent and the two currently most widely used media for adsorption chromatography (alumina and silica). First, the R_f value generally drops as the water content of alumina or silica is decreased. The opposite effect is observed with magnesium hydroxide increases in the R_f being the result of heating the plate at 105°C prior to development. A second apparent anomaly was noted in variable-temperature studies. A large decrease in R_f value on magnesium hydroxide resulted when the development temperature was raised by 10°C, while small increases are the rule with silica and alumina.

Motohashi and Kamata [77] used TLC in the purification of BAc. Twelve components were separated by TLC on a silica gel 60 GF254 with benzene. Furthermore, they recently reported a C18 absorbent for reversed plase TLC analysis of BAc [78].

Boux et al. [79] used TLC kieselgel GF254 in the separation of the metabolites of 7-methyl-BAc and found 7-hydroxymethyl-BAc and 7-carboxylic-BAc.

Table 6 R_f Values of Benz[c]acridines on RP-18 and Silica Gel 60 F_{254}

Compound	RP-18						Silica gel 60 F_{254}
	Methanol	Ethanol	Aceto-nitrile	Acetone	Ethyl acetate	Chloro-form	Benzene
Benz[c]acridine	0.31	0.53	0.33	0.88	0.93t[a]	0.83t	0.72
7-Methylbenz[c]acridine	0.27	0.48	0.18	0.85	0.88t	0.68t	0.71
8-Methylbenz[c]acridine	0.27	0.48	0.38	0.84	0.93t	0.82t	0.70
9-Methylbenz[c]acridine	0.25	0.46	0.36	0.83	0.93t	0.81t	0.70
10-Methylbenz[c]acridine	0.27	0.48	0.23	0.83	0.93t	0.77t	0.92
11-Methylbenz[c]acridine	0.21	0.43	0.32	0.81	0.94	0.91	0.89
5,7-Dimethylbenz[c]acridine	0.22	0.43	0.14	0.80	0.89t	0.56t	0.92
7,9-Dimethylbenz[c]acridine	0.22	0.43	0.13	0.80	0.89t	0.56t	0.61
7,10-Dimethylbenz[c]acridine	0.22	0.43	0.11	0.78	0.85t	0.47t	0.61
7,11-Dimethylbenz[c]acridine	0.17	0.38	0.28	0.77	0.94	0.91	0.86
7,9,10-Trimethylbenz[c]acridine	0.19	0.42	0.07	0.76	0.81t	0.38t	0.83
7,9,11-Trimethylbenz[c]acridine	0.14	0.29	0.23	0.75	0.94	0.91	0.80

[a]t = tailing.
Source: Data from Refs. 77 and 78.

Table 7 R_f Values of Benz[c]acridines on RP-18

Compound	Solvent system									
	I	II	III	IV	V	VI	VII	VIII	IX	X
Benz[c]acridine	0.49	0.50	0.49t[a]	0.36	0.37	0.36	0.51	0.58	0.58t	0.59
7-Methylberz[c]acridine	0.38	0.32	0.35t	0.32	0.32	0.36	0.48	0.54	0.52t	0.55
8-Methylberz[c]acridine	0.47	0.45	0.45t	0.31	0.30	0.33	0.48	0.53	0.56t	0.54
9-Methylbenz[c]acridine	0.44	0.44	0.42t	0.29	0.28	0.30	0.44	0.50	0.52t	0.51
10-Methylbenz[c]acridine	0.42	0.40	0.40t	0.30	0.30	0.34	0.47	0.52	0.54t	0.53
11-Methylbenz[c]acridine	0.45	0.52	0.48	0.24	0.24	0.23	0.40	0.45	0.50	0.48
5,7-Dimethylbenz[c]acridine	0.30	0.28	0.27t	0.26	0.25	0.33	0.43	0.48	0.53t	0.51
7,9-Dimethylbenz[c]acridine	0.30	0.27	0.28t	0.26	0.25	0.39	0.43	0.48	0.53t	0.51
7,10-Dimethylbenz[c]acridine	0.26	0.24	0.25t	0.26	0.25	0.39	0.43	0.48	0.53t	0.53
7,11-Dimethylbenz[c]acridine	0.41	0.48	0.44	0.21	0.19	0.20	0.36	0.41	0.47	0.44
7,9,10-Trimethylbenz[c]acridine	0.20	0.19	0.18t	0.23	0.22	0.41	0.44	0.45	0.52t	0.54
7,9,11-Trimethylbenz[c]acridine	0.36	0.44	0.39	0.17	0.15	0.13	0.31	0.35	0.42	0.38

[a]t = tailing.

Solvent system: I = acetonitrile-acetone (8:2); II = acetonitrile-chloroform (8:2); III = acetonitrile-ethyl acetate (8:2); IV = acetone-water (8:2); V = acetonitrile-acetone-water (7:2:1); VI = acetonitrile-acetone-0.1 M sodium-1-pentanesulfonate (8:2:1); VII = acetonitrile-chloroform-water (8:1:5:0.5); VIII = acetonitrile-chloroform-25% ammonia (8:1.5:0.5); IX = acetonitrile-chloroform-acetic acid (8:1.5:0.5); X = acetonitrile-chlcroform-0.1 M sodium-1-pentanesulfonate (8:1.5:0.5).

Source: Data from Ref. 78.

V. HIGH-PERFORMANCE LIQUID CHROMATOGRAPHY

High-performance liquid chromatography (HPLC) is generally recognized to be superior to PC, TLC, and conventional LC in terms of efficiency and analytical precision. HPLC is the newest, most rapidly developing method of separating BAc and metabolites in biological fluids. It can be used for both qualitative and quantitative purposes. HPLC in its simplest form is based on three elements: (a) a solvent delivery system, (b) injector and column, and (c) detector. Columns and conditions used for the HPLC of BAc and metabolites are presented in Tables 8 and 9.

Dog and Locke [80] reported the HPLC separations of azaarenes using both reversed phase (μBondpak C18) and adsorption (μPorasil) chromatography. Azaarenes with two to five rings are separated within 20 min. A sample recovery is quantitative and permits subsequent UV and fluorescence spectrophotometric identification. The detection limit for most azaarenes is about 1 ng with a 254-nm UV detector.

Colin and coworkers [81] reported some results on the HPLC separation of azaarenes. This work was done for reference compounds ranging from bicyclic molecules (e.g., quinoline) up to tetracyclic ones (e.g., BAc), demonstrating various alkyl substitutions and positions of the nitrogen atom. They utilized both normal phase (LiChrosorb Si 100) and reversed phase (LiChrosorb RP-18 and Hypersil C8) techniques with various mobile phases. The retention patterns are discussed in this chapter.

Chmielowiec [82] discussed a specialty HPLC sorbent based on the organomercuric cation R^-Hg^+ chemically bonded to silica gel. Selectivity of this sorbent for sulfur, oxygen, and nitrogen donor ligands was observed and applied to the separation of aza- and diazaarenes, quinones, furanes, and thiophenes.

Yamauchi and Handa [75] demonstrated the use of reversed phase HPLC coupled with an on-line fluorescence detector for the identification of azaarenes in urban atmospheric particulate matter. Masclet et al. [28] also determined BAc in atmospheric aerosols by reversed phase HPLC utilizing a fluorescence detector.

Snook et al. [55] reported the identification of azaarenes of tobacco smoke by reversed phase HPLC.

Haugen et al. [83] proposed a chromatographic behavior pattern based on 31 kinds of azaarenes separated by cation exchange chromatography. The capacity factor (k') is semilogarithmic and related to the pH.

D'Avila et al. [84] developed hydrophilic complexes with protons and metal ions in the reversed phase liquid chromatography of azaarenes. They recommended the use of complexation with silver for the analysis of azaarenes.

Table 8 HPLC Systems Used for Benz[c]acridine Analysis

Compound	Mode	Column	Mobile phase	Detector	Ref.
Benz[c]acridine	Cation exchange	25 cm × 4.6 mm Partisil 10SCX	40% acetonitrile in 10 mM aqueous sodium citrate	UV, 254 nm	83
	Reversed phase partition	25 cm × 4.6 mm Vydac 5 μm 201TP	65-100% methanol in water	Fluorimeter Ex. λ 366 nm, Em. λ 425 nm	28
		25 cm × 4.9 mm Partisil 10 μm ODS 2	50-90% methanol in water	UV, 254 nm	55
		15 cm × 4 mm LiChrosorb RP-18, 5 μm	60% methanol-water, 50% acetonitrile-water 30% tetrahydrofuran-water, 50% dioxane-water	UV, 254 nm	81
		15 cm × 4 mm Hypersil C8, 5 μm	60% methanol-water, 50% acetonitrile-water 30% tetrahydrofuran-water 50% dioxane-water	UV, 254 nm	81
		30 cm × 4 mm μBondapak C_{18}	20-80% acetonitrile in water	UV, 254 nm	80
		15 cm × 4 mm Hypersil C18	Methanol-water (65:35) (added metal and ammonium acetate)	UV	84
		20 cm × 4.6 mm Silica-ϕ-Hg-Ac	Acetic acid-water-methanol (0.5:50:49.5)	UV	82
	Normal phase partition	15 cm × 4 mm LiChrosorb Si-100	Ethyl acetate-hexane	UV, 254 nm	81
		15 cm × 4 mm LiChrosorb NH_2	Chloroform-hexane	UV, 254 nm	81

Table 8 (Continued)

Compound	Mode	Column	Mobile phase	Detector	Ref.
Benz[c]acridine	Normal phase partition	30 cm × 4 mm µPorasil	1.0% propanol-2 in hexane	UV, 254 nm	80
7-Methylbenz[c]-acridine	Reversed phase partition	25 cm × 4.6 mm Vydac 5 µm 201TP	65-100% methanol in water	Fluorimeter Ex. λ 366 nm, Em. λ 425 nm	28
		15 cm × 4 mm LiChrosorb RP-18 5 µm	60% methanol-water, 50% acetonitrile water 30% tetrahydrofuran-water, 50% dioxane-water	UV, 254 nm	81
		15 cm × 4 mm Hypersil C8, 5 µm	60% methanol-water, 50% acetonitrile-water 30% tetrahydrofuran-water, 50% dioxane-water	UV, 254 nm	81
		15 cm × 4 mm Hypersil C18	Methanol-water (65:35) (added metal and ammonium acetate)	UV	84
		25 cm × 4.6 mm Zorbax ODS	Acetonitrile-water (7:3)	Fluorimeter Ex. λ 285 nm, Em. λ 395 nm	75
		20 cm × 7.7 mm Altex CS C-18, 10 µm	20-100% methanol-water	UV, 254 nm	90
		25 cm × 4 mm LiChrosorb RP-8	Methanol-water gradient	Fluorimeter	91
	Normal phase partition	15 cm × 4 mm LiChrosorb Si-100	Ethyl acetate-hexane	UV, 254 nm	81

Compound	Method	Column	Mobile phase	Detection	Ref.
		15 cm × 4 mm LiChrosorb NH$_2$	Chloroform-hexane	UV, 254 nm	81
		25 cm × 4 mm LiChrosorb Diol	Isooctane-ethanol (97.5:2.5)	UV	85
		25 cm × 4.6 mm Altex silica column, 5 μm	2% or 5% ethanol-cyclohexane	UV, 254 nm	90
7-Methylbenz[c]acridine metabolites	Reversed phase partition	25 cm × 4.6 mm LiChrosorb RP-8	20-70% methanol in water	UV, 254 nm	79, 87, 88, 89, 91
		20 cm × 7.7 mm Altex CS C-18, 10 μm	20-100% methanol-water	UV, 254 nm	90
	Normal phase partition	25 cm × 4.6 mm Altex silica column 5 μm	2% or 5% ethanol-cyclohexane	UV, 254 nm	90
9-Methylbenz[c]-acridine	Normal phase partition	25 cm × 4 mm LiChrosorb Diol	Isooctane-ethanol (97.5:2.5)	UV	85
7,9-Dimethyl-benz[c]acridine	Cation exchange	25 cm × 4.6 mm Partisil 10SCX	40% acetonitrile in 10 mM aqueous sodium citrate	UV, 254 nm	83
	Reversed phase partition	25 cm × 4.6 mm Zorbax ODS	Acetonitrile-water (7:3)	Fluorimeter Ex. λ 285 nm, Em. λ 395 nm	75
7,10-Dimethyl-benz[c]acridine	Cation exchange	25 cm × 4.6 mm Partisil 10SCX	40% acetonitrile in 10 mM aqueous sodium citrate	UV, 254 nm	83
	Reversed phase partition	25 cm × 4.6 mm Zorbax ODS	Acetonitrile-water (7:3)	Fluorimeter Ex. λ 285 nm, Em. λ 395 nm	75
7,11-Dimethyl-benz[c]acridine	Reversed phase partition	25 cm × 4.6 mm Zorbax ODS	Acetonitrile-water (7:3)	Fluorimeter Ex. λ 285 nm, Em. λ 395 nm	75

Table 9 HPLC Retention Data of Benz[c]acridines

Compound	LiChrosorb RP-8 (7 μm) (1 ml/min) A	LiChrosorb RP-18 (10 μm) (1 ml/min)		LiChrosorb RP-18 (5 μm) (0.5 ml/min)		Zorbax C$_{18}$ (1 ml/min)		Zolbax SIL (1.5 ml/min)	LiChrosorb NH$_2$ (5 μm) (1 ml/min)
		B	C	D	E	F	G	H	I	J
Benz[c]acridine	1.00 (10.7 min)	1.00 (13.5 min)	1.00 (7.9 min)	1.00 (9.2 min)	1.00 (15.3 min)	1.00 (19.8 min)	1.00 (7.8 min)	1.00 (10.8 min)	1.00 (4.0 min)	1.00 (17.2 min)
7-Methylbenz[c]acridine	1.24	1.17	1.24	1.20	1.25	1.22	1.33	1.29	2.10	1.13
8-Methylbenz[c]acridine	1.24	1.24	1.24	1.24	1.25	1.23	1.35	1.34	1.11	0.98
9-Methylbenz[c]acridine	1.26	1.26	1.29	1.35	1.32	1.37	1.45	1.48	1.11	1.02
10-Methylbenz[c]acridine	1.22	1.23	1.22	1.30	1.26	1.30	1.36	1.41	1.56	1.06
11-Methylbenz[c]acridine	1.88	1.83	1.79	1.76	1.82	1.77	2.29	2.21	0.61	0.54
5,7-Dimethylbenz[c]acridine	1.65	1.50	1.62	1.54	1.64	1.59	1.90	1.81	2.23	1.12
7,9-Dimethylbenz[c]acridine	1.57	1.47	1.57	1.56	1.59	1.61	1.86	1.82	2.88	1.24
7,10-Dimethylbenz[c]acridine	1.57	1.45	1.57	1.57	1.58	1.61	1.84	1.84	4.70	1.25
7,11-Dimethylbenz[c]acridine	2.34	2.15	2.27	2.16	2.34	2.20	3.19	2.85	0.61	0.56
7,9,10-Trimethylbenz[c]acridine	1.99	1.72	1.97	1.90	1.96	2.01	2.47	2.41	8.13	1.48
7,9,11-Trimethylbenz[c]acridine	3.11	2.78	3.10	2.90	3.20	3.00	4.84	4.28	0.61	0.57

Mobile phase: A = methanol-water (8:2); B = acetonitrile-water (6:4); C = methanol-water (9:1); D = acetonitrile-water (8:2); E = methanol-water (9:1) containing 0.002 M sodium-1-pentanesulfonate; F = acetonitrile-water (8:2) containing 0.002 M sodium-1-pentanesulfonate; G = methanol-water (85:15) containing 0.002 M sodium-1-pentanesulfonate; H = acetonitrile-water (7:3) containing 0.002 M sodium-1-pentanesulfonate; I = ethanol-hexane (5:95); J = hexane. Relative to benz[c]acridine = 1.00.
Source: Data from Ref. 86.

Siouffi et al. [85] reported on the retention data of azaarenes with diol-bonded silica adsorbent. Retention of azaarenes is independent of the basicity but depends greatly on the environment of the nitrogen atom.

Motohashi and Kamata [77,86] recently discussed the normal and reverse phase HPLC behavior of 12 BAc with various mobile phases. In normal phase HPLC and reverse phase HPLC, a steric hindrance between the nitrogen atom and some substituents of the BAc ring apparently influences the retention time of BAc.

Boux et al. [58,79,87-88] described the HPLC separation of metabolites of 7-methyl-BAc. HPLC systems for the metabolites of 7-methyl-BAc used reversed phase columns with methanol-water gradients as eluents. Wright et al. [89] and Ireland et al. [90] also used reversed phase HPLC to identify metabolites of the 7-methyl-BAc in rats.

Gill and Holder [91] reported a sensitive assay for 7-methyl-BAc and their metabolites on reversed phase HPLC (RP-8) using a methanol-H_2O gradient and synchronous luminescence detection of the compound.

VI. GAS-LIQUID CHROMATOGRAPHY

Gas-liquid chromatography (GC) is one of the most widely utilized techniques, either alone or in combination with MS, for the separation and identification of BAc and their metabolites. The attraction of GC lies in its ability to separate and estimate submilligram quantities of complex mixtures. The sensitivity of this technique is almost the same as that of TLC but the selectivity can be enhanced by the use of a nitrogen detector. In this case, GC is more sensitive than TLC. GC involves the same three basic steps as TLC: (a) sample preparation or a prechromatographic extraction step, (b) separation of BAc, and (c) detection. The separation step in GC differs from that in TLC and HPLC in that the specimen is injected into a gas chromatograph and volatilized. GC has the inherent problem that only one class of compounds at a time can be monitored per detector, which makes it a time-consuming method. However, like TLC and HPLC, it permits simultaneous screening for a variety of BAc. Using GC, a single class of compounds may require 30-60 min for the complete screening of BAc. Although its specificity is claimed to be superior to TLC, there are two serious limitations to the GC method: (a) different BAc and their metabolites can have similar retention time and (b) day-to-day reproducibility of GC is less satisfactory with most instruments when operated at high temperatures. Retention times may vary considerably with the condition of the column at such temperatures. Tables 10 and 11 list the more important GC contributions reported in the literature.

Table 10 GLC Retention Data of Benz[c]acridines on Seven Kinds of Stationary Phases

Compound	Stationary phase, column length, and column temperature						
	2% OV-1 2 m 220°C	2% OV-7 2 m 240°C	2% OV-17 2 m 240°C	2% OV-210 2 m 220°C	2% OV-225 2 m 240°C	2% OV-275 1.5 m 220°C	5% XE-60 1 m 240°C
Benz[c]acridine	1.00 (1.7 min)	1.00 (6.5 min)	1.00 (3.0 min)	1.00 (3.7 min)	1.00 (16.2 min)	1.00 (5.7 min)	1.00 (3.9 min)
7-Methylbenz[c]acridine	1.61	1.66	1.70	1.58	1.72	1.66	1.73
8-Methylbenz[c]acridine	1.42	1.41	1.41	1.44	1.45	1.35	1.48
9-Methylbenz[c]acridine	1.35	1.37	1.35	1.39	1.34	1.20	1.39
10-Methylbenz[c]acridine	1.35	1.37	1.36	1.37	1.35	1.22	1.36
11-Methylbenz[c]acridine	1.19	1.15	1.11	1.07	1.04	0.83	1.05
5,7-Dimethylbenz[c]acridine	2.21	2.30	2.31	2.27	2.37	2.10	2.49
7,9-Dimethylbenz[c]acridine	2.13	2.22	2.19	2.22	2.23	1.96	2.32
7,10-Dimethylbenz[c]acridine	2.21	2.23	2.22	2.17	2.26	1.96	2.32
7,11-Dimethylbenz[c]acridine	1.87	1.89	1.84	1.64	1.72	1.32	1.84
7,9,10-Trimethylbenz[c]acridine	3.28	3.47	3.48	3.48	3.57	3.04	3.95
7,9,11-Trimethylbenz[c]acridine	2.47	2.48	2.37	2.15	2.19	1.50	2.45

Relative to benz[c]acridine = 1.00.
Source: Data from Ref. 86.

Table 11 Conditions for the Gas Chromatographic Analysis of Benz[c]acridines

Compound	Stationary phase	Column dimension	Column temperature (°C)	Detection	Ref.
Benz[c]acridine	10% Apiezon L on Chromosorb W	2 m × 0.125 in.	210, 240	FID	21
	XE-60	24 m × 0.4 mm, 70 m × 0.32 mm	200-220	ECD, FID	22
	SE-52	50 m × 0.36 mm	180	ECD, FID	22
	Apiezon L	55 m × 0.30 mm, 50 m × 0.42 mm	220	ECD, FID	22
	Versamid 900	40 m × 0.29 mm	88-201	FID	26
	4% Dexsil 300 on gas chrom Q 100-120	5 m	150-300	MS	27
	Ultra-1	25 m × 0.32 mm	71-300	N-FID	29
	15% Carbowax 20 M on Chromosorb W	5 ft. × 0.125 in.	60-270	FID	30
	SE-54	25 m × 0.32 mm	75-270	N-FID	31
	SE-54	30 m × 0.25 mm	60-290	NPD, MS	34
	SP-2340	20 m × 0.3 mm	100-250	NPD, MS	35
	DB-5 (95% polymethyl + 5% siloxane)	60 m × 0.32 mm	95-270	FID, N-FID	40
	3% Dexsil 400	20 ft × 0.125 in.	100-320	MS	41
	SP-2250 glass SCOT column (50% methyl- + 50% phenylsilicone)	40 m	120-285	FID, AFD, MS	44, 45, 46, 47

Table 11 (Continued)

Compound	Stationary phase	Column dimension	Column temperature (°C)	Detection	Ref.
Benz[c]acridine	DB-5	25 m × 0.25 mm	100-325	FID, ECD, MS	48
	3% Dexsil 300	15 ft × 0.125 in.	100-325	MS	54
	6% OV-17 on Chromosorb G/HP	1.83 m × 3.2 mm	90-250	NPD	55
	3% OV-17 on Chromosorb G		100-250	MS	56
	CP (tm) Sil 5	Capillary	100-260	FID, MS	59
	SE-30 or DB-5	25 m × 0.25 mm	110-260	FID, NPD	60
	OV-101	50 m × 0.31 mm	20-270	FID, MS	83
	2% SE-30 on Chromosorb G	10 ft × 0.125 in.	175-275	FID	92
	2.5% BMBT on Chromosorb W	1.4 m × 2 mm	200-265	FID	93
	SE-52	12 m	50-250	FID	96
	SE-52	Capillary	50-280	FID	98
	SE-54	30 m × 0.26 mm	50-300	MS	99
	SE-52	20 m × 0.30 mm	50-250	FID, NPD, MS	100
Benz[c]acridine metabolites	CP (tm) Sil 5	Capillary	100-260	FID, MS	59
7-Methylbenz[c]acridine	SE-54	25 m × 0.32 mm	75-270	N-FID	31

Compound	Column	Dimensions	Temperature	Detector	Ref.
	DB 5 (95% polymethyl- + 5% phenyl siloxane)	60 m × 0.32 mm	95–270	FID, N-FID	40
	3% OV-101 on gas chrom Q	1 m	200–280	MS	79
	3% OV-17 on gas chrom Q	1 m	180–260	MS	87
	3% OV-17		180–260	MS	90
	2.5% BMBT on Chromosorb W	1.4 m × 2 mm	200–265	FID	93
	SP-2340	40 m × 0.3 mm	238–260		97
7-Methylbenz[c]acridine metabolites	3% OV-101 on gas chrom Q	1 m	200–280	MS	79
	3% OV-17 on gas chrom Q	1 m	180–260	MS	87
	3% OV-17		180–260	MS	90
10-Methylbenz[c]acridine	XE-60	24 m × 0.4 mm, 70 m × 0.32 mm	200–220	ECD, FID	22
	SE-52	50 m × 0.36 mm	180	ECD, FID	22
	Apiezon L	55 m × 0.30 mm, 50 m × 0.42 mm	220	ECD, FID	22
	Versamid 900	40 m × 0.29 mm	88–201	FID	26
1,10-Dimethylbenz[c]-acridine	XE-60	24 m × 0.4 mm, 70 m × 0.32 mm	200–220	ECD, FID	22
	SE-52	50 m × 0.36 mm	180	ECD, FID	22
	Apiezon L	55 m × 0.30 mm, 50 m × 0.42 mm	220	ECD, FID	22

Table 11 (Continued)

Compound	Stationary phase	Column dimension	Column dimension (°C)	Detection	Ref.
2,10-Dimethylbenz[c]-acridine	XE-60	24 m × 0.4 mm, 70 m × 0.32 mm	200-220	ECD, FID	22
	SE-52	50 m × 0.36 mm	180	ECD, FID	22
	Apiezon L	55 m × 0.30 mm, 50 m × 0.42 mm	220	ECD, FID	22
7,9-Dimethylbenz[c]-acridine	XE-60	24 m × 0.4 mm, 70 m × 0.32 mm	200-220	ECD, FID	22
	SE-52	50 m × 0.36 mm	180	ECD, FID	22
	Apiezon L	55 m × 0.30 mm, 50 m × 0.42 mm	220	ECD, FID	22
	OV-101	50 m × 0.31 mm	20-270	FID, MS	83
	3% Dexsil 400 on Chromosorb W	12 ft × 0.125 in.	100-320	FID, NPD	94
	SE-54	30 m × 0.26 mm	50-300	MS	99

8,10-Dimethylbenz[c]-acridine	XE-60	24 m × 0.4 mm, 70 m × 0.32 mm	200–220	ECD, FID	22
	SE-52	50 m × 0.36 mm	180	ECD, FID	22
	Apiezon L	55 m × 0.30 mm, 50 m × 0.42 mm	220	ECD, FID	22
7,10-Dimethylbenz[c]-acridine	XE-60	24 m × 0.4 mm, 70 m × 0.32 mm	200–220	ECD, FID	22
	SE-52	50 m × 0.36 mm	180	ECD, FID	22
	Apiezon L	55 m × 0.30 mm, 50 m × 0.42 mm	220	ECD, FID	22
	OV-101	50 m × 0.32 mm	20–270	FID, MS	83

The first report where GC was used for BAc analysis was by Sawicki et al. [21]. The BAc separation was carried out on a stainless steel tube packed with Chromosorb W (60-80 mesh) coated with 10% Apiezon L. In addition, the techniques of high-temperature GC, mixed-absorbent two-dimensional TLC, and spectrophotofluorometry were combined in a coordinated study of the BAc of the urban atmosphere [21].

Searl et al. [92] studied analytical methods for arenes in coke oven effluents by the combined use of GC and UV. Although other techniques are involved, the method is designated as a GC-UV procedure. In practice, samples are collected on a filter and the filter is extracted with cyclohexane. An internal standard is added and a portion of the extract injected into a gas chromatograph for separation into fractions that are trapped. UV adsorption spectra of selected fractions provide a quantitative measurement of the arenes. BAc have not been observed.

Schmeltz et al. [30] used a GC equipped with a flame ionization detector to isolate characteristic products from pyrolysis of azaarenes.

Rubin and Bayne [94] studied optimum operating conditions for the rubidium (metal) bead-type nitrogen-phosphorus (NP) gas chromatographic detector to detect small quantities of nitrogen compounds in an arene matrix.

One method for the analysis of BAc of tobacco smoke using GC was described by Snook et al. [54,55]. They also applied the NP detector to this study.

Cautreels and Cauwenberghe [27] described the determination of BAc in airborne particulate matter by GC and GC/MS.

Schultz et al. [56] developed GC and GC/MS methods to obtain specific azaarenes in coal liquefaction products. Guerin et al. [41] also described a method for the determination of BAc in petroleum substituents using GC and GC/MS.

Pailer and Holozek [93] used nematic liquid crystals as a stationary phase for the gas chromatographic separation of azaarenes. Good separation and high symmetry of the peaks as well as the relative independence of the amount of stationary phase used are the advantages of this method.

Motohashi and Kamata [77] reported the retention times of 12 BAc on GC using a glass column packed with Dexsil 400. They examined the GC behavior of 12 BAc on various types of silicone stationary phases [86]. It was concluded that the elution of BAc on nonpolar stationary phases follows the order of increasing number of carbon atoms, but on polar stationary phases the prediction of the retention sequence is very difficult because of steric hindrance between the nitrogen atoms and some substituents of the BAc ring.

Excellent separations between closely related BAc isomers were obtained by combination of a nonpolar (OV-1) and polar (OV-275) stationary phase.

Boux et al. [79,87] and Ireland et al. [90] used GC-MS for the identification of metabolites of 7-methyl-BAc in rats.

One method for improving the separation efficiency in GC is to use columns of small diameter. In recent years the advances of capillary packed columns have been demonstrated and capillary GC is now a powerful technique for the analysis of complex mixtures of azaarenes. The application of capillary columns for the analysis of BAc was first reported by Alberini et al. [22]. Their separation of BAc in atmospheric dusts uses capillary columns with silicon rubber SE-52, XE-60, or Apiezon L as liquid phases.

Brocco et al. [26] developed and used GC with a high-efficiency glass capillary column coated with Versamid 900 for the separation of heterocyclenes in atmospheric particulates.

Nielsen et al. [29] also reported the determination of basic azaarenes in airborne particulate matter by capillary GC with an NP detector.

Grimer et al. described the determination of azaarenes in high-protein foods [40,60,95] and in cigarette smoke [31]. Azaarenes were analyzed by capillary GC and measured by comparison to corresponding peak areas of an internal standard.

Lee et al. [96] reported the retention indexes of arenes (containing benz[c]acridine) on an SE-52 capillary column.

Ignatiadis et al. [97] described the reproducible methods for the preparation of nonpolar, medium polar, and polar capillary columns with suitable wall surface deactivation for the analysis of azaarenes.

Schmitter et al. [98] investigated the retention behavior of azaarenes on several stationary phases of capillary columns. Steric hindrance in the vicinity of the nitrogen atom was found to drastically influence the retention of azaarenes. The most useful and complementary selection of stationary phases was OV-73, OV-61, and SP-2340. Examples of separations of complex azaarene mixtures extracted from crude oils were discussed.

A coupled capillary column GC/MS system is an important tool for both qualitative and quantitative analysis. Several workers have described the use of capillary column GC and GC/MS for the identification of nitrogen compounds in coal tar products [44-48,99,100], coal conversion oil [83], synthetic fuels [101], and sediments [34,35].

A capillary column GC/MS was also developed by Jacob et al. [59] for the identification of biological samples.

REFERENCES

1. A. Lacassagne, N. P. Buu-Hoi, R. Daudel, and F. Zajdela, The relation between carcinogenic activity and the physical and chemical properties of angular benzacridines. Adv. Cancer Res. 4, 315-369 (1956).
2. A. Lacassagne, N. P. Buu-Hoi, F. Zajdela, N. P. Giao, P. Jacquingnon, and M. Dufour, Effect of the nature and number of substituents on the carcinogenic activity of benz[a]- and benz[c]acridines. C. R. Acad. Sci. 267, 981-983 (1968).
3. C. Liebermann, P. Lazar, I. Chouroulinkov, and M. Guerin, Efficacy of rapid skin tests for classification of substituted angular benzacridines according to their carcinogenic activity. C. R. Soc. Biol. 162, 835-838 (1968).
4. R. L. Chang, W. Levin, A. W. Wood, S. Kumar, H. Yagi, D. M. Jerina, R. E. Lehr, and H. Conney, Tumorigenicity of dihydrodiols and diol-epoxides of benz[c]acridine in newborn mice. Cancer Res. 44, 5161-5164 (1984).
5. B. A. Walker, E. G. Rogan, and N. H. Cromwell, Mutagenicity of selected functionalized benz[c]acridines and a benz[a]phenazine in the Salmonella typhimurium/microsome assay. Anticancer Res. 4, 399-402 (1984).
6. S. H. Epstein, M. Small, H. L. Falk, and N. Mantel, On the association between photodynamic and carcinogenic activities in polycyclic compounds. Cancer Res. 24, 855-862 (1964).
7. G. Klopman, M. R. Frierson, and H. S. Rosenkranz, Computer analysis of toxicological data bases: Mutagenicity of aromatic amines in Salmonella tester strains. Environ. Mutagen 7, 625-644 (1985).
8. H. R. Glatt, H. Schwind, F. Zajdela, A. Croisy, P. C. Jacquignon, and F. Oesch, Mutagenicity of 43 structurally related heterocycle compounds and its relationship to their carcinogenicity. Mut. Res. 66, 307-328 (1979).
9. A. W. Wood, R. L. Chang, W. Levin, D. E. Ryan, P. E. Thomas, R. E. Lehr, S. Kumar, M. Schaefer-Ridder, U. Engelhardt, H. Yagi, D. M. Jerina, and A. H. Conney, Mutagenicity of diol-epoxides and tetrahydroepoxides of benz[a]acridine and benz[c]acridine in bacteria and in mammalian cells. Cancer Res. 43, 1656-1662 (1983).
10. W. Levin, A. W. Wood, R. L. Chang, S. Kumar, H. Yagi, D. M. Jerina, R. E. Lehr, and A. H. Conney, Tumor-initiating activity of benz[c]acridine and twelve of its derivatives on mouse skin. Cancer Res. 43, 4625-4628 (1983).
11. D. M. DeMarini, H. N. Pham, A. J. Kalz, and H. E. Brockman, Relationships between structures and mutagenic potencies of 16 heterocyclic nitrogen mustards (ICR compounds) in Salmonella typhimurium. Mut. Res. 136, 185-199 (1984).

12. J. H. Gill, A. M. Bonin, E. Podobna, R. S. U. Baker, C. C. Duke, C. A. Rosario, A. J. Ryan, and G. M. Holder, 7-Methylbenz[c]acridine: Mutagenicity of some of its metabolites and derivatives, and the identification of trans-7-methylbenz[c]acridine-3,4-dihydrodiol as a microsomal metabolite. Carcinogenesis 7, 23-31 (1986).
13. R. L. Chang, W. Levin, A. W. Wood, N. Shirai, A. J. Ryan, C. C. Duke, D. M. Jerina, G. M. Holder, and A. H. Conney, High tumorigenicity of the 3,4-dihydrodiol of 7-methylbenz[c]acridine on mouse skin and in newborn mice. Cancer Res. 46, 4552-4555 (1986).
14. R. S. U. Baker, A. M. Bonin, I. Stupans, and G. M. Holder, Comparison of rat and guinea pig as sources of the S9 fraction in the Salmonella mammalian microsome mutagenicity test. Mut. Res. 71, 43-52 (1980).
15. I. Niculescu-Duvaz, T. Creascu, M. Tugulea, A. Croisy, and P. C. Jacquignon, A quantitative structure-activity analysis of the mutagenic and carcinogenic action of 43 structurally related compounds. Carcinogenesis 2, 269-275 (1981).
16. A. E. Freeman, E. K. Weisberger, J. H. Weisburger, R. G. Wolford, J. M. Maryak, and R. J. Huebner, Transformation of cell cultures as an indication of the carcinogenic potential of chemicals. J. Natl. Cancer Inst. 51, 799-807 (1973).
17. P. Markovits, J. Coppey, D. Papadopoulo, A. Mazabraud, and M. Hubert-Habart, Malignant transformation of hamster embryo cells in tissue culture by 7-10-dimethylbenz[c]acridine. Int. J. Cancer 14, 215-225 (1974).
18. D. Papadopoulo, S. Levy, V. Poirer, C. Pene, P. Markovits, and M. Hubert-Habart, Effects of several dimethylbenzacridines on secondary hamster embryo cells: Neoplastic transformation. Eur. J. Cancer 17, 179-186 (1981).
19. R. P. Deutsch-Wenzel, H. Brune, and G. Grimmer, Experimental studies on the carcinogenicity of five nitrogen containing polycyclic aromatic compounds directly injected into rat lungs. Cancer Res. 20, 97-101 (1983).
20. E. Sawicki, S. P. Mcpherson, T. W. Stanley, J. Meeker, and W. C. Elbert, Quantitative composition of the urban atmosphere in terms of polynuclear aza heterocyclic compounds and aliphatic and polynuclear aromatic hydrocarbons. J. Air Water Poll. 9, 515-524 (1965).
21. E. Sawicki, T. W. Stanley, S. P. Mcpherson, and M. Morgan, Use of gas-liquid and thin-layer chromatography in characterising air pollutants by fluorometry. Talanta 13, 619-629 (1966).
22. G. Alberini, V. Cantuti, and G. P. Cartoni, Gas chromatography of heterocyclic nitrogen compounds and their evaluation in atmospheric dusts. Gas Chromatogr. 1966 Int. Symp. Anal. Instrum. Div. Inst. Soc. Am. 6, 258-271 (1967).

23. E. Sawicki, Airborne carcinogens and allied compounds. Arch. Environ. Health 14, 46-53 (1967).
24. E. Sawicki, Fluorescence analysis in air pollution research. Talanta 16, 1231-1266 (1969).
25. T. W. Stanley, M. J. Morgan, and E. M. Grisby, Application of rapid thin-layer chromatographic procedure to the determination of benz[a]pyrene, benz[c]acridines, and 7H-benz[d,e]anthracene-7-one in air particulates from many American cities. Environ. Sci. Technol. 2, 699-702 (1968).
26. D. Brocco, A. Cimino, and M. Possanzini, Determination of aza-heterocyclic compounds in atmospheric dust by a combination of thin-layer and gas chromatography. J. Chromatogr. 84, 371-377 (1973).
27. W. Cautreels and K. V. Cauwenberghe, Determination of organic compounds in airborne particulate matter by gas chromatography-mass spectrometry. Atm. Environ. 10, 447-457 (1976).
28. P. Masclet, M. A. Bresson, S. Beyne, and G. Mouvier, Fast determination without preseparation of polycyclic aromatic nitrogen derivatives in atmospheric aerosols. Analusis 13, 401-405 (1985).
29. T. Nielsen and P. Clausen, Determination of basic azaarenes and polynuclear aromatic hydrocarbons in airborne particulate matter by gas chromatography. Anal. Chim. Acta 187, 223-231 (1986).
30. I. Schmeltz, W. S. Schlotzhauer, and E. B. Higman, Characteristic products from pyrolysis of nitrogenous organic substances. Beitr. Tabakforsch. Intern. 6, 134-138 (1972).
31. G. Grimmer, K.-W. Naujack, and G. Dettbarn, Gas chromatographic determination of polycyclic aromatic hydrocarbons, azaarenes, aromatic amines in the particle and vapor phase of mainstream and sidestream smoke of cigarettes. Toxicol. Lett. 35, 117-124 (1987).
32. E. Sawicki, J. E. Meeker, and M. Morgan, Polynuclear aza compounds in automotive exhaust. Arch. Environ. Health 11, 773-775 (1965).
33. T. Handa, T. Yamauchi, K. Sawai, T. Yamamura, Y. Koseki, and T. Ishii, In situ emission levels of carcinogenic and mutagenic compounds from diesel and gasoline engine vehicles on an expressway. Environ. Sci. Technol. 18, 895-902 (1984).
34. C. A. Krone, D. G. Burrows, D. W. Brown, P. A. Robisch, A. J. Friedman, and D. C. Malins, Nitrogen-containing aromatic compounds in sediments from a polluted harbor in Puget Sound. Environ. Sci. Technol. 20, 1144-1150 (1986).
35. S. G. Wakeham, Azaarenes in recent lake sediments. Environ. Sci. Technol. 13, 1118-1123 (1979).

36. E. Sawicki, T. W. Stanley, and W. C. Elbert, Characterization of polynuclear aza heterocyclic hydrocarbons separated by column and thin-layer chromatography from air pollution source particulates. J. Chromatogr. 18, 512-519 (1965).
37. E. Sawicki, J. E. Meeker, and M. J. Morgan, The quantitative composition of air pollution source effluents in terms of aza heterocyclic compounds and polynuclear aromatic hydrocarbons. J. Air Water Poll. 9, 291-298 (1965).
38. E. Sawicki, T. W. Stanley, and W. C. Elbert, Comparison of fluorimetric methods of assay for benz[c]acridine and benz[h]quinoline in urban atmospheres and air pollution source effluents. J. Chromatogr. 26, 72-78 (1967).
39. J. F. Mckay, J. H. Weber, and D. R. Latham, Characterization of nitrogen bases in high-boiling petroleum. Anal. Chem. 48, 891-898 (1976).
40. G. Grimmer and K.-W. Naujack, Determination of basic nitrogen-containing polycyclic aromatic compounds (azaarenes) in petroleum and petroleum products. Fresenius Z. Anal. Chem. 321, 27-31 (1985).
41. M. R. Guerin, C.-H. Ho, T. K. Rao, B. R. Clatk, and J. L. Epler, Polycyclic aromatic primary amines as determinant chemical mutagens in petroleum substitutes. Environ. Res. 23, 42-53 (1980).
42. R. A. Pelroy and M. R. Petersen, Use of Ames test in evaluation of shale oil fractions. Environ. Health Pers. 30, 191-203 (1979).
43. R. A. Pelroy and M. R. Petersen, Mutagenicity of shale oil compounds. Environ. Sci. Res. 15, 463-475 (1979).
44. P. Burchill, A. A. Herod, J. P. Mahon, and E. Pritchard, Comparison of methods for the isolation of basic nitrogen compounds from coal tars. J. Chromatogr. 265, 223-238 (1983).
45. P. Burchill, A. A. Herod, J. P. Mahon, and E. Pritchard, The class separation of nitrogen compounds in coal tars by liquid chromatography on a polar bonded-phase silica. J. Chromatogr. 281, 109-124 (1983).
46. P. Burchill, A. A. Herod, and E. Pritchard, Investigation of nitrogen compounds in coal tar products. I. Unfractionated materials. Fuel 62, 11-19 (1983).
47. P. Burchill, A. A. Herod, and E. Pritchard, Estimation of basic nitrogen compounds in some coal liquefaction products. J. Chromatogr. 246, 271-295 (1982).
48. D. W. Later, R. A. Pelroy, D. D. Mahlum, C. W. Wright, M. L. Lee, W. C. Weimer, and B. W. Wilson, Aromatic hydrocarbons and related nitrogen-containing heteroatomic species in products from coal liquefaction processes. Polynucl. Aromat. Hydrocarbons, 1982 Int. Symp., 7th, 1983, pp. 771-783, edited by M. Cook and A. J. Dennis, Battelle Press, Columbus, Ohio.

49. J. R. Kershaw, Fluorescence spectroscopic analysis of coal-derived liquids. Determination of polycyclic aromatic hydrocarbon ring systems and identification of basic nitrogen heterocycles. Fuel 62, 1430-1435 (1983).
50. E. Sawicki, T. W. Stanley, and W. C. Elbert, The application of thin-layer chromatographic and spectral procedures to the analysis of aza heterocyclic hydrocarbons in complex mixtures. Occ. Health Rev. 16, 8-16 (1964).
51. E. Sawicki, J. E. Meeker, and M. Morgan, Column chromatographic separation of basic polynuclear aromatic compounds from complex mixtures. J. Chromatogr. 17, 252-256 (1965).
52. E. Sawicki, H. Johnson, and K. Kosinski, Chromatographic separation and spectral analysis of polynuclear aromatic amines and heterocyclic imines. Microchem. J. 10, 72-102 (1966).
53. C. R. Engel and E. Sawicki, A superior thin-layer chromatographic procedure for the separation of aza arenes and its application to air pollution. J. Chromatogr. 31, 109-119 (1967).
54. M. E. Snook, Nitrogen analogues of polynuclear aromatic hydrocarbons in tobacco smoke. Carcinogenesis 3, 203-215 (1978).
55. M. E. Snook, P. J. Fortson, and O. T. Cbortyk, Isolation and identification of aza-arenes of tobacco smoke. Beitr. Tabakforsch. Intern. 11, 67-78 (1981).
56. J. L. Shultz, C. M. White, F. K. Schweighardt, and A. G. Sharkey, Characterization of the heterocyclic components in coal liquefaction products. Part I: Nitrogen compounds, 1-25, Pittsburgh, PA: United States Energy Research and Development Administration, Pittsburgh Energy Research Center, 1977.
57. H.-J. Klimisch and A. Beiss, Separation of N-heteropolycyclic aromatic hydrocarbons from the separation by ion-exchange chromatography and fractionation of cigarette smoke condensate J. Chromatogr. 128, 117-124 (1976).
58. L. J. Boux and G. M. Holder, The activation and DNA binding of 7-methylbenz[c]acridine catalysed by mouse liver microsomes. Cancer Lett. 25, 333-342 (1985).
59. J. Jacob, A. Schmoldt, W. Kohbrok, G. Raab, and G. Grimmer, On the metabolic activation of benz[a]acridine and benz[c]acridine by rat liver and lung microsomes. Cancer Lett. 16, 297-306 (1982).
60. G. Grimmer and K.-W. Naujack, Gas chromatographic profile analysis of basic nitrogen-containing aromatic compounds (aza-arenes) in high protein foods. J. Assoc. Off. Anal. Chem. 69, 537-541 (1986).
61. E. Sawicki and J. D. Pfaff, Analysis for aromatic compounds on paper and thin-layer chromatograms by spectrophotophorimetry: Application to air pollution. Anal. Chim. Acta 32, 521-534 (1965).

62. A. M. Luly and K. Sakodynsky, A paper chromatographic study of aza-heterocyclic hydrocarbons using aqueous solvents. J. Chromatogr. 19, 624-629 (1965).
63. M. Lederer and G. Roch, Paper chromatography of some aza heterocyclic hydrocarbons. J. Chromatogr. 31, 618-627 (1967).
64. S. Caroli and M. Lederer, Paper chromatography of azaheterocyclic hydrocarbons. J. Chromatogr. 37, 333-340 (1968).
65. E. Sawicki, T. W. Stanley, J. D. Pfaff, and W. C. Elbert, Thin-layer chromatographic separation and analysis of polynuclear aza heterocyclic compounds. Anal. Chim. Acta 31, 359-375 (1964).
66. E. Sawicki, T. W. Stanley, and H. Johnson, Direct spectrophotofluorometric analysis of aromatic compounds on thin-layer chromatograms. Microchem. J. 8, 257-284 (1964).
67. E. Sawicki and H. Johnson, Thin-layer chromatographic characterization tests for basic polynuclear compounds. Application to air pollution. Mikrochim. Ichnoanal. Acta (2-4), 435-450, (1964).
68. E. Sawicki, W. C. Elbert, and T. W. Stanley, The fluorescence-quenching effect in thin-layer chromatography of polynuclear aromatic hydrocarbons and their aza analogs. J. Chromatogr. 17, 120-126 (1965).
69. E. Sawicki, T. W. Stanley, and W. C. Elbert, Direct fluorometric scanning of thin-layer chromatograms and its application to air pollution studies. J. Chromatogr. 20, 348-353 (1965).
70. E. Sawicki, T. W. Stanley, and W. C. Elbert, Simple fluorimetric method of assay for benz[c]acridine and other aza heterocyclic compounds in urban atmospheres and air pollution source effluents. Am. Chem. Soc., Div. Water Air Waste Chem., Preprints 6, 111-116 (1966).
71. L. H. Klemm, C. E. Klopfenstein, and H. P. Kelly, Thin-layer chromatography of azines and aromatic nitrogen heterocycles on alumina. J. Chromatogr. 23, 428-445 (1966).
72. T. P. Wang and A. Brockhaus, The content of tar compound benz[c]acridine in the air of two cities in north-Rhinewestphalia. Mikrochim. Acta 2, 55-60 (1976).
73. T. W. Stanley, J. E. Meeker, and M. J. Morgan, Extraction of organics from airborne particulates. Effect of various solvents and conditions on the recovery of benz[a]pyrene, benz[c]acridine, and 7H-benz[de]anthracene-7-one. Environ. Sci. Technol. 1, 927-931 (1967).
74. H. Matsushita, Y. Esumi, and K. Yamada, Identification of polynuclear hydrocarbons in air pollutants. Bunseki Kagaku 19, 951-966 (1970).
75. T. Yamauchi and T. Handa, Characterization of aza heterocyclic hydrocarbons in urban atmospheric particulate matter. Environ. Sci. Technol. 21, 1177-1181 (1987).

76. L. K. Keefer and D. E. Johnson, Magnesium hydroxide as a thin-layer chromatographic adsorbent. II. A unique system for separating polynuclear aza aromatic compounds. J. Chromatogr. 47, 20-26 (1970).
77. N. Motohashi and K. Kamata, A convenient method of the purification for carcinogenic and noncarcinogenic methyl-substituted benz[c]acridines. Yakugaku Zasshi 103, 795-799 (1983).
78. K. Kamata and N. Motohashi, Separation of methyl-substituted benz[c]acridines by reversed-phase high-performance thin-layer chromatography. J. Chromatogr. 396, 437-440 (1987).
79. L. J. Boux, C. M. Ireland, D. J. Wright, G. M. Holder, and A. J. Ryan, Thin-layer chromatographic and high-performance liquid chromatographic separation of metabolites of the weak carcinogen, 7-methylbenz[c]acridine. J. Chromatogr. 227, 149-157 (1982).
80. M. Dog and D. C. Locke, Separation of aza-arenes by high-pressure liquid chromatography. J. Chromatogr. Sci. 15, 32-35 (1977).
81. H. Colin, J.-M. Schmitter, and G. Guiochon, Liquid chromatography of azaarenes. Anal. Chem. 53, 625-631 (1981).
82. J. Chmielowiec, Organomercuric bonded phase for high performance liquid chromatographic separations of pai-electron-, sulfur-, oxygen-, and nitrogen-containing compounds. J. Chromatogr. Sci. 19, 296-307 (1981).
83. D. A. Haugen, M. J. Peak, and K. M. Suhrbier, Isolation of mutagenic aromatic amines from a coal conversion oil by cation exchange chromatography. Anal. Chem. 54, 32-37 (1982).
84. L. A. D'Avila, H. Colin, and G. Guiochon, Hydrophilic complexes with protons and metallic ions in the reversed-phase liquid chromatography of azaarenes. Anal. Chem. 55, 1019-1024 (1983).
85. A. M. Siouffi, M. Righezza, and G. Guiochon, Separation of aromatic compounds by liquid chromatography on diol-bonded phase columns. J. Chromatogr. 368, 189-202 (1986).
86. K. Kamata and N. Motohashi, Separation of methyl-substituted benz[c]acridines by gas-liquid chromatography and high-performance liquid chromatography. J. Chromatogr. 319, 331-340 (1985).
87. L. J. Boux, C. D. Duke, G. M. Holder, C. M. Ireland, and A. J. Ryan, Metabolism of 7-methylbenz[c]acridine: Comparison of rat liver and lung microsomal preparations and identification of some minor metabolites. Carcinogenesis 4, 1429-1435 (1983).
88. L. J. Box and G. M. Holder, The metabolism of the carcinogen 7-methylbenz[c]acridine by hepatocytes isolated from untreated and induced rats. Xenobiotica 15, 11-20 (1985).

89. D. J. Wright, H. K. Robinson, G. M. Holder, A. J. Ryan, A. M. Bonin, and R. S. U. Baker, Metabolism of the carcinogen 7-methylbenz[c]acridine in the rat. Xenobiotica 15, 825-834 (1985).
90. C. M. Ireland, H. T. A. Cheung, A. J. Ryan, and G. M. Holder, Rat liver microsomal metabolites of 7-methylbenz[c]acridine. Chem.-Biol. Interact. 40, 305-318 (1982).
91. J. H. Gill and G. M. Holder, Application of synchronous luminescence to the separate determination of CO chromatographing metabolites of the carcinogen, 7-methylbenz[c]acridine. J. Pharm. Biomed. Anal. 4, 31-36 (1986).
92. T. D. Searl, F. J. Cassidy, W. H. King, and R. A. Brown, An analytical method for polynuclear aromatic compounds in coke oven effluents by combined use of gas chromatography and ultraviolet absorption spectrometry. Anal. Chem. 42, 954-957 (1970).
93. M. Pailer and V. Hlozek, Use of a nematic liquid crystal for the gas chromatographic separation of aza-heterocyclic compounds. J. Chromatogr. 128, 163-165 (1976).
94. I. B. Rubin and C. K. Bayne, Statistical designs for the optimization of the nitrogen-phosphorus gas chromatographic detector response. Anal. Chem. 51, 541-546 (1979).
95. G. Grimmer, Recommended method for the gas chromatographic profile analysis of basic N-containing aromatic compounds (azaarenes) in high protein foods. Pure Appl. Chem. 55, 2067-2071 (1983).
96. M. L. Lee, D. L. Vassilaros, C. M. White, and M. Novotny, Retention indices for programmed-temperature capillary-column gas chromatography of polycyclic aromatic hydrocarbons. Anal. Chem. 51, 768-773 (1979).
97. I. Ignatiadis, J. M. Schmitter, and G. Guiochon, Capillary gas chromatography of azaarenes. I. Preparation of columns. J. Chromatogr. 246, 23-36 (1982).
98. J. M. Schmitter, I. Ignatiadis, and G. Guiochon, Capillary gas chromatography of azaarenes. II. Application to petroleum nitrogen bases. J. Chromatogr. 248, 203-216 (1982).
99. C. E. Rostad, W. E. Pereira, and M. F. Hult, Partitioning studies of coal-tar constituents in a two-phase contaminated ground-water system. Chemosphere 14, 1023-1036 (1985).
100. J. C. Lauer, H. D. H. Valles, and D. Cagiant, Improved characterization of coal tar distillation cuts (200-500°C). 2. Capillary GC-MS determination of neutral and basic nitrogen aromatic compounds and hydroxylated aromatic compounds. Fuel 67, 1446-1455 (1988).
101. D. W. Later, M. L. Lee, K. D. Bartle, R. C. Kong, and D. L. Vassilaros, Chemical class separation and characterization of organic compounds in synthetic fuels. Anal. Chem. 53, 1612-1620 (1981).

Index

A

Absorption spectroscopy for studying complex boronate formations, 316-319
Acetanilides, 198, 215
Acetophenones, 206
Acrylic acid esters, 196, 221, 223
3-Acyloxy-1,4-benzodiazepines, 219
Adonispigments, 213
Affinity chromatography:
  boronate ligands used for, 303-308
  using immobilized boronates (frontal analysis), 316
Alcohols:
  as colored derivatives, 193
  dimethylaminobenzeneazobenzoates of, 215
  3,5 dinitrobenzoates of, 191, 192, 195, 197
  isoprenoid, 214

[Alcohols]
  phenylazobenzoates of, 209
  terpene, 213, 214, 216
  wax, 212
Aliphatic acids, 196
  as derivatives, 192
Aliphatic amines, 219
Alkaloids, 186, 187, 188, 196, 215
  rauwolfia, 182, 185, 194, 196
Alkoxyphenylcarbamic acid, 190
Alkoxyphenylcarbamic acid esters, 189
O-Alkyl-O-arylphenylphosphonothioates, 210
Alkylated phenols, 201
Alkylmercuridithizonates, 215
Alkylphenols, 206
n-Alkylphenyl ketones, 221
S-Alkylphenylthiouronium picrates, 214
Alkyl sulfoxides, 201
Alkyltrityl amines, 214
ω-Amino acid derivatives, 219

383

Amino acids, 195, 222
Aminoacyl transfer RNA, boronate affinity chromatography for, 321-324
Aminophenols, 221
Androgen esters, 216
Androgens, 205
Anesthetics, local, 187, 188, 201
Aniline derivatives, 223
Anion-active tensides, 200
Anthraquinone derivatives, 193
Antibiotics, 204
  oligosaccharide, 205
  polyenic, 205
Antiinflammatory drugs, 221
Antimalarial sulfonamides, 220
Antimycin A, 222
Aqueous size exclusion chromatography, 119, 151
  commercial packings for, 121-126
  electrostatic interactions in, 126-134
  hydrophobic interactions in, 134-136
  of micelles, 145-147
  of proteins, 143-145
  universal calibration for, 136-142
Aristolochic acids, 184
Aromatic acids as 2,4-dinitrobenzyl esters, 193
Aromatic amines, 190, 196
Aromatic polyamines, 203
Arylacetic acids, 207
Arylaliphatic acids, 206, 207
3-Aryl-n-butyric acids, 206
Arylpropionic acids, 206
4-Arylthiazoles, 191
Azaarenes, silicic acid chromatography of, 341, 342
Azo dyes, 200
  phenols as, 183

B

Bacteriochlorophylls, 202
Barbiturates, 186, 187, 202, 220
Benz(c)acridines:
  conditions for gas chromatographic analysis of, 365, 367-371

[Benz(c)acridines]
  GLC retention data on seven kinds of stationary phases for, 365, 366
  R_f values of
    on RP-18, 351, 359
    on RP-18 and silica gel 60 F_{254}, 351, 358
  TLC separation of, 351, 352-356
Benzenesulfonamidopyrimidines, 217
1,4-Benzodiazepines, 187, 190, 200, 201, 203, 206, 208
Benzoic acids, substituted, 202
Benzophenones, 208, 221
Benzoquinones, chlorinated, 195
Benzoylarylaliphatic acids, 207
Biospecific affinity chromatography, 4
Bisdichloroacetamides, 205
Bisoxirane method for activation of carbohydrate matrices, 311-312
Boronate affinity chromatography, 293-335
  affinity matrices used for immobilization of different ligands, 312-314
  boronate complex formation, 295-300
    chemistry of interaction, 295-298
    mechanisms of interaction, 298-300
  boronate ligands used for affinity chromatography, 303-308
  chemistry of activation and coupling of ligands, 308-312
    bisoxirane method for activation of carbohydrate matrices, 311-312
    carbodiimide ligand coupling method, 308-309
    cyanogen bromide agarose activation method, 310-311
    periodate oxidation dextrose activation method, 311
  conditions to enhance complex formations, 300-303

[Boronate affinity chromatography]
  nature of boronate ligand, 301
  nature of cis-diols and bond distances, 302-303
  reaction environment, 300-301
  examples of applications of, 321-329
  large biomolecules, 321-329
  small biomolecules, 321
  methods for studying complex formations, 314-321
  absorption spectroscopy, 316-319
  affinity chromatography using immobilized boronates (frontal analysis), 316
  ion exchange chromatography, 315
  optical rotation, 321
  proton and ^{11}B-NMR, 319-320
Bruceolides, 196

C

Cannabidiol, 192
Cannabinoids, 187, 193
Cannabinols, 191, 193
Cannabis, 192, 193
Capsaicinoids, 217
Carbamates, 185
Carbanilic acid, 189
Carbanilic acid esters, 188
Carbazole halogen derivatives, 223
Carbodiimide ligand coupling method, 308-309
Carboxamide derivatives, 221
Carcinogenic benz(c)acridine determination, 337-381
  column chromatography for, 341-344
  gas-liquid chromatography for, 365-373
  high-performance liquid chromatography for, 360-365
  paper chromatography for, 344-350

[Carcinogenic benz(c)acridine determination]
  thin-layer chromatography for, 350-359
Cardenolide glycosides, 182, 186
Cardenolides, 199, 206
Cardiac glycosides, 187
Cardioactive glycosides, 189
Carotene aldehydes, 210
Carotenoids, 211
Cephalosporins, 205
Chemistry of activation and coupling of boronate ligands, 308-312
  bisoxirane method for activation of carbohydrate matrices, 311-312
  carbodiimide ligand coupling method, 308-309
  cyanogen bromide agarose activation method, 310-311
  periodate oxidation dextrose activation method, 311
Chlorinated benzoquinones, 195
Chlorinated biphenyls, 218
Chlorinated pesticides, 191, 192, 216
Chlorinated phenols, 194
Chloroplast pigments, 202
Chlorophenols, 201
Chlorophyll derivatives, 202, 217, 224
Chlorophylls, 203
Cholesterol, 209
Cholesterol esters, 211
Chromatographic systems using impregnated layers, 182-224, 225
Coil planet centrifuge (CPC):
  cross-axis, 285-287
  multilayer, 274 276
  for peptide purification horizontal flow-through, 268-273
  operation of, 265-267
Colchicine alkaloids, 207
Color couplers, 191
Column chromatography (CC)
  for carcinogenic benz(c)-acridine determination, 341-344

Commercial packing available for aqueous size exclusion chromatography, 121-126
Competitive equilibrium isotherms, determination of, 73-74
Corticoids, 184
Corticosteroids, 182, 183, 197, 205
Countercurrent chromatography (CCC) for peptide purification, 253-292
  early instrumentation for countercurrent chromatography of peptides, 255-262
    countercurrent chromatography, 259-262
    countercurrent distribution, 257-259
  methods for, 265-268
    analysis of the countercurrent chromatographic separation, 267-268
    operation of the coil planet centrifuge, 265-267
  multicoil countercurrent chromatograph, 276-285
    capability, 283-285
    operation, 276-283
  multilayer coil planet centrifuge, 274-276
  potential instruments for, 285-289
    cross-axis coil planet centrifuge, 285-287
    "foam" chromatography, 287-289
  theory of CCC, 262-264
    partition coefficient relationships, 262-264
    resolution, 264
  types of peptides separated by CCC, 289
  use of horizontal flow-through coil planet centrifuge, 268-273
Cross-axis coil planet centrifuge, 285-287
Crotonolactones, 190
Crown ethers, 222

Cucurbitacin, 185
Cyanogen bromide agarose activation method, 310-311
2-Cyanomethylbenzimidazoles, 219

D

Deamosterol, 209
Dermorphine oligopeptides, 207
N,N-Dialkyldithioxamides, 208
Dian, 184, 192
Dicarboxylic, 197
Dicarboxylic acids, 199
Diethylphenyl phosphates, 200, 207
Digilanides, 189
Digitalis cardenolides, 183, 189
Digitoxin, 185, 187, 197
Diglycerides, 208
Digoxin, 185, 188, 197
Dihydrodigoxin, 188
Dimethylaminobenzeneazobenzoates of alcohols, 215
4-Dimethylamino-3,5-dinitrobenzamides, 199
Dimethylaminoethylalkoxycarbanilate chlorides, 190
3,5-Dinitrobenzoates of alcohols, 191, 192, 195, 197, 215
Dinitrooctylphenyl cocotonate, 222
Dinitrophenols, 201, 214
2,4-Dinitrophenylamines, 199
2,4-Dinitrophenylhydrazones, 198, 209, 221
  of aldehydes, 195, 197, 199, 204, 212, 224
  of aliphatic ozo compounds, 191, 195, 197, 199
  of aromatic aldehydes, 193, 198
  of benzophenones, 192
  of carbonyls, 199
  of ketones, 197, 212
2,4-Dinitrophenylosazones of α-dicarbonyls, 213
Dioleylphosphatidylcholine, 222
Diphenhydramine, 217
Dirac problem, 2

Index

Displacement chromatography:
 analytical applications of, 101–107
 preparative applications of, 98–101
 separation by, 94–98
 for three-component/multicomponent problems, 91–94
 for two-component problems, 69–70
Disulfides, 217
Dothistromin, 220

E

Ecdysteroids, 221, 224
Electrostatic interactions in aqueous size exclusion chromatography, 126–134
Elution mode for three-component/multicomponent problems, 91
Epoxy acids, 208
Equilibrium equation, 6–7
Equilibrium isotherms:
 competitive, determination of, 73–74
 for multicomponent samples, 10–13
Ergot alkaloids, 182, 183, 184, 185, 186, 187, 188, 190
 hydrogenated, 189, 201
Erythromycins, 185

F

Fat-soluble food dyes, 213, 215
Fat-soluble vitamins, 198, 211, 218
Fatty acid esters, 212, 213
Fatty acid methyl esters, 204, 210, 213
Fatty acids, 203, 204, 208, 209, 212, 214
 free, 209
Fatty alcohols, 204, 208
Fatty aldehydes, 209
Fatty hydroxy acids, 210
Fatty ketones, 209

Fatty oils, 223
"Foam" chromatography, 287–289
Food dyes, fat-soluble, 213, 215
Free fatty acids, 209
Frontal analysis:
 single-component, 37
 two-component, 66–69
Furil oximes, 187
Furocoumarins, 183, 189

G

Gas liquid chromatography (GLC) for carcinogenic benz(c)-acridine determination, 365–373
Gel permeation chromatography (GPC), 119, 120
Gitoxin, 187
Glucocorticoids, 184
Glyceollin isomers, 190
Glycerides, 210, 212, 215
Glycosylated hemoglobin, boronate affinity chromatography for, 325–329
Gradient elution:
 separation of a binary mixture in, 108–109
 two-component problems and, 62–66
Guanidinoalkanesulfonic acids, 217

H

Herbicides:
 phenylurea, 189, 197
 triazine, 189, 197, 216, 221
Heterocyclic bases, 196
Hexachlorobenzene, 193
High-performance liquid chromatography (HPLC) for carcinogenic benz(c)acridine determination, 360–365
Horizontal flow-through coil planet centrifuge, 268–273
Hydrogenated ergot alkaloids, 189, 201
Hydrophobic interaction chromatography (HIC), 4

Hydrophobic interactions in aqueous size exclusion chromatography, 134-136
Hydroxamic acids, 202
Hydroxyanthroquinones, 193
2-Hydroxybenzophenones, 202
Hydroxycinnamic acids, 220
4-Hydroxycoumarin derivatives, 201
Hypoglycemic sulfonamides, 220

I

Ideal chromatography model, 13
  elution mode and
    single-component problems, 15-18
    two-component problems, 44-49
  numerical analysis of, 30-33
Immobilized boronates, affinity chromatography using, 316-319
Impregnated layers:
  chromatographic systems using, 182-224, 225
  slurry compositions used for preparation of, 161, 162-163
Indanols, 186, 191, 193, 201
Instrumentation, early, for countercurrent chromatography of peptides, 255-262
Iodinated aromatic acids, 207
Ion exchange chromatography (IXC), 4
  for studying complex boronate formations, 315
Isonicotinic acid hydrazide-type compounds, 218
Isoprenoid alcohols, 214
Isoprenoid quinones, 213, 216
Isotherm parameters, determination of, for single-component problems, 37-43

K

Ketokarotenoids, 213

Kinetic chromatography models, 14
  elution mode and
    single-component problems, 34-37
    two-component problems, 57-62

L

Lanatosides, 182, 183
Ligands, boronate, chemistry of activation and coupling of, 308-312
Linear chromatography, single-component problems and, 18-20
Liquid-liquid chromatography (LLC), 4
Liquid-solid chromatography (LSC), 4
Lobelia alkaloids, 182
Local anesthetics, 187, 188, 201
Lutein esters, 212

M

Mass balance equation, 5, 7-8
Menaquinones, 215
Metallocarboranes, 223
Methacrylic acid esters, 221
Methyl esters, 203
Micelles, size exclusion chromatography of, 145-147
Mobile phase:
  in partition thin-layer chromatography, 168-176
  for nonaqueous polar stationary phases, 170-171
  for unpolar stationary phases, 171-176
  relationship between concentration in the stationary phase and, 8-9
Monocarboxylic acid, 199
Morphine, 182
Multicoil countercurrent chromatograph (MC-CCC), 276-285
  capability of, 283-285

Index

[Multicoil countercurrent chromatograph (MC-CCC)]
  operation of, 278-283
Multicomponent problems, 80-109
  analytical applications of displacement chromatography, 101-107
  displacement chromatography and, 91-94
  elution mode and, 91
  preparative applications of displacement chromatography, 98-101
  separation of binary mixture in a gradient elution, 108-109
  system peaks, 107-108
  theory of separations by displacement chromatography, 94-98
Multicomponent samples, basic equations and models for, 10-13
Multilayer coil planet centrifuge (MLCPC), 274-276

N

Naphthalenes:
  derivatives, 220
  polychlorinated, 218
Naphthols, 206, 222
Naphthylamines, 196
Natural quinones, 214
5-Nitroimidazoles, 207, 209, 220, 223
Nitroimidiazothiazoles, 207
Nitrophenols, 186, 190, 194, 196, 200, 201, 224
Nonequilibrium equation, 6-7
Nonlinear chromatography, 2-118
  basic equations and models, 7-14
    ideal model, 13
    initial and boundary conditions, 9-10
    kinetic models, 14
    mass balance equation, 7-8
    multicomponent case and the equilibrium isotherms, 10-13

[Nonlinear chromatography]
  relationship between concentrations in the two phases of the columns, 8-9
  semi-ideal model, 13-14
  results and discussion, 14-109
    single-component problems, 14-43
    three-component and multicomponent problems, 89-109
    two-component problems, 43-89
Normal phase chromatography (NPC), 4

O

Oligopeptides, 208
Oligosaccharide antibiotics, 205
Optical rotation for studying complex boronate formations, 321
Organic acids, 223
Organic stationary liquids, thin-layer chromatography using, 153-252
  applications, 179-225
    analytical separations, 179
    lipophilicity determination and QSAR studies, 179-225
  partition solvent system for, 154-179
    amount of stationary phase, 165-168
    choice of the stationary phase, 160-161
    impregnation procedure, 161-165
    mobile phase, 168-176
    the solute, 176-178
    support, 156-160
Organic thiols and thioethers, 210
Organomercuridithizonates, 216
Oxyethylated nonylphenol, 220

P

Palmitic acid, 206
Paper chromatography (PC) for carcinogenic benz(c)acridine determination, 344-350
Partition solvent system for thin layers with organic stationary liquids, 154-179
Penicillins, 198, 200, 205, 206
Peptide purification, countercurrent chromatography for, 253-292
  early instrumentation for countercurrent chromatography of peptides, 255-262
  countercurrent chromatography, 259-262
  countercurrent distribution, 257-259
  methods for, 265-268
    analysis of the countercurrent chromatographic separation, 267-268
    operation of the coil planet centrifuge, 265-267
  multicoil countercurrent chromatograph, 276-285
    capability, 283-285
    operation, 278-283
  multilayer coil planet centrifuge, 274-276
  potential instruments for, 285-289
    cross-axis coil planet centrifuge, 285-287
    "foam" chromatography, 287-289
  theory of countercurrent chromatography, 262-264
    partition coefficient relationship, 262-264
    resolution, 264
  types of peptides separated by CCC, 289
  use of horizontal flow-through coil planet centrifuge, 268-273
Periodate oxidation dextrose activation method, 311
Pesticides, 191, 204, 215, 220
  chlorinated, 191, 192, 216

Phenolic acids, 198, 219
Phenolic antioxidants, 204, 211
Phenols, 183, 184, 185, 186, 187, 190, 191, 192, 193, 194, 201, 205, 206
  alkylated, 201
  as azodyes, 183
  chlorinated, 194
  as derivatives, 186, 188, 221, 222
  oxidation products of, 183
  polyhydric, 185, 194
Phenothiazines, 201, 205
Phenylazobenzoates of alcohols, 209
Phenylcarbamic acid esters, 190
Phenylhydrazones, substituted, 216
Phenylurea herbicides, 189, 197
Phospholipid adducts, 223
Phosphorus pesticides, 212
Plant phloroglucinol butanones, 211
Plasma proteins, boronate affinity chromatography for, 325-329
Plastic pigments, 211
Plastide pigments, 203
Plastoquinone, 212
Polyalcohols, 194
Polychlorinated biphenyls, 215, 217, 218, 219
Polychlorinated naphthalenes, 218
Polycyclic hydrocarbons, 191, 192
Polyenic antibiotics, 205
Polyhydric phenols, 185, 194
Polymeric packings for aqueous size exclusion chromatography, 121, 124-125
Polyprenols, 215, 218, 219
Polyunsaturated esters, 219
Porphyrin esters, 204
Preparative chromatography, optimization of, for two-component problems, 78-83
Prostaglandins, 207
Proteins, size exclusion chromatography of, 143-145

Index

Proton and ^{11}B-NMR for studying complex boronate formations, 319-320
Pyrazine carbothioamide derivatives, 219

Q

Quaternary ammonium compounds, 221
QSAR (quantitative structure activity relationships) studies, thin layers with organic stationary phases for, 154, 156, 160, 165, 168, 179-225

R

Rauwolfia, 182
Rauwolfia alkaloids, 182, 185, 194, 196
Resins, 212
Reversed phase chromatography (RPC), 4
Riemann problem, 2
Rifamycin derivatives, 205
Rifamycins, 205

S

Salicylamide, 185
Semi-ideal chromatography model, 13-14
  elution mode and
    single-component problems, 20-30
    two-component problems, 49-57
  numerical analysis of, 30-33
Siliceous packings for aqueous size exclusion chromatography, 121, 122-123
Silicic acid chromatography of azaarenes, 341, 342
Single-component problems, 14-43
  determination for isotherm parameters, 37-43

[Single-component problems]
  elution mode and ideal model, 15-18
  elution mode and kinetic models, 34-37
  elution mode and semi-ideal model, 20-30
  frontal mode, 37
  linear chromatography, 18-20
  numerical analysis of the ideal and semi-ideal models, 30-33
Size exclusion chromatography (SEC), 4
  aqueous, 119-151
    commercial packings for, 121-126
    electrostatic interactions in, 126-134
    hydrophobic interactions in, 134-136
    of micelles, 145-147
    of proteins, 143-145
    universal calibration for, 136-142
Softeners, 197
Specialized transfer RNAs, boronate affinity chromatography for, 325
Stationary phase, relationship between concentration in mobile phase and, 8-9
Steroidal sapogenins, 182
Steroids, 182, 183, 195, 196, 205, 209, 211, 215
Sterol acetates, 209
Sterol esters, 213
Sterols, 195, 208, 209, 220
Substituted benzoic acid, 202
Substituted phenylhydrazones, 216
Sugar acetate, 195
Sugars, 194
Sulfonamides, 200, 206, 209
  antimalarial, 220
Surfactants, 200
System peaks:
  with pure-component sample, 70-73
  three-component/multicomponent problems and, 107-108

T

Terepene alcohols, 213, 214, 216
Testosterone derivatives, 188, 205
Testosterone esters, 205
Tetracyclines, 199
Thin-layer chromatography (TLC) for carcinogenic benz(c)-acridine determination, 350-359
Thin layers with organic stationary phases, 153-252
 applications, 179-225
  analytical separations, 179
  lipophilicity determination and QSAR studies, 179-225
 partition solvent system for, 154-179
  amount of stationary phase, 165-168
  choice of the stationary phase, 160-161
  impregnation procedure, 161-165
  mobile phase, 168-176
  the solute, 176-178
  support, 156-160
Thiobenzamides, 193, 200
Thiocarbamic acid derivatives, 184, 213
Thiohydrazides, 208
Thiols, 210
Thionophosphates, 220
Thiophosphate pesticides, 191
Thioureas, 217
Three-component problems, 89-109
 analytical applications of displacement chromatography, 101-107
 displacement chromatography, 91-94
 elution mode, 91
 preparative applications of displacement chromatography, 98-101
 separation of a binary mixture in a gradient elution, 108-109
 system peaks, 107-108

[Three-component problems]
 theory of separation by displacement chromatography, 94-98
Tranquilizers, 218
Triazine herbicides, 189, 197, 216, 221
Triazines, 221
s-Triazine derivatives, 221
Triglycerides, 203, 208, 209, 210, 211, 212, 214, 216
Two-component problems, 43-89
 determination of competitive equilibrium isotherms, 73-74
 displacement chromatography, 69-70
 elution mode and ideal model, 44-49
 elution mode and kinetic models, 57-62
 elution mode and semi-ideal model, 49-57
 frontal analysis, 66-69
 gradient elution, 62-66
 optimization of preparative chromatography, 75-83
 other applications, 84-89
 system peak with a pure-component sample, 70-73

U

Ubiquinones, 211, 212, 213
Universal calibrations for aqueous size exclusion chromatography, 136-142
Urea, 185

V

Vitamin A, 198
Vitamin A compounds, 223
Vitamin D, 198
Vitamin E, 211, 213
Vitamin K, 211, 212
Vitamin K derivatives, 219
Vitamins, fat-soluble, 198, 211, 218

W

Wax alcohols, 212
Waxes, 210

X

Xanthates, 217
Xanthone derivatives, 207